Circuit Analysis and Solid State Electronic Devices

Circuit Analysis and Solid State Electronic Devices

Dr C P Vardhani
Assistant Professor
Department of Physics
Osmania University College for Women
Koti, Hyderabad, A.P.

Publishing Globally
NEW AGE INTERNATIONAL (P) LIMITED, PUBLISHERS
New Delhi • Bangalore • Chennai • Cochin • Guwahati
Hyderabad • Kolkata • Lucknow • Mumbai
Visit us at **www.newagepublishers.com**

Published by New Age International (P) Ltd., Publishers
First Edition: 2013

BRANCHES

- **Bangalore** 37/10, 8th Cross (Near Hanuman Temple), Azad Nagar, Chamarajpet, Bangalore-560 018
  Tel.: (080) 26756823, Telefax: 26756820, **E-mail: bangalore@newagepublishers.com**
- **Chennai** 26, Damodaran Street, T. Nagar, Chennai-600 017
  Tel.: (044) 24353401, Telefax: 24351463, **E-mail: chennai@newagepublishers.com**
- **Cochin** CC-39/1016, Carrier Station Road, Ernakulam South, Cochin-682 016
  Tel.: (0484) 2377004, Telefax: 4051303, **E-mail: cochin@newagepublishers.com**
- **Guwahati** Hemsen Complex, Mohd. Shah Road, Paltan Bazar, Near Starline Hotel, Guwahati-781 008
  Tel.: (0361) 2513881, Telefax: 2543669, **E-mail: guwahati@newagepublishers.com**
- **Hyderabad** 105, 1st Floor, Madhiray Kaveri Tower, 3-2-19, Azam Jahi Road, Nimboliadda, Hyderabad-500 027
  Tel.: (040) 24652456, Telefax: 24652457, **E-mail: hyderabad@newagepublishers.com**
- **Kolkata** RDB Chambers (Formerly Lotus Cinema) 106A, 1st Floor, S.N. Banerjee Road, Kolkata-700 014
  Tel.: (033) 22273773, Telefax: 22275247, **E-mail: kolkata@newagepublishers.com**
- **Lucknow** 16-A, Jopling Road, Lucknow-226 001
  Tel.: (0522) 2209578, 4045297, Telefax: 2204098, **E-mail: lucknow@newagepublishers.com**
- **Mumbai** 142C, Victor House, Ground Floor, N.M. Joshi Marg, Lower Parel, Mumbai-400 013
  Tel.: (022) 24927869, Telefax: 24915415, **E-mail: mumbai@newagepublishers.com**
- **New Delhi** 22, Golden House, Daryaganj, New Delhi-110 002
  Tel.: (011) 23262368, 23262370, Telefax: 43551305, **E-mail: sales@newagepublishers.com**

ISBN: 978-81-224-3448-4

₹ 199.00

C-6480-08-12

Printed in India at Print 'O' Pack, Delhi.

Typeset at Excellent Graphics, Delhi.

PUBLISHING FOR ONE WORLD

NEW AGE INTERNATIONAL (P) LIMITED, PUBLISHERS
7/30 A, Daryaganj, New Delhi-110002
Visit us at **www.newagepublishers.com**

Knowledge is a Pyramid,
To build it upward, it needs solid foundation

Dedicated

To my family members

and

The loving memory of my father

Late C.S. Rao

and

My Mother

Late C. Subbayamma

and

To all my Students

Preface

This book provides an extensive coverage of Circuit Analysis and Solid State Electronic Devices for B.Sc. I year (Electronics) students of common core syllabus of AP universities. It contains comprehensive up-to-date text for B.Sc. (Electronics) I year course. The designed consideration and performance characteristics of Electronic Devices and Circuit Analysis have been discussed at length.

The written text is supplemented with suitable diagrams and large number of worked examples for reinforcing the key aspects. Each chapter has a good number of problems at the end. This will help the students to widen the horizons of the knowledge who want to get into more details or test their grasp on the subject matter.

I am highly indebted to THE MISSILE MAN Dr. Prahlada, former Distinguished Scientist, Chief Controller R&D, SI, DRDO, New Delhi, and present Vice Chancellor, Defence Institute of Advanced Technology (DIET), PUNE, for his inspiration, valuable suggestions which gave boost to my intuition in writing this book successfully.

Special thanks are also due to Prof. K. Narasimha Reddy, Vice-Chancellor, Mahatma Gandhi University, Nalgonda, for his encouragement and constant support.

Last but not least, I thank all my family members—brothers, sister and friends, students, and my well wishers for their constant support.

I also thank the New Age International Publishers for printing and bringing out this book successfully.

Dr. C.P. Vardhani

Syllabus

Paper-I Circuit Analysis and Electronic Devices (120 Hours)

Unit – I (30 Hours)

AC Fundamentals : The Sine wave – Average and RMS values – The J operator – Phasor diagram – Phasor representation of sinusoidal currents and voltages – Polar and rectangular forms of complex numbers – Complex impedance and admittance – circuit analysis using complex number representation.

Passive networks : Concept of voltage and current sources – KVL and KCL – Application to simple circuits (AC and DC) consisting of resistors and sources (one or two) – Solution of simple networks using Node and Mesh analysis.

Network theorems (DC and AC) : Superposition Theorem – Thevenin's Theorem – Norton's Theorem – Maximum power transfer Theorem – Millman Theorem – Reciprocity Theorem – Application to simple networks.

Unit – II (30 Hours)

Cathode Ray Oscilloscope – Cathode ray tube (CRT) and its working – electron gun – focussing – defection sensitivity – fluorescent screen.

RC and RL Circuits : Transient response of RL and RC circuits with step input – time constants. Frequency response of RC and RL circuits – Types of Filters : Low pass filter – High pass filter – frequency response – Passive differentiating and integrating circuits.

Resonance : Series resonance and parallel resonance RLC circuits – Resonant frequency – Q factor – Band width – Selectivity.

Unit – III (30 Hours)

PN Junction : Depletion region – Junction capacitance – Diode equation (no derivation) – Effect of temperature on reverse saturation current. Construction, working, V-I characteristics and simple applications of

(*i*) Junction diode; (*ii*) Zener diode; (*iii*) Tunnel diode and (*iv*) Varactor diode.

Bipolar Junction Transistor (BJT) : PNP and NPN transistors – current components in BJT – BJT static characteristics (Input and Output) – Early effect – CB, CC, CE configurations (cut off, active and saturation regions) CE configuration as two port network – h-parameters – h-parameter equivalent circuit. Experimental arrangement to study input and output characteristics of BJT in CE configuration. Determination of h-parameters from the characteristics. Biasing and load line analysis – Transistor biasing Fixed bias and self bias arrangement.

Unit – IV (30 Hours)

Field Effect Transistor (FET) : Structure and working of JFET and MOSFET – output and transfer characteristics – Experimental arrangements for studying the characteristics and to determine FET parameters. Application of FET as voltage variable resistor and MOSFET as a switch – Advantages of FET over transistor.

Uni Junction Transistor (UJT) : Structure and working of UJT – Characteristics. Application of UJT as a relaxation oscillator.

Silicon Controlled Rectifier (SCR) : Structure and working of SCR. Two transistor representation, characteristics of SCR. Experimental set up to study the SCR characteristics. Application of SCR for power control.

Photo Electric Devices : Structure and operation of LDR, Photo voltaic cell, Photo diode, Photo transistors and LED.

(Note: Solving related problems in all the Units)

Practicals Paper – I (90 hours – 30 sessions)

Circuit Analysis and Electronic Devices Lab

1. Measurement of peak voltage, frequency and phase using CRO.
2. Thevenin's theorem – verification.
3. Norton's theorem – verification.
4. Maximum power transfer theorem – verification.
5. CR and LR circuits – Frequency response (Low pass and High Pass).
6. CR and LR circuits – Differentiation and integration – tracing of waveforms.
7. LCR – Series resonance circuit – Frequency response – Determination of fo, Q and band width.
8. To draw volt-ampere characteristics of Junction diode and determine the cut-in voltage, forward and reverse resistances.
9. Zener diode V-I Characteristics—Determination of Zener breakdown voltage.
10. Voltage regulator using Zener diode.
11. BJT input and output characteristics (CE configuration) and determination of 'h' parameters.
12. FET – Characteristics and determination of FET parameters.
13. UJT – (*i*) V-I Characteristics, (*ii*) Relaxation oscillator
14. LDR/Photo diode/Photo transistor/Solar Cell-characteristics.
15. SCR Volt-ampere characteristics.

Contents

UNIT–II

UNIT–III

UNIT–I

1. A.C. Fundamentals
2. Passive Networks
3. Network Theorems (D.C. and A.C.)

CHAPTER 1

A.C. Fundamentals

1.1 WHAT IS ALTERNATING CURRENT (A.C.)?

Alternating current is the current which constantly changes in amplitude, and which reverses direction at regular intervals. We know that direct current flows only in one direction, and that the amplitude of current is determined by the number of electrons flowing past a point in a circuit in one second. If, for example, a coulomb of electrons moves past a point in a wire in one second and all of the electrons are moving in the same direction, the amplitude of direct current in the wire is one ampere. Similarly, if half a coulomb of electrons moves in one direction past a point in the wire in half a second, then reverses direction and moves past the same point in the opposite direction during the next half-second, a total of one coulomb of electrons passes the point in one second. The amplitude of the alternating current is one ampere.

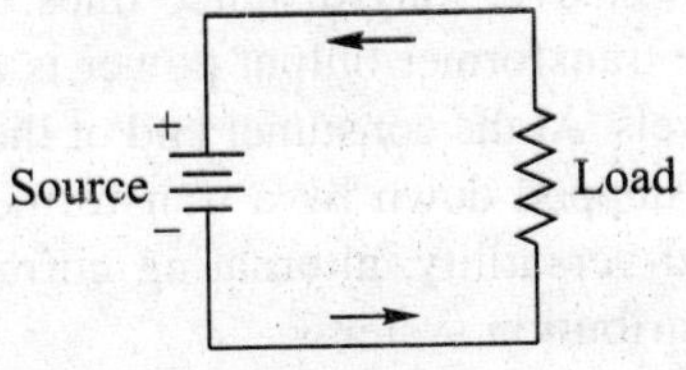

Fig. 1.1

1.2 ALTERNATING CURRENT VS. DIRECT CURRENT

The figure shown above shows the schematic diagram of a very basic D.C. circuit. It consists of nothing more than a source (a producer of electrical energy) and a load (whatever is to be powered by that electrical energy). The source can be any electrical source: a chemical battery, an electronic power supply, a mechanical generator, or any other possible continuous source of electrical energy. For simplicity, we represent the source in this figure as a battery.

At the same time, the load can be any electrical load: a light bulb, electronic clock or watch, electronic instrument, or anything else that must be driven by a continuous source of electricity. The figure here represents the load as a simple resistor.

Regardless of the specific source and load in this circuit, electrons leave the negative terminal of the source, travel through the circuit in the direction shown by the arrows, and eventually return to the positive terminal of the source. This action continues for as long as a complete electrical circuit exists.

Now consider the same circuit with a single change, as shown in the Fig. 1.2. This time, the energy source is constantly changing.

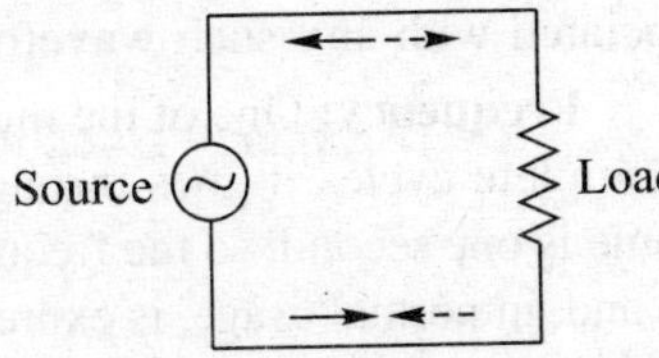

Fig. 1.2

It begins by building up a voltage which is positive on top and negative on the bottom, and therefore pushes electrons through the circuit in the direction shown by the solid arrows. However, then the source voltage starts to fall off, and eventually reverse polarity. Now current will still flow through the circuit, but this time in the direction shown by the dotted arrows. This cycle repeats itself endlessly, and as a result the current through the circuit reverses direction repeatedly. This is known as an alternating current.

1.3 DISADVANTAGES OF D.C. COMPARED TO A.C.

Disadvantage of the direct-current system becomes evident when the direct current (I) from the generating station must be transmitted a long distance over wires to the consumer. When this happens, a large amount of power is lost due to the resistance (*R*) of the wire. The power loss is equal to I^2R. However, this loss can be greatly reduced if the power is transmitted over the lines at a very high voltage level and a low current level. This is not a practical solution to the power loss in the d.c. system since the load would then have to be operated at a dangerously high voltage. Because of the disadvantages related to transmitting and using direct current, practically all modern commercial electric power companies generate and distribute alternating current. (a.c.).

Unlike direct voltages, alternating voltages can be stepped up or down in amplitude by a device called a transformer. Use of the transformer permits efficient transmission of electrical power over long-distance lines. At the electrical power station, the transformer output power is at high voltage and low current levels. At the consumer end of the transmission lines, the voltage is stepped down by a transformer to the value required by the load. Due to its inherent advantages and versatility, alternating current has replaced direct current in all but a few commercial power distribution systems.

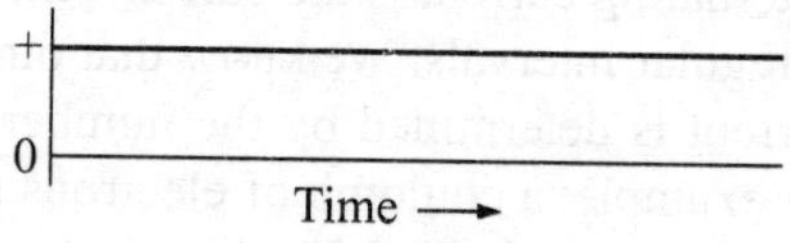

Fig. 1.3

1.4 PROPERTIES OF ALTERNATING CURRENT

A D.C. power source, such as a battery, outputs a constant voltage over time, as depicted in Fig. 1.4. Of course, once the chemicals in the battery have completed their reaction, the battery will be exhausted and cannot develop any output voltage. But until that happens, the output voltage to the right will remain essentially constant. The same is true for any other source of D.C. electricity: the output voltage remains constant over time.

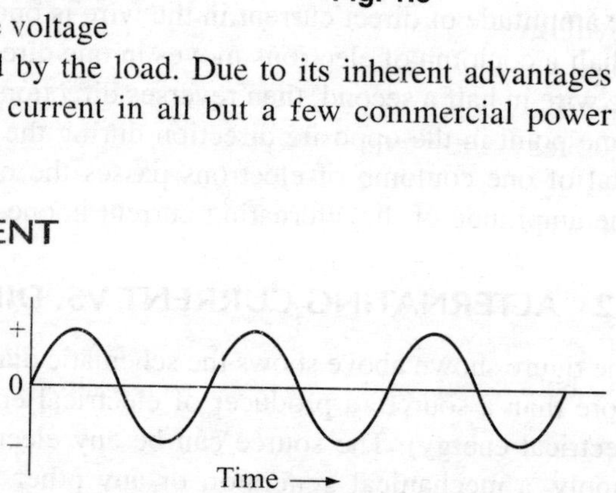

Fig. 1.4

By contrast, an A.C. source of electrical power changes constantly in amplitude and regularly here. Because the changes are so regular, alternating voltage and current have a number of properties associated with any such waveform. These basic properties include the following list:

Frequency: One of the most important properties of any regular waveform identifies the number of complete cycles it goes through in a fixed period of time. For standard measurements, the period of time is one second, so the frequency of the wave is commonly measured in cycles per second (cycles/sec) and, in normal usage, is expressed in units of Hertz (Hz). It is represented in mathematical equations by the letter '*f*'.

Period: Sometimes we need to know the amount of time required to complete one cycle of the waveform, rather than the number of cycles per second of time. This is logically the reciprocal of frequency. Thus, period is the time duration of one cycle of the waveform, and is measured in seconds/ cycle.

Wavelength: Because an A.C. wave moves physically as well as changing in time, sometimes we need to know how far it moves in one cycle of the wave, rather than how long that cycle takes to complete. This of course depends on how fast the wave is moving as well. The Greek letter (lambda) is used to represent wavelength in mathematical expressions. And, $\lambda = c/f$. As shown in the figure to the above, wavelength can be measured from any part of one cycle to the equivalent point in the next cycle. Wavelength is very similar to period as discussed above, except that wavelength is measured in distance per cycle while period is measured in time per cycle.

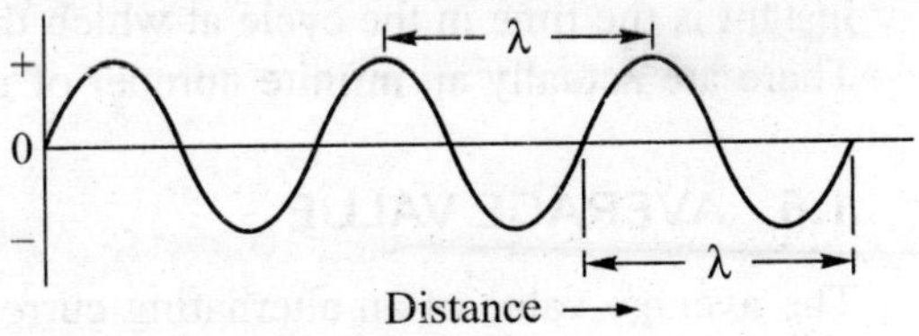

Fig. 1.5

Amplitude: Mathematically, the amplitude of a sine wave is the value of that sine wave at its peak. This is the maximum value, positive or negative, that it can attain. However, when we speak of an A.C. power system, it is more useful to refer to the effective voltage or current.

1.5 THE SINE WAVE

In discussing alternating current and voltage, you will often find it necessary to express the current and voltage in terms of maximum or peak values, peak-to-peak values, effective values, average values, or instantaneous values. Each of these values has a different meaning and is used to describe a different amount of current or voltage.

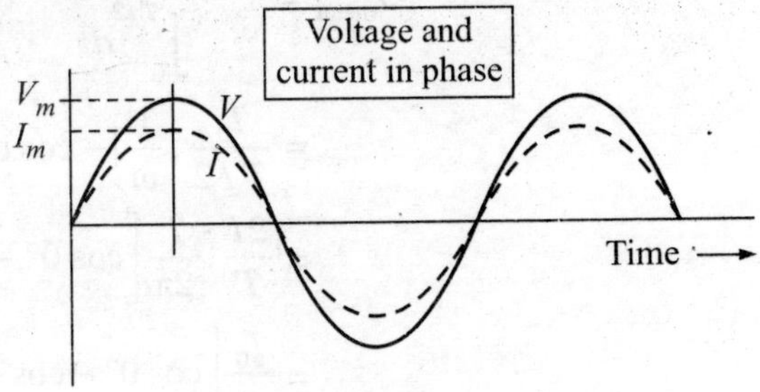

Fig. 1.6

Peak Value[Ip]

Refer to figure, it is the maximum value of voltage [V_p] or Current [I_p]. The peak value applies to both positive and negative values of the cycle.

Peak-Peak value [Ip-p]

During each complete cycle of ac there are always two maximum or peak values, one for the positive half-cycle and the other for the negative half-cycle. The difference between the peak positive value and the peak negative value is called the peak-to-peak value of the sine wave. This value is twice the maximum or peak value of the sine wave and is sometimes used for measurement of ac voltages.

Note the difference between peak and peak-to-peak values in figure below. Usually alternating voltage and current are expressed in effective values rather than in peak-to-peak values.

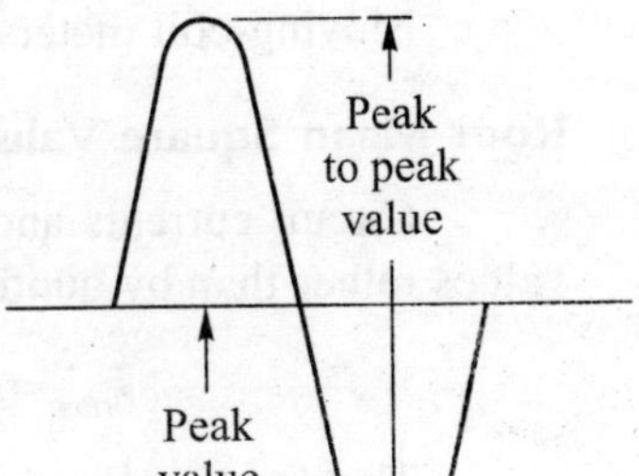

Fig. 1.7

Instantaneous Value

The instantaneous value of an alternating voltage or current is the value of voltage or current at one particular instant. The value may be zero if the particular instant is the time in the cycle at which the polarity of the voltage is changing. It may also be the same as the peak value, if the selected instant is the time in the cycle at which the voltage or current stops increasing and starts decreasing. There are actually an infinite number of instantaneous values between zero and the peak value.

1.6 AVERAGE VALUE

The average value of an alternating current or voltage is the average of all the instantaneous values during one alternation. Since the voltage increases from zero to peak value and decreases back to zero during one alternation, the average value must be some value between those two limits.

Derivation of Average Value of Current [I_{AV}]

Average value derivation: The average value of A.C. is the average over one complete cycle and is clearly zero, because there are alternately equal positive and negative half cycles.

Alternating current is represented as $I = I_0 \sin \omega t$

$$I_{mean} = \frac{\int_0^{T/2} I_0 \sin \omega t \, dt}{\int_0^{T/2} dt}$$

$$= \frac{I_0}{T/2} \cdot \frac{1}{\omega} [-\cos \omega t]_0^{T/2}$$

$$= \frac{2I}{T} \cdot \frac{T}{2\pi} \left[\cos 0° - \cos \frac{\omega T}{2} \right]$$

$$= \frac{I_0}{\pi} \left[\cos 0° - \cos \frac{\omega}{2} \cdot \frac{2\pi}{\omega} \right]$$

$$= \frac{2I_0}{\pi} = \frac{2}{\pi} \times \text{Peak value of current}$$

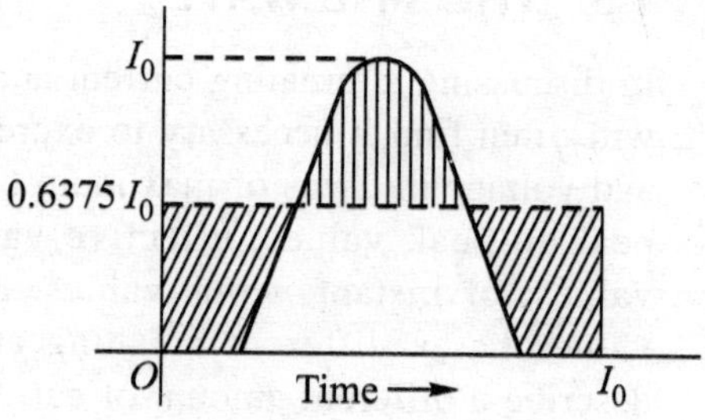

Fig. 1.8

Similarly, $E_{mean} = \dfrac{2E_0}{\pi} = \dfrac{2}{\pi} \times$ Peak value of voltage

Moving-coil meters respond to the average value of the current through them.

Root Mean Square Value

Circuit currents and voltage in A.C. circuits are generally stated as root-mean-square or rms values rather than by quoting the maximum values. The root-mean-square for a current is defined by

$$I_{rms} = \sqrt{(I^2)_{avg}}$$

That is, you take the square of the current and average it, then take the square root. When this process is carried out for a sinusoidal current

$$[I_m^2 \sin^2 \omega t]_{avg} = \frac{I_m^2}{2} \quad \text{so} \quad I_{rms} = \sqrt{(I^2)_{avg}} = \frac{I_m}{\sqrt{2}}$$

Since the A.C. voltage is also sinusoidal, the form of the rms voltage is the same. These rms values are just the effective value needed in the expression for average power to put the A.C. power in the same form as the expression for D.C. power in a resistor. In a resistor where the power factor is equal to 1.

$$P_{avg} = \frac{V_m I_m}{2} \cos\phi = V_{rms} I_{rms} \cos\phi = V_{rms} I_{rms}, \text{ for a resistor } R$$

$$\textbf{Form Factor} = \frac{V_{rms}}{V_{ave}} \text{ or } \frac{I_{rms}}{I_{ave}} = 1.11 \text{ (approx.)}$$

Derivation of RMS Value

RMS value of current is defined as $I_{rms} = \sqrt{\bar{I}^2}$. The average value of I^2 over one complete cycle is given by

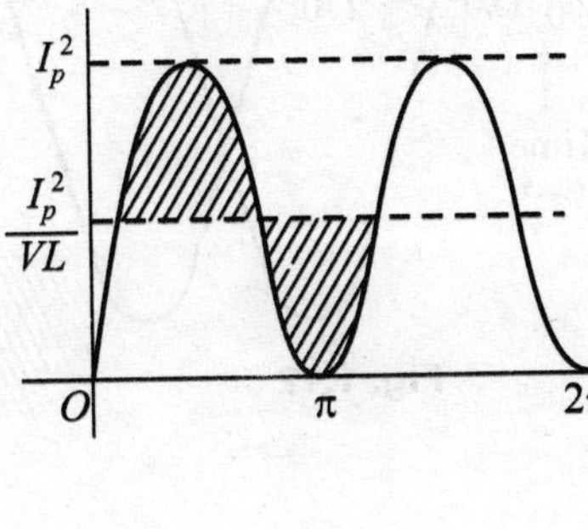

Fig. 1.9

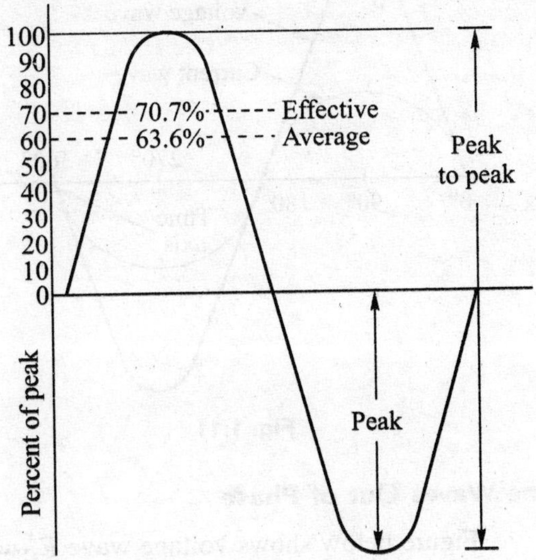

Fig. 1.10

$$\bar{I}^2 = \frac{\int_0^T I^2 dt}{\int_0^T dt} = \frac{1}{T}\int_0^T I_0^2 \sin^2 \omega t\, dt$$

$$\bar{I}^2 = \frac{I_0^2}{T}\int_0^T \left(\frac{1-\cos 2\omega t}{2}\right) dt$$

$$= \frac{I_0^2}{2T}\left[t - \frac{\sin 2\omega t}{2\omega}\right]_0^T = \frac{I_0^2}{2T}[T-0]$$

$$\bar{I}^2 = \frac{I_0^2}{2}$$

Thus the root mean square value of an alternating current is $I_{rms} = \sqrt{\bar{I}^2} = \sqrt{\frac{I_0^2}{2}} = \frac{I_0}{\sqrt{2}}$

Similarly the RMS value of an a.c. voltage is

$$V_{rms} = \frac{V_0}{\sqrt{2}} \text{ or } \frac{E_0}{\sqrt{2}}.$$

The rms value is the effective value required in power calculations. The rms value of a sine-wave current produces the same heating effect in a resistor as an identical d.c. current.

1.7 SINE WAVES IN PHASE

When a sine wave of voltage is applied to a resistance, the resulting current is also a sine wave. This follows Ohm's law which states that current is directly proportional to the applied voltage. To be in phase, the two sine waves must go through their maximum and minimum points at the same time and in the same direction.

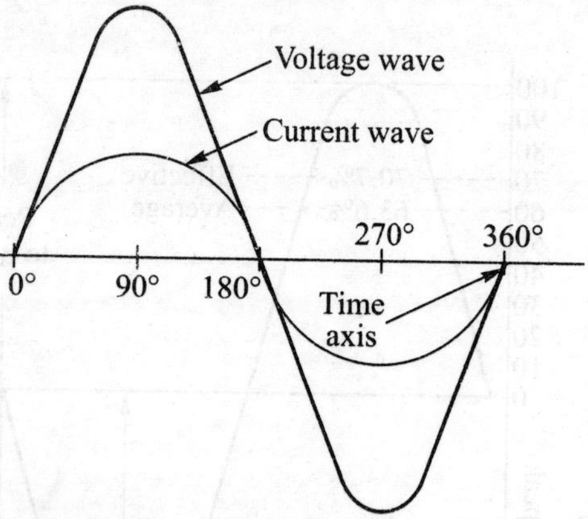

Fig. 1.11

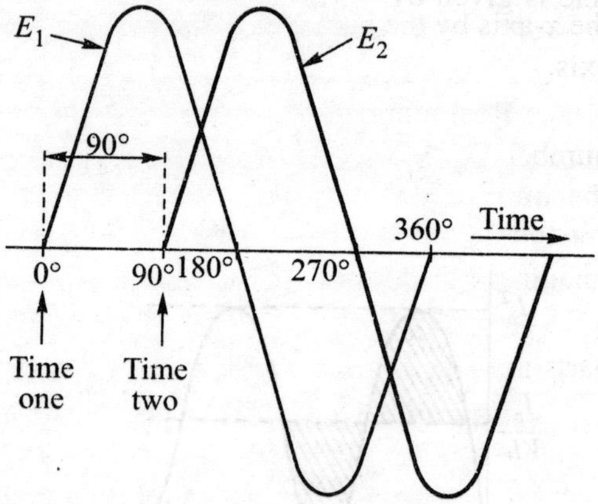

Fig. 1.12

Sine Waves Out of Phase

Figure below shows voltage wave E_1 which is considered to start at 0° (time one). As voltage wave E_1 reaches its positive peak, voltage wave E_2 starts its rise (time two). Since these voltage waves do not go through their maximum and minimum points at the same instant of time, a phase difference exists between the two waves. The two waves are said to be out of phase. For the two waves in figure, the phase difference is 90°.

1.8 THE OPERATOR J

Many problems in A.C. get simplified by the use of operator $j = \sqrt{-1}$, which is an imaginary quantity. Multiplication of a given vector quantity $\bar{A}$ by j or $\sqrt{-1}$ denotes the rotation of the vector through 90° in the anticlockwise direction (*i.e.* from OX to OY in Fig. 1.13). Similarly a multiplication by $-j$ represents the rotation of the vector through 90° in the clockwise direction (*i.e.*, form OX to OY'). Multiplication by $j^2 = -1$ means rotation by 90° and a further rotation by 90° *i.e.* the application of j^2 to a vector rotates it through 180° (form OX to OX'). Thus

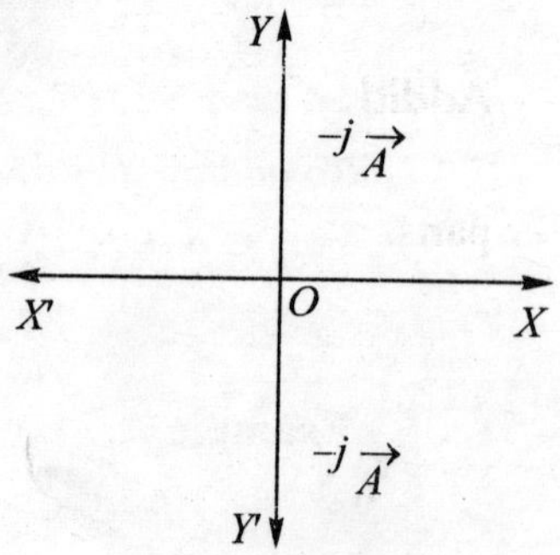

Fig. 1.13

$$j^2 A = -\bar{A}$$

Which is a vector equal in magnitude but opposite in direction to $\bar{A}$.

1.9 COMPLEX NUMBERS

The standard symbol for the set of all complex numbers is *C*.

For example, the equation $z = x + yj$ is to be understood as saying that the complex number z is the sum of the real number x and the real number y times i. In general, the x part of a complex number $z = x + yj$ is called the real part of z, while y is called the imaginary part of z. (Sometimes yj is called the imaginary part.)

When we use the *xy*-plane for the complex plane *C*, we'll call the *x*-axis by the name real axis, and the *y*-axis we'll call the imaginary axis.

Real numbers are to be considered as special cases of complex numbers; they're just the numbers $x + yj$ when y is 0, that is, they're the numbers on the real axis. For instance, the real number 2 is $2 + 0i$. The numbers on the imaginary axis are sometimes called purely imaginary numbers.

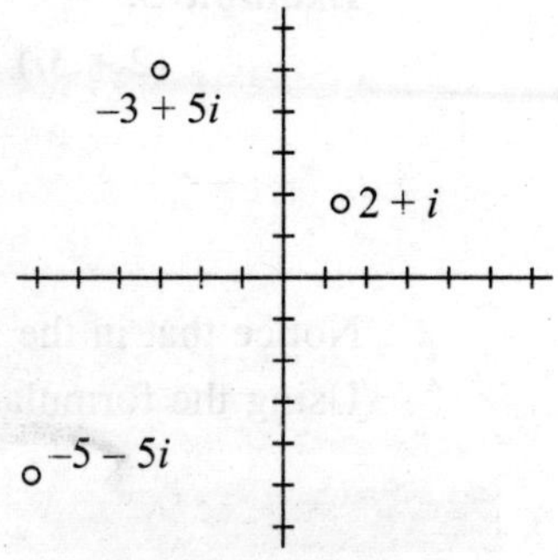

Fig. 1.14

Two complex numbers are equal if and only if their real parts are equal and their imaginary parts are equal. *i.e.*, $a + bi = c + di$ if and only if $a = c$ and $b = d$.

Example 1.

$$2 - 5i$$
$$6 + 4i$$
$$0 + 2i = 2i$$
$$4 + 0i = 4.$$

1.10 COMPLEX ARITHMETIC

When a number system is extended the arithmetic operations must be defined for the new numbers, and the important properties of the operations should still hold. For example, addition of whole numbers is commutative. This means that we can change the order in which two whole numbers are added and the sum is the same: $3 + 5 = 8$ and $5 + 3 = 8$.

We need to define the four arithmetic operations on complex numbers.

Addition and Subtraction

To add or subtract two complex numbers, you add or subtract the real parts and the imaginary parts.

$$(a + jb) + (c + jd) = (a + c) + j\,(b + d)$$
$$(a + jb) - (c + jd) = (a - c) + j\,(b - d)$$

Example 2.

$$(3 - 5j) + (6 + 7j) = (3 + 6) + (-5 + 7)j = 9 + 2j$$
$$(3 - 5j) + (6 + 7j) = (3 - 6) + (-5 - 7)j = -3 - 12j.$$

Multiplication

The formula for multiplying two complex numbers is

$$(a + bj) * (c + dj) = (ac - bd) + (ad + bc)j.$$

You do not have to memorize this formula, because you can arrive at the same result by treating the complex numbers like expressions with a variable, multiply them as usual, then simplify. The only difference is that powers of i do simplify, while powers of x do not.

Example 3.

$$\begin{aligned}(2 + 3i)(4 + 7i) &= 2*4 + 2*7i + 4*3i + 3*7*i2 \\ &= 8 + 14i + 12i + 21*(-1) \\ &= (8 - 21) + (14 + 12)i \\ &= -13 + 26i.\end{aligned}$$

Notice that in the second line of the example, the i^2 has been replaced by -1. Using the formula for multiplication, we would have gone directly to the third line.

EXERCISE

Perform the following operations:

(a) $(-3 + 4i) + (2 - 5i)$ (b) $3i - (2 - 4i)$

(c) $(2 - 7i)(3 - 4i)$ (d) $(1 + i) + (2 - 3i)$.

Division

Definition: The conjugate (or complex conjugate) of the complex number $a + bi$ is $a - bi$. Conjugates are important because of the fact that a complex number times its conjugate is real; *i.e.*, its imaginary part is zero.

$$(a + bi)(a - bi) = (a^2 + b^2) + 0i = a^2 + b^2.$$

Example 4.

Number	*Conjugate*	*Product*
$2 + 3i$	$2 - 3i$	$4 + 9 = 13$
$3 - 5i$	$3 + 5i$	$9 + 25 = 34$
$4i$	$-4i$	16

Suppose we want to do the division problem $(3 + 2i) \div (2 + 5i)$. First, we want to rewrite this as a fractional expression $\dfrac{3+2i}{2+5i}$.

Even though we have not defined division, it must satisfy the properties of ordinary division. So, a number divided by itself will be 1, where 1 is the multiplicative identity; *i.e.*, 1 times any number is that number.

So, when we multiply $\dfrac{3+2i}{2+5i}$ by $\dfrac{3-5i}{2-5i}$, we are multiplying by 1 and the number is not changed.

Notice that the quotient on the right consists of the conjugate of the denominator over itself. This choice was made so that when we multiply the two denominators, the result is a real number. Here is the complete division problem, with the result written in standard form.

$$\frac{3+2i}{2+5i} = \frac{3+2i}{2+5i} \times \frac{2-5i}{2-5i}$$

$$= \frac{(3+2i)(2-5i)}{(2+5i)(2-5i)}$$

$$= \frac{16-11i}{29} = \frac{16}{29} - \frac{11}{29} i.$$

1.11 PHASORS

In an a.c. circuit, the e.m.f. or current vary sinusoidally wih time and may be mathematically represented as

$$E = E_0 \sin \omega t$$

and $$I = I_0 \sin (\omega t \pm \theta)$$

Where θ is the phase angle between alternating e.m.f. and current.

Displacement of S.H.M. also varies sinusoidally with time *i.e.*

$$Y = A \sin \omega t$$

And its instantaneous value is equal to the projection of the amplitude A on Y-axis. Therefore, instantaneous values of alternating e.m.f. (E) and current (I) may be considered as the projections of e.m.f. amplitude (E_0) and current amplitude (I_0) respectively. The quantities, such as alternating e.m.f. and alternating current are called phasor. Thus a phasor is a quantity which varies sinusoidally with time and represented as the projection of rotating vector.

1.12 PHASOR DIAGRAM

The generator at the power station which produces our A.C. mains rotates through 360 degrees to produce one cycle of the sine wave form which makes up the supply.

In the next diagram there are two sine waves.

They are out of phase because they do not start from zero at the same time.

To be in phase they must start at the same time.

The waveform A starts before B and is LEADING by 90 degrees.

Waveform B is LAGGING A by 90 degrees.

The next left hand diagram, known as a PHASOR DIAGRAM, shows this in another way.

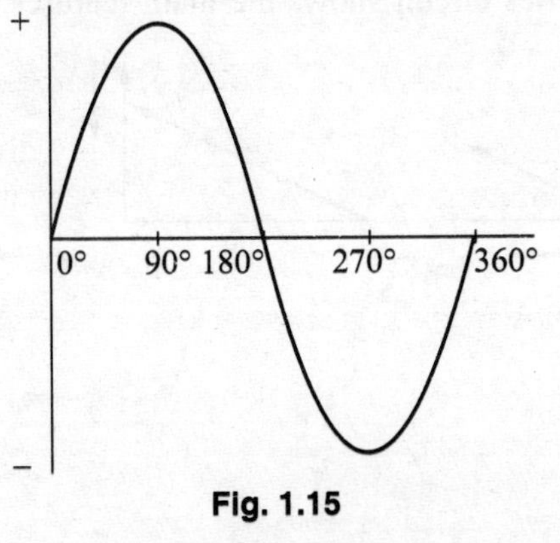

Fig. 1.15

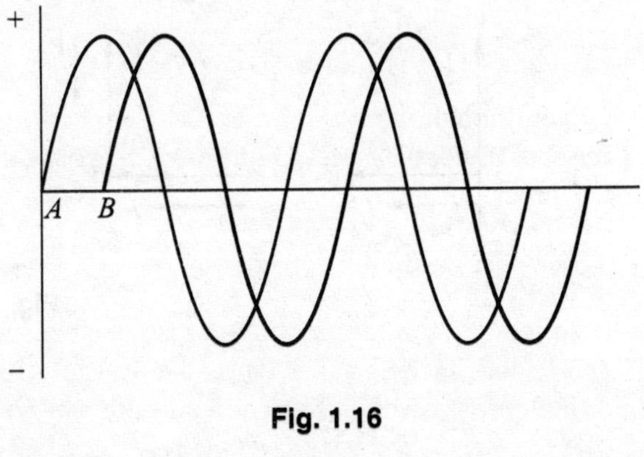

Fig. 1.16

The phasors are rotating anticlockwise as indicated by the arrowed circle.

A is leading *B* by 90 degrees.

The length of the phasors is determined by the amplitude of the voltages *A* and *B*.

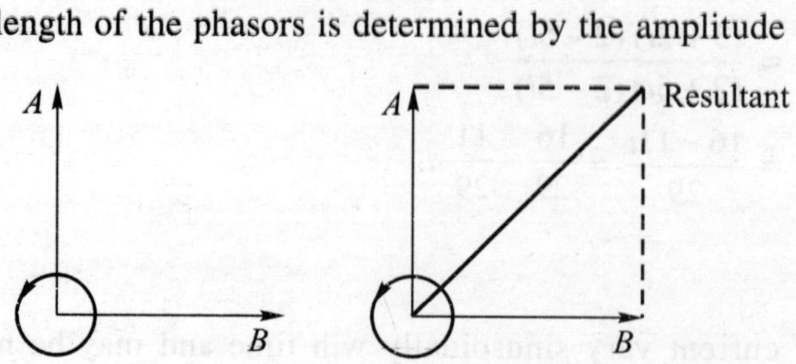

$$\text{Resultant} = \sqrt{A^2 + B^2}$$

Fig. 1.17

Since the voltages are of the same value then their phasors are of the same length. If voltage *A* was half the voltage of *B* then its phasor would be half the length of *B*.

The voltages *A* and *B* cannot be added together directly to find the resulting voltage, because they are not in phase.

The result of the two voltages can be found by completing the phasor diagram as shown on the right.

The resulting voltage is slightly greater in amplitude than *A* or *B* and leads *B* by 45 degrees and lags *A* by 45 degrees.

Since the two voltages are 90 degrees apart, then the resultant can be found by using Pythagoras, as shown.

Fig. 1.18

Fig. 1.19

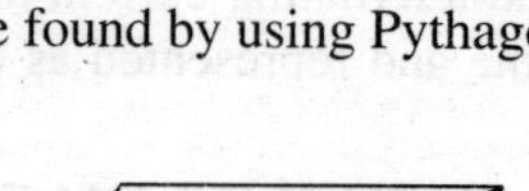

Fig. 1.20

It is sometimes helpful to treat the phase as if it defines a vector in a plane. The usual reference for zero phase is taken to be the positive *x*-axis and is associated with the resistor since the voltage and current associated with the resistor are in phase. The length of the phasor is proportional to the magnitude of the quantity represented, and its angle represents its phase relative to that of the current through the resistor. The phasor diagram for the RLC series circuit shows the main features.

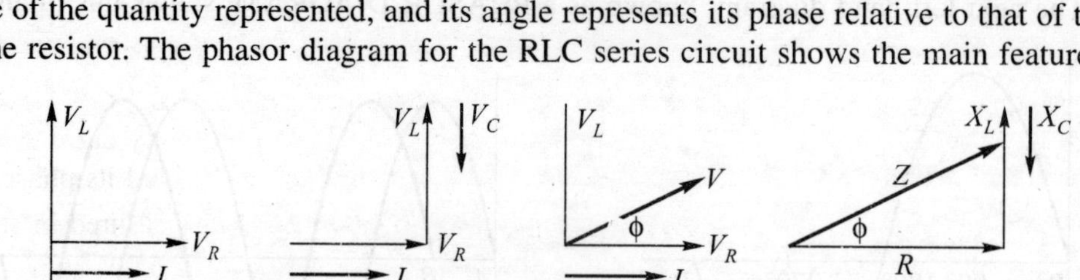

Fig. 1.21

$$V = \sqrt{V_R^2 + (V_L - V_C)^2} \qquad Z = \sqrt{R^2 + (X_L - X_C)^2}$$

$$\phi = \tan^{-1}\frac{V_L - V_C}{V_R} \qquad \phi = \tan^{-1}\frac{X_L - X_C}{R}$$

Note that the phase angle, the difference in phase between the voltage and the current in an A.C. circuit, is the phase angle associated with the impedance Z of the circuit.

1.13 PHASOR REPRESENTATION OF SINUSOIDAL CURRENTS AND VOLTAGES

A phase vector ("phasor") is a representation of a sine wave whose amplitude (A), phase (θ), and frequency (ω) are time-invariant. It is a subset of a more general concept called analytic representation. Phasors reduce the dependencies on these parameters to three independent factors, thereby simplifying certain kinds of calculations. In older texts, a phasor is also referred to as a sinor.

An alternating current or voltage is graphically represented by a line fixed at one end and rotating anticlockwise with a constant angular velocity ω radians/sec. The rotating line is called a phasor. The instantaneous value of the voltage or current is given by the length of the phasor and the instantaneous phase by the angle which the phasor makes with reference line. The phase angle is taken positive in anticlockwise direction and negative in clockwise direction.

The figure shown below shows the phasor representation of a sinusoidal voltage. Here OA represents phasor. Its projection on the vertical axis will give the instantaneous value of the e.m.f., *i.e.*,

$$OB = OA \sin \omega t$$

or

$$E = E_o \sin \omega t.$$

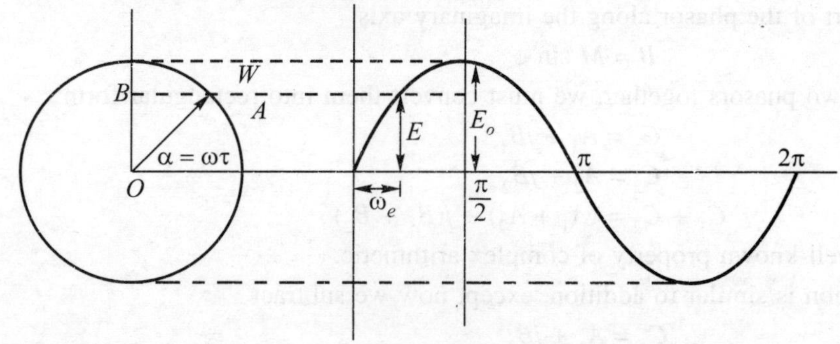

Fig. 1.22

The fraction of a cycle or time period that has elapsed since an alternating current or voltage last passed a given reference point, which is generally the starting point, is called its phase.

Phase of the alternating current or voltage may be expressed in time measured in seconds or fraction of a time period or the angle expressed in the degree or radians.

The phase difference between two alternating quantities is more important than their absolute phases. If two alternating current or voltages act simultaneously in the same circuit, they may do so in such a manner that their peak values do not occur at the same time. The time interval between two positive peak values of a.c. current or voltage is known as the phase difference. The quantity which reaches its maximum value earlier as compared to the other quantity is said to lead on the other. Similarly, a quantity which reaches its maximum value later than the other, is said to lag behind of the other.

1.14 PHASOR ALGEBRA–POLAR AND RECTANGULAR FORMS OF COMPLEX NUMBERS

1. Phasors have two components, the magnitude (M) and the phase angle (ϕ). Phasors are related to sinusoids through our cosine convention:

$$C = M\angle\phi = M \cos(t\omega + \phi)$$

Remember, there are 3 forms to phasors:

$$C = M\angle\phi \text{ “Phasor Form”}$$

$$C = A + jB \text{ “Rectangular Form”}$$

$$C = Me^{j\phi} \text{ “Exponential Form”}$$

Phasor and Exponential forms are identical and are also referred to as polar form.

2. When working with phasors it is often necessary to convert between rectangular and polar form. To convert from rectangular form to polar form:

$$M = \sqrt{A^2 + B^2}$$

$$\phi = \text{arc tan}\left(\frac{B}{A}\right)$$

To convert from polar to rectangular form:

A is the part of the phasor along the real axis

$$A = M \cos \phi$$

B is the part of the phasor along the imaginary axis

$$B = M \sin \phi$$

3. To add two phasors together, we must convert them into rectangular form:

$$C_1 = A_1 + jB_1$$

$$C_2 = A_2 + jB_2$$

$$C_1 + C_2 = (A_1 + A_2) + j(B_1 + B_2)$$

This is a well-known property of complex arithmetic.

4. Subtraction is similar to addition, except now we subtract

$$C_1 = A_1 + jB_1$$

$$C_2 = A_2 + jB_2$$

$$C_1 - C_2 = (A_1 - A_2) + j(B_1 - B_2)$$

5. To multiply two phasors, we should first convert them to polar form to make things simpler. The product in polar form is simply the product of their magnitudes, and the phase is the sum of their phases.

$$C_1 = M_1\angle\phi_1$$

$$C_2 = M_2\angle\phi_2$$

$$C = M_1 * M_2\angle\phi_1 + \phi_2$$

Keep in mind that in polar form, phasors are exponential quantities with a magnitude (M), and an argument (ϕ). Multiplying two exponentials together forces us to multiply the magnitudes, and add the exponents.

6. Division is similar to multiplication, except now we divide the magnitudes, and subtract the phases.

$$C_1 = M_1\angle\phi_1$$
$$C_1 = M_2\angle\phi_2$$
$$C = \frac{M_1}{M_2}\angle\phi_1 - \phi_2$$

7. An important relationship that is worth understanding is the inversion property of phasors:

$$C = M_C\angle 0 = -M_C\angle 180$$

On the normal Cartesian plane, for instance, the negative *X*-axis is 180 degrees around from the positive *X*-axis. By using that fact on an imaginary axis, we can see that the Negative Real axis is facing in the exact opposite direction from the Positive Real axis, and therefore is 180 degrees apart.

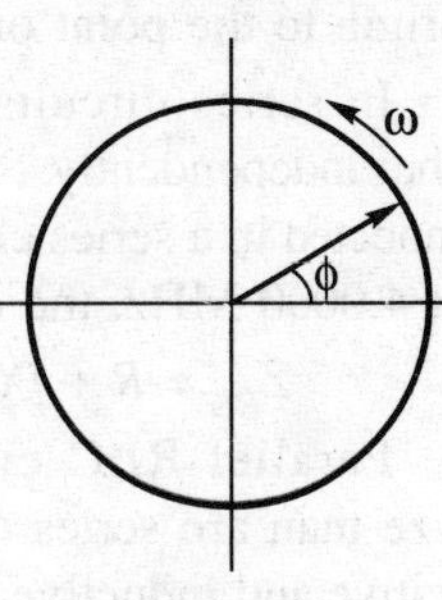

Fig. 1.23

8. Similar to the inversion property is the complex conjugation property of phasors. Complex conjugation is denoted with an asterisk above the phasor to be conjugated. Since phasors can be graphed on the Real-Imaginary plane, a 90 degree phasor is a purely imaginary number, and a – 90 degree phasor is its complex conjugate:

$$C = M\angle 90°$$
$$C^* = M\angle -90° = M\angle 270°$$

Essentially, this holds true for phasors with all angles: the sign of the angle is reversed to produce the complex conjugate of the phasor in polar notation. In general, for polar notation, we have:

$$C = M\angle\phi$$
$$C^* = M\angle -\phi$$

In rectangular form, we can express complex conjugation as:

$$C = A + jB$$
$$C^* = A - jB$$

Notice the only difference in the complex conjugate of *C* is the sign change of the imaginary part.

1.15 COMPLEX IMPEDANCE AND ADMITTANCE

Impedance, denoted *Z*, is an expression of the opposition that an electronic component, circuit, or system offers to alternating and/or direct electric current. Impedance is a vector (two-dimensional) quantity consisting of two independent scalar (one-dimensional) phenomena: resistance and reactance.

Resistance, denoted *R*, is a measure of the extent to which a substance opposes the movement of electrons among its atoms. The more easily the atoms give up and/or accept electrons, the lower the resistance, which is expressed in positive real number ohms. Resistance is observed with alternating current (A.C.) and also with direct current (D.C.). Examples of materials with low resistance, known as electrical conductors, include copper, silver, and gold. High-resistance substances are called insulators or dielectrics.

Reactance, denoted *X*, is an expression of the extent to which an electronic component, circuit, or system stores and releases energy as the current and voltage fluctuate with each A.C. cycle. Reactance is expressed in imaginary number ohms. It is observed for A.C., but not for DC. When A.C. passes through a component that contains reactance, energy might be stored and released in the form of a magnetic field, in which case the reactance is inductive (denoted $+jX_L$); or energy might be stored and released in the form of an electric field, in which case the reactance is capacitive (denoted $-jX_C$). Reactance is conventionally multiplied by the positive square root of -1, which is the unit imaginary number called the j operator, to express Z as a complex number of the form $R + jX_L$ (when the net reactance is inductive) or $R - jX_C$ (when the net reactance is capacitive).

The illustration shows a coordinate plane modified to denote complex-number impedances. Resistance appears on the horizontal axis, moving toward the right. (The left-hand half of this coordinate plane is not normally used because negative resistances are not encountered in common practice). Inductive reactance appears on the positive imaginary axis, moving upward. Capacitive reactance is depicted on the negative imaginary axis, moving downward. As an example, a complex impedance consisting of 4 ohms of resistance and $+j5$ ohms of inductive reactance is denoted as a vector from the origin to the point on the plane corresponding to $4 + j5$.

In series circuits, resistances and reactances add together independently. Suppose a resistance of 100.00 ohms is connected in a series circuit with an inductance of 10.000? H. At 4.0000 MHz, the complex impedance is:

$$Z_{RL} = R + jX_L = 100.00 + j251.33$$

Parallel RLC circuits are more complicated to analyze than are series circuits. To calculate the effects of capacitive and inductive reactance in parallel, the quantities are converted to inductive susceptance and capacitive susceptance. Susceptance is the reciprocal of reactance. Susceptance combines with conductance, which is the reciprocal of resistance, to form complex admittance, which is the reciprocal of complex impedance. Entire volumes have been devoted to the theoretical and practical aspects of resistance, conductance, reactance, susceptance, impedance, and admittance. An intermediate electronics text or reference book is recommended for further study.

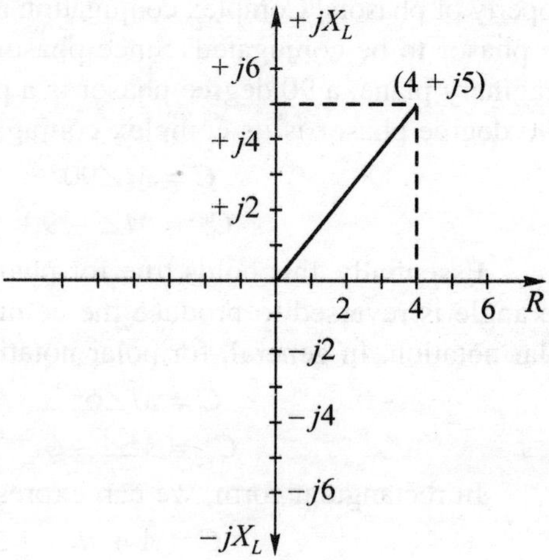

Fig. 1.24

1.16 CIRCUIT ANALYSIS USING COMPLEX NUMBER REPRESENTATION

Consider that an alternating e.m.f. $E = E_0 \sin \omega t$ is applied across an inductive coil L. Since the voltage across the inductor ($V_L = \omega L I_0$) leads the current I passing through it by 90°, hence by the use of operator 'j' we may write

$$V_L = j\omega \, LI_0$$

Because multiplication by 'j' means rotation by a positive (or anticlockwise) angle of 90°. Thus the inductive reactance ωL can be completely given by the term $j\omega L$.

Similarly, the capacitive reactance $\dfrac{1}{\omega C}$ can be completely given by

$$-j/\omega C = 1/j\omega C$$

Because the voltage across the capacitor lags the current passing through it by 90°.

If we have a resistance R in series with an inductance L, then

$$\text{Impedance} = Z = R + j\omega L$$

Which is known as complex or vector impedance. Its Modulus is

$$|Z| = \sqrt{R^2 + \omega^2 L^2}$$

and the phase angle is $\theta = \tan^{-1}(\omega L/R)$.

Similarly, the vector impedance of a resistance R in series with a capacitance C is given by

$$Z = R + 1/j\omega C$$

In LCR circuit, the vector impedance is given by

$$Z = R + j[\omega L^{-1}/j\omega C]$$

A.C. network analysis can be simplified by expressing sinusoidal voltage and currents by complex number representation as following:

$$E = E_0 \sin \omega t$$
$$I = I_0 \sin (\omega t \pm \theta)$$

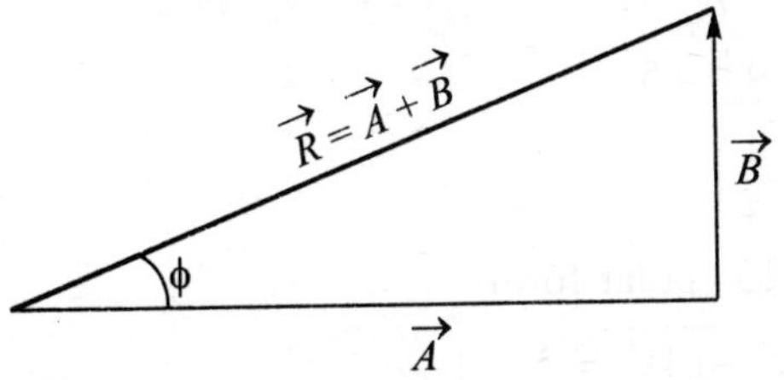

Fig. 1.25. Circuit analysis using complex representation.

SOLVED EXAMPLES

Example 1: *50 Ω resistor is connected to 220 volts a.c. supply. Calculate R.M.S. value of current.*

Solution: Here $V_{rms} = 220$ volts, $R = 50\ \Omega$

$$\therefore \quad I_{RMS} = \frac{V_{rms}}{R} = \frac{220}{50} = 4.4 \text{ amperes.}$$

Example 2: *Express the following complex numbers into rectangular form:*

(a) $20\angle 30°$ *(b)* $20\angle 60°$ *(c)* $20\angle 94°$

(d) $10\angle -116°$ *(e)* $10\angle -200°$ *(f)* $10\angle -267.3°$

Solution: (*a*)

$$20\angle 30° = 20 \cos 30° = j20 \sin 30°$$
$$= 20 \times 0.866 + j20 \times 0.5$$
$$= 17.32 + 10j \text{ (rectangular form)}$$

(*b*)

$$20\angle 60° = 20 \cos 60° + j \sin 60° \times 20$$
$$= 20 \times 0.5 + j\ 20 \times \sqrt{\frac{3}{2}}$$
$$= 5 \times 10\sqrt{3}\,j \text{ (rectangular form)}$$

(*c*) $20\angle 94° = 20 \cos 94° + j \sin 94° \times 20$
$= -20 \times 0.069 + 20j \times 0.997$
$= -1.38 + 19.94\,j$ (rectangular form)

(*d*) $10\angle -116° = 10 \cos 116° - j10 \sin 116°$
$= 10 \times 0.44 - j10 \times 0.898$
$= 0.44 - j8.98$ (rectangular form)

(*e*) $10\angle -200° = 10 \cos 200° - j10 \sin 200°$
$= -10 \times 0.94 + j10 \times 0.34$
$= -94 + j3.4$

(*f*) $10\angle 267.3° = 10 \cos 267.3° + j10 \sin 267.3°$
$= -10 \times 0.047 - j10 \times 0.998$
$= -0.47 - j9.98.$

Example 3: *Explain the following complex numbers into polar form:*
(i) 3 + j4 *(ii) – 3 + j4* *(iii) – 3 – j4.*

Solution: (*i*) $3 + j4 = \sqrt{3^2 + 4^2} = 5$

$\angle\theta = \tan^{-1}\left(\frac{4}{3}\right)$

$\therefore$ $3 + j4 = 5\angle 53.13°$ polar form

(*ii*) $-3 + j4 = \sqrt{(-3)^2 + (4)^2} = 5$

$\angle\theta_{ref} = -\tan^{-1}\left(\frac{4}{3}\right) = -53.13°$

$\angle\theta_{ref}$ is not the actual angle.

$\theta = 180° - 53.13° = 126.87°$

$-3 + j4 = 5\angle 126.87°$

(*iii*) $-3 - j4 = \sqrt{(-3)^2 + (-4)^2} = 5$

$\angle\theta_{ref} = \tan^{-1}\left(\frac{4}{3}\right) = 53.13°$

Actual angle $\theta = 180 + \theta_{ref} = 180 + 53.13 = 233.13°$

$-3 - j4 = 5\angle 233.13°.$

Example 4: *Find the sum or difference of the following complex numbers:*
(i) 10∠45° + (6 + j8) *(ii) 10∠45° – (8 + j6)* *(iii) (– 3 + j4) – (3 – j4).*

Solution: (*i*) $10\angle 45° + (6 + j8) = (10 \cos 45° + j10 \sin 45°) + (6 + j8)$
$= (10 \cos 45° + 6) + j(10 \sin 45° + 8)$
$= (10 \times 0.707 + 6) + j(10 \times 0.701 + 8)$
$= (13.07 + j15.07)$

(*ii*) $10\angle 45° - (8 + j6) = (10 \cos 45° + j10 \sin 45°) - (8 + j6)$
$= (10 \times 0.707 - 8) + j(10 \sin 45° - 6)$
$= -0.93 + j(7.07 - 6) = (-0.93 + j1.07)$

(*iii*) $(-3 + j4) - (3 - j4) = -3 - 3 + j(4 + 4) = -6 + j8.$

Example 5: *Find the product of 10∠45° × 10∠– 45°.*

Solution: (10∠45°)(10∠45°) = 10 × 10∠45 – 45° = 100∠0°.

Example 6: *Evaluate* $\dfrac{12.5 + j8.66}{(-4 - j3)}$.

Solution: $$\frac{12.5 + j8.66}{(-4 - j3)} = \frac{\sqrt{(12.5)^2 + (8.66)^2}\ \angle\theta_1 = \tan^{-1}\left(\frac{8.66}{12.51}\right)}{\sqrt{(-4)^2 + (-3)^2}\ \angle\theta_{ref} = \tan^{-1}(3/4)}$$

$$= \frac{15.2\angle\theta_1 = 4.6°}{5\angle\theta_{ref} = 36.87} = \frac{15.2\angle 4.6°}{5\angle 180° + 36.87°} = 3.04 \angle -212.3°.$$

Example 7: *Find the rms and maximum value of the voltage function.*

Solution: $$V_{rms} = \left[\left(\frac{100}{\sqrt{2}}\right)^2 + \left[\frac{-173}{\sqrt{2}}\right]^2\right]^{1/2} = \sqrt{\frac{10^4}{2} + \frac{(173)^2}{2}}$$

$$= \sqrt{\frac{39929}{2}} = 141.29 \text{ volts}$$

and $$V_{peak} = V_{Max} = \sqrt{2}\ V_{rms} = 1.414 \times 141.29 = 200 \text{ volts.}$$

Example 8: *Calculate the effective value of current in a circuit when i(t) = (10 + 10 sin ωt) mA.*

Solution: ∴ $$I_{rms} = \sqrt{10^2 + \left(\frac{10}{\sqrt{2}}\right)^2} = \sqrt{100 + 50} = 12.2 \text{ mA.}$$

Example 9: *An alternating current is represented by i = 120 sin (57t + π/2). Calculate (i) Frequency; (ii) time period; (iii) peak value; (iv) mean value; (v) rms value.*

Solution: Given equation: $i = 120 \sin (57t + \pi/2)$

Compare this with standard equation

$$i = i_0 \sin (\omega t + \theta)$$

Now

(*i*) $\omega = 2\pi f = 57$

$$\text{Frequency} = f = \frac{57}{2 \times 3.14} = \frac{57}{6.28} = 9.07 \text{ Hz}$$

(*ii*) $$\text{Time period} = T = \frac{2\pi}{\omega} = \frac{2 \times 3.14}{57} = 0.110 \text{ sec}$$

(*iii*) Peak value of current (I_0) = 120 amp

(*iv*) $$\text{Mean value} = I_{Mean} = \frac{2i_0}{\pi} = \frac{2 \times 120}{3.14} = 76.43 \text{ Amp}$$

(*v*) $$\text{R.m.s. value} = I_{rms} = \frac{i_0}{\sqrt{2}} = \frac{120}{1.414} = 84.86 \text{ amp.}$$

Example 10: *Find peak, average and r.m.s. of an a.c. current represented by* $i = 28.4 \sin 628\, t$?

Solution: Given, $i = 28.4 \sin 628\, t$

Peak value (i_0) $= 28.4$ A

Average value $= \dfrac{2i_0}{\pi} = \dfrac{2 \times 28.4}{\pi} = 18.089$ Amp

R.m.s. value $= \dfrac{i_0}{\sqrt{2}} = \dfrac{28.4}{1.414} = 20.08$ Amp

Frequency $= f = \dfrac{\omega}{2\pi} = \dfrac{628}{2\pi} = 100$ Hz.

Example 11: *If A.C. supply voltage is 220 volt, then find the average e.m.f. during a positive half cycle.*

Solution: If E_0 is the peak value of the e.m.f., then

$$E_{rms} = \frac{E_0}{\sqrt{2}}$$

$$E_0 = \sqrt{2} \times E_{rms} = \sqrt{2} \times 220 = 311 \text{ Volt}$$

$$E_{av} = \frac{2E_0}{\pi} = \frac{2 \times 311}{3.14} = 198 \text{ Volt.}$$

Example 12: *If A.C. supply voltage is 120 volt then find its form factor.*

Solution: $E_{rms} = \dfrac{E_0}{\sqrt{2}} = E_0 = \sqrt{2}\ E_{rms}$

$$= \sqrt{2} \times 120 = 169.68 \text{ Volts}$$

$$E_{av} = \frac{2E_0}{\pi} = \frac{2 \times 169.68}{3.14} = 108.07 \text{ Volts}$$

Form Factor $= \dfrac{E_{rms}}{E_{av}} = \dfrac{169.68}{108.07} = 1.15.$

EXERCISES

1. Express the following complex numbers into polar form:
   (*i*) $-10 - j10$ (*ii*) $-10 + j10$
   (*iii*) $3 - j4$.
2. Express the following complex numbers into rectangular form:
   (*i*) $20\angle 67°$ (*ii*) $5\angle -263°$
   (*iii*) $10\angle 94°$.
3. Find the sum (or) difference of the following complex numbers:
   (*i*) $10\angle 250° + (3 - j4)$ (*ii*) $(-6 - j8) - (4 + j2)$
   (*iii*) $10\angle 26° - 5\angle 240°$.
4. Find the product of $10\angle -120° \times 5 \times \angle 260°$.
5. Find the product of $(-3 + j4)\ (10\angle 60°)$.
6. Find the product of $(3 + j4)\ (3 - j4)$.

7. Evaluate:

(i) $\frac{10\angle -60^2}{5\angle 120°}$ (ii) $\frac{10\angle 30^2}{5\angle 180°}$ (iii) $\frac{10-20}{(-5-j10)}$.

8. Find the r.m.s. and maximum value of the voltage functions.

$$V(t) = 200 \cos \omega t - 178 \sin \omega t \text{ volts}$$

and $$V(t) = 200 \sin \left(5\omega t + \frac{\pi}{2}\right) \text{ volts.}$$

9. Find the r.m.s. value of current given by

$$i = 10 + 100 \sin 100\, t.$$

10. An alternating current is represented by $i = 220 \sin (157t + \pi/2)$. Calculate:

(i) Frequency (ii) Time period (iii) Peak value

(iv) Mean value (v) r.m.s. value.

11. The alternating current is represented by $i = 10 \sin \left(\omega t + \frac{\pi}{6}\right)$. Find the its r.m.s. value, average value and frequency and initial phase.

12. An A.C. e.m.f. is given by $E = 8 \sin \omega t + 6 \sin 2\, \omega t$ volts. Calculate its r.m.s. value.

QUESTIONS

A. Short answer type:

1. Define R.M.S. value, power factor and form factor.
2. Explain the significance of operator *j*.
3. Describe Wattless current.
4. Compare A.C. with direct current.
5. Explain the terms frequency, phase, peak, average and R.M.S. value of A.C. voltage.
6. What is an alternating voltage? Explain it.
7. Explain the term power factor. What is its significance?
8. Derive the expression of average value of A.C. voltage.
9. Derive the expression of R.M.S value of A.C. voltage.
10. Describe complex impedance and admittance.
11. Explain the phasor representation of sinusoidal currents and voltages.
12. Describe the terms period, wavelength and phase angle for an a.c. waveform.

B. Long answer type:

1. What do you understand by alternating current? Explain its amplitude, period, frequency.
2. Define complex number. Explain how complex numbers can be used to express sinusoidal quantities.
3. Describe operator *J*. Draw the phasor diagram. Explain the phasor representation of A.C. voltages and currents.
4. How do you generate sinusoidal voltages and currents? Discuss phasor representation of sinusoidal voltages and currents.
5. What is alternating current? How does it differ from direct current? Derive the expression for average and virtual values of A.C. Define Form Factor. Give its significance.
6. Illustrate by the graph the difference between an alternating current and a direct current. Use the graph to explain frequency, cycle, peak and R.M.S. values of current.

7. Derive an expression for average and R.M.S. values of A.C. current. Under what condition will the current be Wattless?
8. Explain operator '*j*' and phasor representation of sinusoidal voltages. Explain peak, average and r.m.s. values of an alternating voltage and derive relation between them.
9. Explain the period, wavelength and phase angle of an A.C. waveform.
10. Describe the procedure of circuit analysis using complex number representation.
11. Describe complex impedance and admittance. Analyse phasor algebra.
12. Define root mean square value of current and average value of current and derive expressions for them.
13. What is a complex number? Explain complex number representation for circuit analysis.
14. Explain the polar and rectangular forms of complex numbers. Explain R.M.S. and average values of A.C. current and A.C. voltage.

CHAPTER 2

Passive Networks

2.1 INTRODUCTION

An arrangement of passive and/or active circuit elements which form closed paths is called a network. If a network contains energy sources as well as other circuit elements, it is called an active network while a network containing circuit elements without any energy source is known as the passive network.

The following are the terms related to the networks:

1. **Node:** A node of a network is an equipotential surface at which two or more circuit elements are joined.
2. **Junction:** A junction is that point in a network where three or more elements are joined.
3. **Branch:** Any group of elements in series, having two terminals in the network is called a branch.
4. **Loop:** Any closed path in a network is called a loop.
5. **Mesh:** The space that the loop encloses is called a mesh.
6. **Tree:** A tree is defined as that part of a network composed of those branches between junction points which can be drawn without forming a closed path.

A network is said to be linear when the current in all branches is directly proportional to the driving voltage. If relation between voltage and current in any branch of network is nonlinear, the network is said to be non-linear.

If the inductors, resistors and capacitors are not electrically separated in a network, it is called a distributed network and if, the inductors, resistors and capacitors are physically separated, it is called a lumped network.

2.2 CONCEPT OF VOLTAGE AND CURRENT SOURCES

Devices that supply electrical energy are called energy sources. They may be classified as

1. Voltage sources
2. Current sources

Real voltage sources can be represented as ideal voltage sources in series with a resistance *r*, the ideal voltage source having zero resistance. Real current sources can be represented as ideal current sources in parallel with a resistance *r*, the ideal current source having infinite resistance.

Electrical sources, in practice, include things like: Batteries, the ever-present wall plug, power supplies, and many other kinds of electrical devices that deliver electrical energy.

1. Voltage sources

An independent voltage source maintains a voltage (fixed or varying with time) which is not affected by any other quantity. A voltage source is any device or system that produces an electromotive force between its terminals or derives a secondary voltage from a primary source of the electromotive force. An ideal voltage source is a theoretical circuit component that supplies a fixed potential difference across its terminals that is completely independent of the current it supplies.

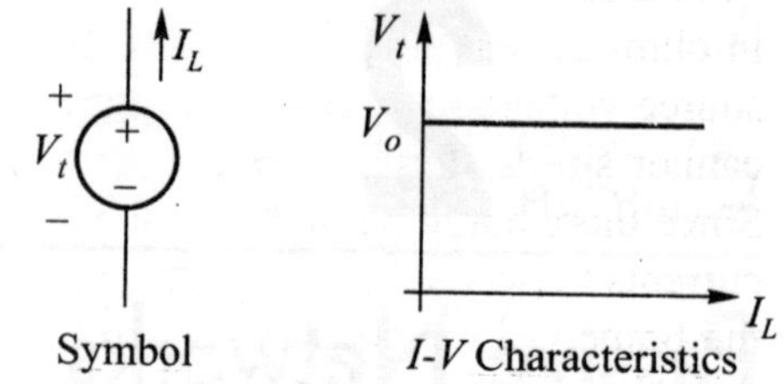

Fig. 2.1

2. Current sources

An independent current source maintains a current (fixed or time-varying) which is unaffected by any other quantity. A current source is an electrical or electronic device that delivers or absorbs electric current. Current sources can be theoretical or practical. A current source is the dual of a voltage source. An ideal current source is a conceptual source used in network theory and analysis that delivers or absorbs electrical energy such that the electrical current is independent of the voltage across its terminals.

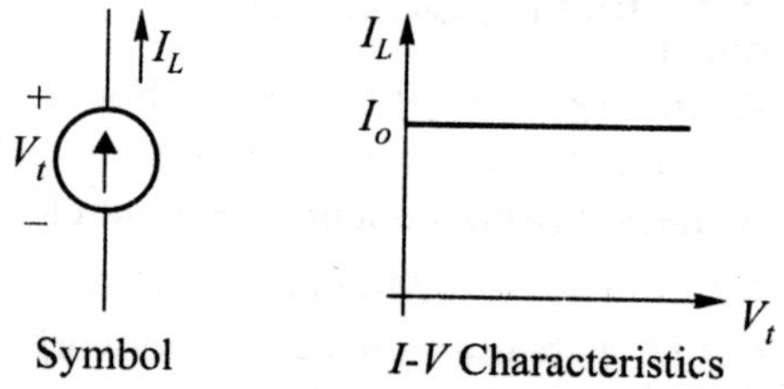

Fig. 2.2

Dependant sources

A dependant voltage or current source is one which depends on some other quantity in the circuit. It may depend on either voltage or current. These are explained as following.

1. Voltage Controlled Dependent Voltage Source

In the above figure, we find the dependent source whose positive terminal is designated as "n^+" and whose negative terminal is designated as "n^-" The controlling voltage is a branch voltage at some other circuit location. In this case, the positive terminal of the controlling branch is designated as "nc^+" while the negative terminal is designated as "nc^-". The "gain" of the dependent voltage source is a, a dimensionless quantity. For example, if V_x happened to be 16.0 volts while $a = 4$, then node "n^+" would be at 64.0 volts higher potential than node "n^-".

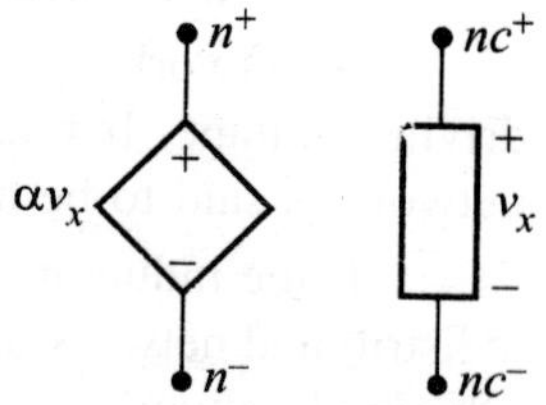

Fig. 2.3

2. Voltage Controlled Dependent Current Source

In the above voltage-controlled dependent current source a current equal to γ times v_x flows from node "n^+" through the source and out node "n^-". γ is called the transconductance and has the dimensions of siemens (inverse ohms).

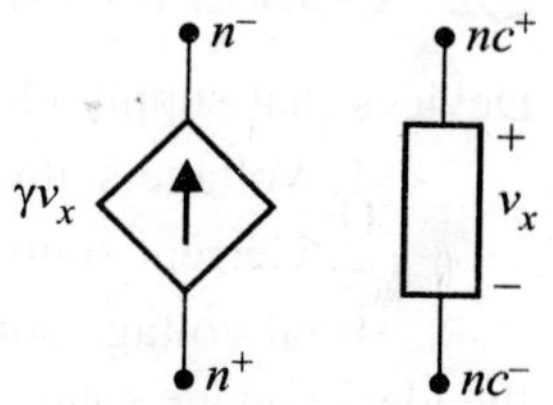

Fig. 2.4

3. Current Controlled Dependent Voltage Source

The current-controlled dependent voltage source as shown above, produces a voltage proportional to the current, i_x, in a different branch of the network. The transresistance, ρ, in ohms is multiplied by i_x in amps to produce the dependent source voltage in volts. Unlike the two previous examples, we cannot simply designate the controlling branch by its nodes. Since there could be multiple branches carrying very different currents between any pair of nodes, we must explicitly identify the branch of the controlling current.

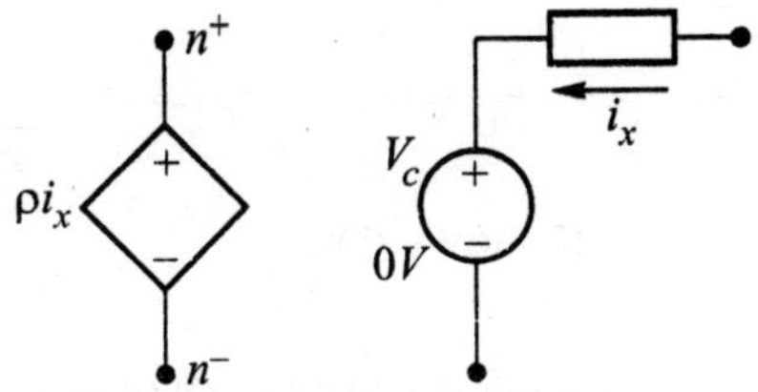

Fig. 2.5

4. Current Controlled Dependent Current Source

The current-controlled dependent current source produces a current proportional to the controlling current, i_x, flowing in a different branch. The current gain, β, is dimensionless. Designating the control scheme is similar to setting up the current-controlled dependent voltage source previously discussed. We must use a voltage source connected in series with the controlling element so that the controlling current enters the positive terminal of the independent voltage source used as an ammeter. If no voltage source is needed for its voltage, we use a zero—valued voltage source as shown in the figure.

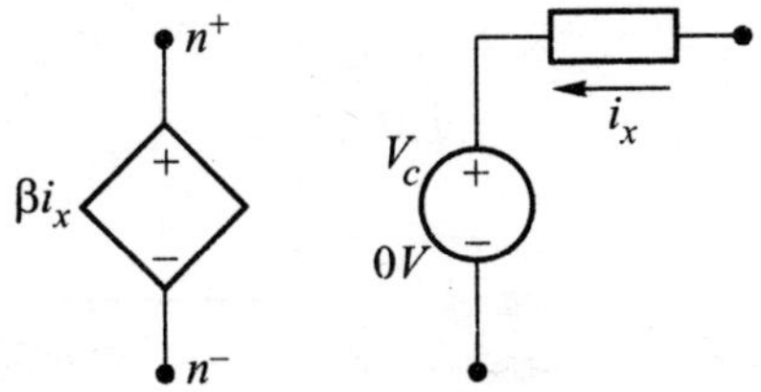

Fig. 2.6

2.3 CONVERSION OF VOLTAGE SOURCES INTO CURRENT SOURCES

The following procedure shows the conversion between voltage sources and current sources.

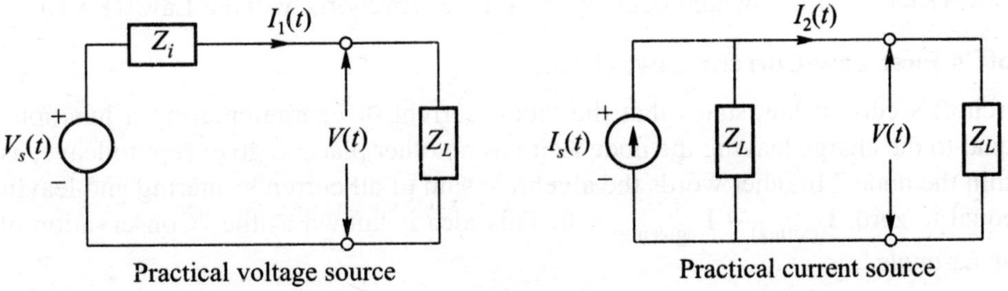

Fig. 2.7

The load current $I_1(t)$ from the voltage source

$$I_1(t) = \frac{V_s(t)}{Z_1 + Z_2} \qquad \text{...(1)}$$

The load current $I_2(t)$ from the current source is

$$I_1(t) = \frac{I_s(t) Z_2}{Z_2 + Z_2} \qquad \text{...(2)}$$

If the two sources are equivalent to each other

$$I_2 = I_1$$

$$\frac{V_s(t)}{Z_1 + Z_2} = \frac{I_s(t)Z_2}{Z_2 + Z_2}$$

It shows that

$\Rightarrow$ $$V_s(t) = I_s(t)Z_2 \quad \text{and} \quad Z_1 = Z_2$$

Thus a constant voltage source $V_s(t)$ with an internal impedance Z may be converted into an equivalent constant current source of current $I_s(t) = \frac{V_s(t)}{Z}$ in parallel with impedance.

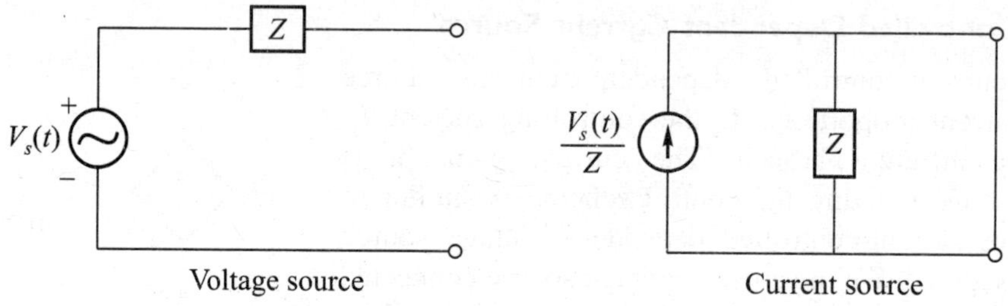

Fig. 2.8

2.4 KIRCHOFF'S CIRCUIT LAWS

Sometimes in complex circuits such as bridge or *T* networks, we cannot use simply Ohm's Law alone to find the voltages or currents circulating within a circuit. For these types of calculations we need certain rules which allow us to obtain the circuit equations and for this we can use "Kirchoff's Circuit Laws".

In 1845, a German physicist, Gustav Kirchoff, developed a pair of rules or laws which deal with the conservation of current and energy in electrical circuits, one which deals with current, Kirchoffs Current Law, (KCL) and one which deals with voltage, Kirchoffs Voltage Law, (KVL).

1. Kirchoff's First Law-Current Law, (KCL)

Kirchoff's current law states that the "total current or charge entering a junction or node is exactly equal to the charge leaving the node as it has no other place to go except to leave, as no charge is lost within the node." In other words the algebraic sum of all currents entering and leaving the node must be equal to zero, $\mathbf{I}_{\mathbf{(exiting)}} + \mathbf{I}_{\mathbf{(entering)}} = \mathbf{0}$. This idea is known as the "Conservation of Charge".

For Example,

Here, the 3 currents entering the node, I_1, I_2, I_3 are positive in value and the 2 currents leaving the node, I_4 and I_5 are negative in value. Then this means we can also rewrite the equation as;

$$I_1 + I_2 + I_3 - I_4 - I_5 = 0$$

In other words, Kirchoff's first law states "when steady current flows in an electrical circuit, there is no accumulation of charge at any point or junction.

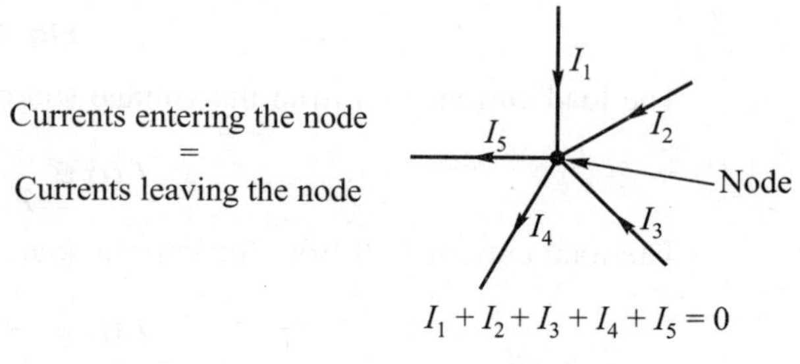

Fig. 2.9

2. Kirchoff's Second Law-Voltage law, (KVL)

Kirchoff's voltage or loop law states that "in any closed loop network the total voltage around the loop is equal to the sum of all the voltage drops within the loop" which is also equal to zero. In other words the algebraic sum of all voltages within the loop must be equal to zero. This idea is known as the "Conservation of Energy".

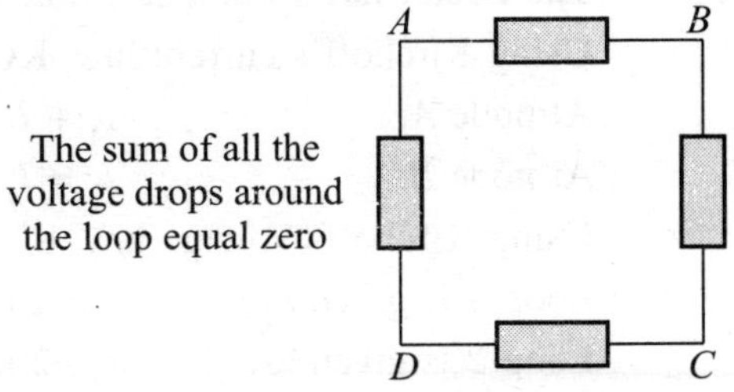

Fig. 2.10

For example,

$$V_{AB} + V_{BC} + V_{CD} + V_{DA} = 0$$

Starting at any point in the loop continue in the same direction noting the direction of all the voltage drops, either positive or negative, and returning back to the same starting point. It is important to maintain the same direction either clockwise or anticlockwise or the final voltage sum will not be equal to zero.

KVL can also be stated as "The algebraic sum of the product of the current and resistance in any closed loop of a closed circuit is equal to the algebraic sum of electromotive forces (e.m.f.) acting in that loop.

$$\Sigma IR = 0.$$

2.5 APPLICATION TO SIMPLE NETWORKS CONSISTING OF RESISTORS WITH A.C. AND D.C. SOURCES

Application of Kirchoff's Laws

These two laws enable the currents and voltages in a circuit to be found, *i.e.*, the circuit is said to be analyzed, and the basic procedure is as follows:

1. Assume all voltage sources and resistances are given. (If not label them V_1, V_2 ..., R_1, R_2 etc.)
2. Each branch with a branch currents. (I_1, I_2, I_3 etc.)
3. Kirchoff's first law equations for each node.
4. Kirchoff's second law equations for each of the independent loops of the circuit.
5. Linear simultaneous equations as required to find the unknown currents.

Example

Find the current flowing in the 40 Ω resistor, R_3

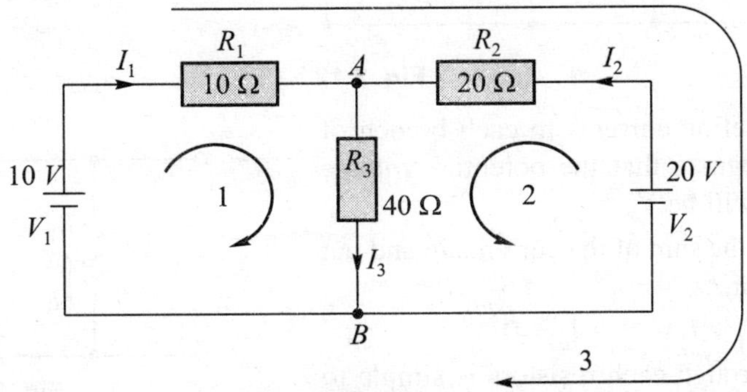

Fig. 2.11

The circuit has 3 branches, 2 nodes (A and B) and 2 independent loops.

Using Kirchoff's current law, KCL equations are;

At node A: $I_1 + I_2 = I_3$

At node B: $I_3 = I_1 + I_2$

Using Kirchoffs voltage law, KVL equations are;

Loop 1 is given as: $10 = R_1 \times I_1 + R_3 \times I_3 = 10I_1 + 40I_3$

Loop 2 is given as: $20 = R_2 \times I_2 + R_3 \times I_3 = 20I_2 + 40I_3$

Loop 3 is given as: $10 - 20 = 10I_1 - 20I_2$

As I_3 is the sum of $I_1 + I_2$ we can rewrite the equations as;

Eq. No. 1: $10 = 10I_1 + 40\,(I_1 + I_2) = 50I_1 + 40I_2$

Eq. No. 2: $20 = 20I_1 + 40\,(I_1 + I_2) = 40I_1 + 60I_2$

We now have two "Simultaneous Equations" that can be reduced to give us the value of both I_1 and I_2.

Substitution of I_1 in terms of I_2 gives I_1 as $-$ 0.143 Amps

Substitution of I_2 in terms of I_1 gives I_2 as + 0.429 Amps

As: $I_3 = I_1 + I_2$ The current flowing in resistor R_3 is given as:

$-0.143 + 0.429 = 0.286$ Amps

and the voltage across the resistor R_3 is given as: $0.286 \times 40 = 11.44$ volts

The negative sign for I_1 means that the direction of current flow initially chosen was wrong, but nevertheless still valid. In fact, the 20 V battery is charging the 10 V battery.

Some more simple applications of Kirchoff's laws

1. Parallel Resistors

Let's consider one on the most common electrical calculations—that for resistors in parallel. We want to find the equivalent resistance for the network of resistors shown.

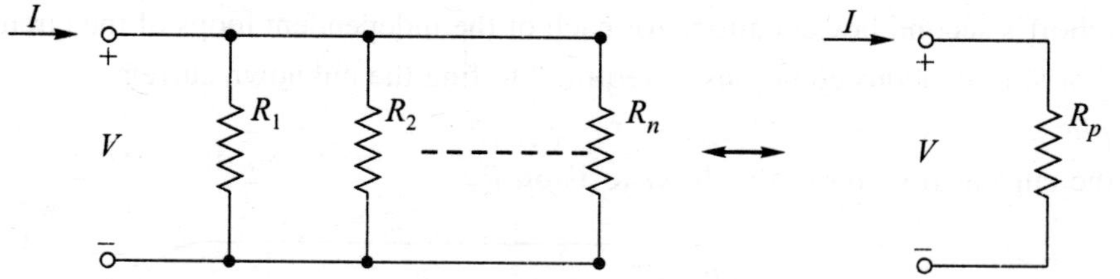

Fig. 2.12

First we can define currents in each branch of the circuit. Also recognize that the potential voltage across reach resistor will be V.

Now, consider the sum of the currents in and out of the upper conductor,

$$\Sigma I = I - I_1 - I_2 - \ldots - I_n = 0$$

The current through each resistors is simple to calculate, so if we add the current through the resistors, and then relate the expression to R_p, the equation becomes,

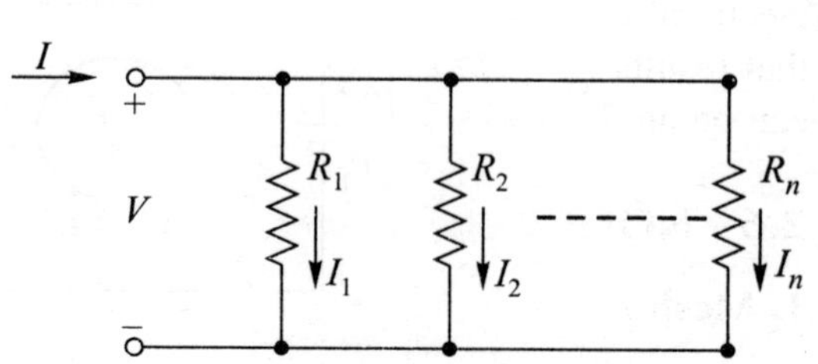

Fig. 2.13

$$\therefore \quad I - \frac{V}{R_1} - \frac{V}{R_2} \ldots - \frac{V}{R_x} = 0$$

$$\therefore \quad \frac{1}{V} = \left(\frac{1}{R_1} + \frac{1}{R_2} + \ldots + \frac{1}{R_x}\right) = \frac{1}{R_F}$$

$$\therefore \quad \boxed{R_F = \frac{1}{\frac{1}{R_1} + \frac{1}{R_2} + \ldots + \frac{1}{R_X}}}$$

2. Series Resistors

Now consider another problem with series resistors. We can use Kirchoff's voltage law to sum the voltages in the circuit loop. In this case the input voltage is a voltage rise, and the resistors are voltage drops (the signs will be opposite)

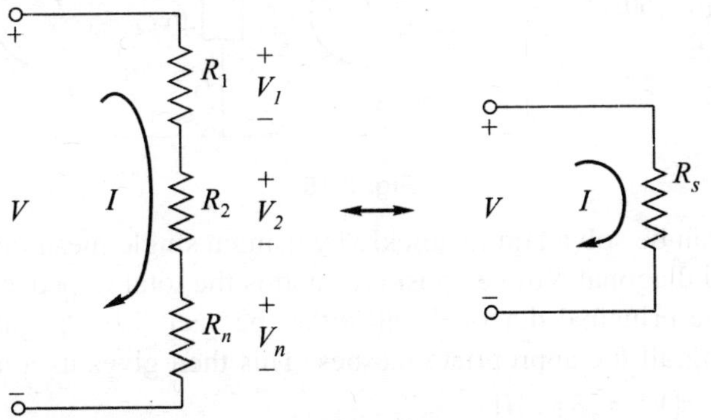

Fig. 2.14

First, sum the voltages about the loop,

$$\Sigma V = -V + V_1 + V_2 + \ldots + V_X = 0$$

Next, use ohms law to replace voltages with the current, and then relate the values to the equivalent resistor R_s.

$$\therefore \quad -V + IR_1 + IR_2 + \ldots + IR_X = 0$$

$$\therefore \quad \frac{V}{I} = \boxed{R_1 + R_2 + \ldots + R_X = R_2}$$

Till now we saw that complex circuits such as bridge or T networks can be solved using "Kirchoff's Circuit Laws". While Kirchoff's Laws give us the basic method for analyzing any complex electrical circuit, these are ways of improving upon this method by using Mesh or Nodal Analysis that results in a lessening of the maths involved and when large networks are involved this reduction can be an advantage.

2.6 NODE AND MESH ANALYSIS

1. Mesh Analysis

A more easier method of solving the about circuit is by using Mesh Analysis or Loop Analysis which is also sometimes called "Maxwell's Circulating Currents" method. Instead of labelling the branch currents we need to label each "closed loop" with a circulating current. As a general rule of thumb, only label inside loop in a clockwise direction with circulating currents as the aim is to cover all the elements of the circuit at least once. Any required branch current may be found from the appropriate loop or mesh currents as before using Kirchoff's method.

For example: $i_1 = I_1$, $i_2 = I_2$ and $I_3 = I_1 - I_2$

We now write Kirchoff's second voltage law equations in the same way as before to solve them but the advantage of this method is that it ensures that the information obtained from the circuit equations is the minimum required to solve the circuit as the information is more general and can easily be put into a matrix form.

For example, consider the circuit from the previous section.

$$10 = 50I_1 - 40I_2$$
$$-20 = -40I_1 + 60I_2$$
$$\text{or } [V] = [R][I]$$

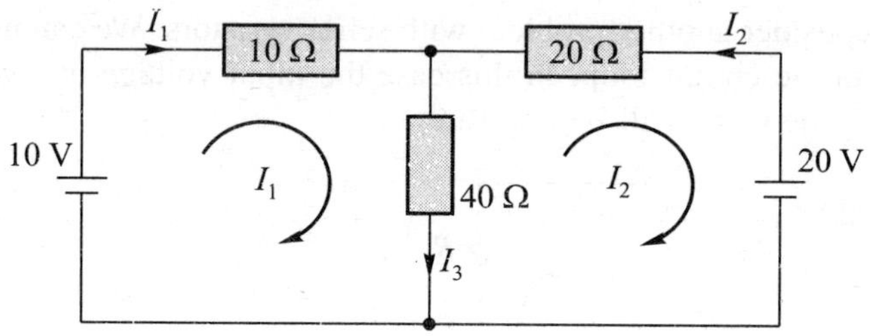

Fig. 2.15

These equations can be solved quite quickly by using a single mesh impedance matrix *Z*. Each element ON the principal diagonal will be "positive" and is the total impedance of each mesh. Where as, each element OFF the principal diagonal will either be "zero" or "negative" and represents the circuit element connecting all the appropriate meshes. This then gives us a matrix of:

$$[V] = [R] \,.\, [I]$$

$$\begin{bmatrix} 10 \\ 20 \end{bmatrix} = \begin{bmatrix} 10+40 & -40 \\ -40 & 60 \end{bmatrix} \cdot \begin{bmatrix} I_1 \\ I_2 \end{bmatrix}$$

where [*V*] gives the total battery voltage for loop 1 and then loop 2.

[*I*] states the names of the loop currents which we are trying to find.

[*R*] is called the resistance matrix.

and this gives I_1 as – 0.143 Amps and I_2 as + 0.429 Amps

As: $I_3 = I_1 + I_2$

The current I_3 is therefore given as: – 0.143 + 0.429 = 0.286 Amps

which is the same value of 0.286 amps, we found using Kirchoff's circuit laws in the previous section.

2. Nodal Analysis

As well as using Mesh Analysis to solve complex circuits it is also possible to use Nodal Analysis methods. Nodal Analysis complements Mesh Analysis in that it is equally powerful, based on the same concepts of matrix analysis. Nodal analysis uses the "Nodal" equations of Kirchoff's First Law to describe the circuit. If there are "*N*" nodes in the circuit there will be "*N* – 1" independent nodal equations and these alone are sufficient to describe and hence solve the circuit.

At each node point write down Kirchoff's first law equations, "currents entering the node are exactly equal to the currents leaving the node" then express each current in terms of the voltage across the branch. For N nodes, one node will be the reference node and all other voltages will be measured with respect to it.

For example, consider the circuit from the previous section.

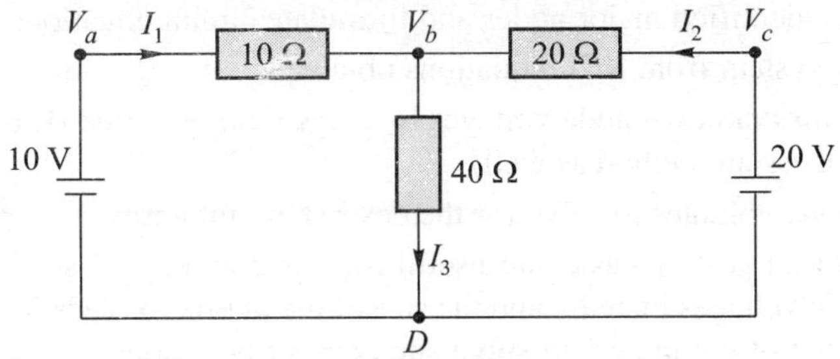

Fig. 2.16

In the above circuit, node D is chosen as the reference node and the other three nodes are assumed to have voltages, V_a, V_b and V_c with respect to node D. For example;

$$\frac{(V_a - V_b)}{10} + \frac{(V_c - V_b)}{20} = \frac{V_b}{40}$$

As $V_a = 10V$ and $V_c = 20V$, V_b can be easily found by:

$$1 + \frac{V_b}{10} + 1 - \frac{V_b}{20} = \frac{V_b}{40}$$

$$2 = V_b\left(\frac{1}{40} + \frac{1}{20} + \frac{1}{10}\right)$$

$$V_b = \frac{80}{7} V$$

$$\therefore \qquad I_3 = \frac{2}{7} \text{ or } 0.286 \text{ Amps.}$$

From both Mesh and Nodal Analysis methods we have looked at so far, this is the simplest method of solving this particular circuit. Generally, Nodal analysis is more appropriate when there are a larger number of current sources around. The network is then defined as: $[I] = [Y]\,[V]$ where $[I]$ are the driving current sources, $[V]$ are the nodal voltages to be found and $[Y]$ is the admittance matrix of the network which operates on $[V]$ to give $[I]$.

2.7 SOLUTION OF NETWORKS USING NODE AND MESH ANALYSIS

1. Nodal Analysis

The following is a general procedure for using Nodal Analysis method to solve electric circuit problems. The aim of this algorithm is to develop a matrix system from equations found by applying KCL at the major nodes in an electric circuit. Cramer's rule is then used to solve the unknown major node voltages.

Once the Node voltages are solved, normal circuit analysis methods (Ohm's law; Voltage and Current Divider principles etc...) can then be used to find whatever circuit entity.

Steps of Nodal Analysis

1. Choose a reference node. (Rule of thumb: take Node with most branches connecting to it)
2. Identify and Number major nodes. (Usually 2 or 3 major Nodes)
3. Apply KCL to identified major nodes and formulate circuit equations.
4. Create Matrix system from KCL equations obtained.
5. Solve Matrix for unknown node voltages by using Cramer's rule (It is simpler although you can still use Gaussian method as well)
6. Use solved Node voltages to solve for the desired circuit entity.

The above algorithm is very basic and useful for 2 × 2 and 3 × 3 size matrices. Generally as the number of major node voltages increase and the size of the matrix exceeds 3 × 3, numerical methods are employed with the aid of computers to solve such circuit networks.

Let's try an example to illustrate the above nodal analysis algorithm.

Example 1. *Consider figure with the following parameters:*

$V_1 = 15\ V$

$V_2 = 7\ V$

$R_1 = 2\ \Omega$

$R_2 = 20\ \Omega$

$R_3 = 10\ \Omega$

$R_4 = 5\ \Omega$

$R_5 = 2\ \Omega$

$R_6 = 2\ \Omega$

Find current through R_3 using Nodal Analysis method.

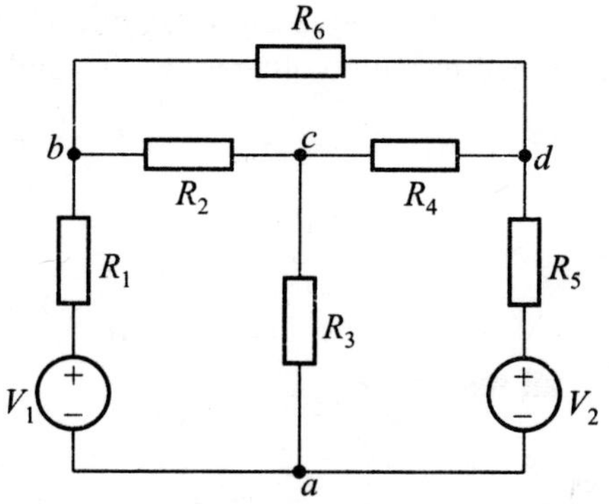

Fig. 2.17

Solution. Above figure shows Voltages at Nodes *a, b, c* and *d*.

We use node *a* as common node (ground if you like). Thus $V_a = 0$ V as we did previously.

Now that we have labelled the currents flowing in this circuit using passive sign convention, and have identified Nodes *b, c* and *d* as major nodes, we proceed as follows:

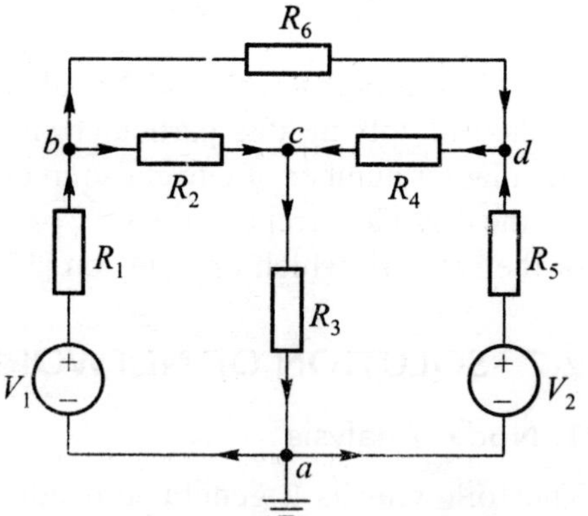

Voltage at nodes

Fig. 2.18

KCL @ Node b:

$$i_1 = i_2 + i_6$$

Thus by applying Ohm's law to above equation, we get

$$\frac{V_1 - V_b}{R_1} = \frac{V_b - V_c}{R_2} + \frac{V_b - V_d}{R_6}$$

Therefore,

$$V_b\left(\frac{1}{R_1} + \frac{1}{R_2} + \frac{1}{R_6}\right) - V_c\left(\frac{1}{R_2}\right) - V_d\left(\frac{1}{R_6}\right) = \frac{V_1}{R_1} \qquad ...(1)$$

KCL @ Node c:

$$i_3 = i_2 + i_4$$

Thus by applying Ohm's law to above equation, we get

$$\frac{V_c}{R_3} = \frac{V_b - V_c}{R_2} + \frac{V_d - V_c}{R_4}$$

Therefore,

$$V_b\left(-\frac{1}{R_2}\right) + V_c\left(\frac{1}{R_2} + \frac{1}{R_3} + \frac{1}{R_4}\right) + V_d\left(-\frac{1}{R_4}\right) = 0 \qquad ...(2)$$

KCL @ Node d:

$$i_4 = i_5 + i_6$$

Thus by applying Ohm's law to above equation, we get

$$\frac{V_d - V_c}{R_4} = \frac{V_b - V_d}{R_6} + \frac{V_2 - V_d}{R_5}$$

Therefore,

$$V_b\left(\frac{1}{R_6}\right) + V_c\left(\frac{1}{R_4}\right) + V_d\left(-\frac{1}{R_4} - \frac{1}{R_5} - \frac{1}{R_6}\right) = \frac{V_2}{R_5} \qquad ...(3)$$

The next step in this algorithm is to construct a matrix. In order to do that easily we substitute all resistances in above equations (1), (2) and (3) with their equivalent Admittances as follows:

$$G_1 = \frac{1}{R_1}$$

etc. thus equations (1), (2) and (3) will be re-written as follows:

$$V_b(G_1 + G_2 + G_6) + V_c(-G_2) + V_d(-G_6) = (V_1 \times G_1) \qquad ...(1)$$

$$V_b(-G_2) + V_c(G_2 + G_3 + G_4) + V_d(-G_4) = 0 \qquad ...(2)$$

$$V_b(G_6) + V_c(G_4) + V_d(-G_4 - G_5 - G_6) = (V_2 \times G_5) \qquad ...(3)$$

Now we can create a matrix with the above equations as follows:

$$\begin{bmatrix} (G_1 + G_2 + G_6) & -(G_2) & -(G_6) \\ (-G_2) & (G_2 + G_3 + G_4) & (-G_4) \\ (G_6) & (G_4) & -(G_4 + G_5 + G_6) \end{bmatrix} \cdot \begin{bmatrix} V_b \\ V_C \\ V_d \end{bmatrix} = \begin{pmatrix} (V_1 \times G_1) \\ 0 \\ (V_2 \times G_5) \end{pmatrix}$$

The following matrix is the above with values substituted:

$$A \,.\, \vec{X} = \vec{Y} \rightarrow \begin{bmatrix} 1.05 & -0.05 & -0.5 \\ -0.05 & 0.35 & -0.2 \\ 0.5 & 0.2 & -1.2 \end{bmatrix} \cdot \begin{bmatrix} V_b \\ V_c \\ V_d \end{bmatrix} = \begin{bmatrix} 7.5 \\ 0 \\ 3.5 \end{bmatrix}$$

Now that we have arranged equations (1), (2) and (3) into a matrix we need to get Determinants of the General matrix, and Determinants of alterations of the general matrix as follows:

Solving determinants of:

Matrix A: General matrix A from KCL equations

Matrix A_1: General Matrix A with Column 1 substituted by $\vec{Y}$.

Matrix A_2: General Matrix A with Column 2 substituted by $\vec{Y}$.

Matrix A_3: General Matrix A with Column 3 substituted by $\vec{Y}$.

As follows:

$$\det \begin{bmatrix} 1.05 & -0.05 & -0.5 \\ -0.05 & 0.35 & -0.2 \\ 0.5 & 0.2 & -1.2 \end{bmatrix} = \det A$$

$$= -0.3$$

$$\det \begin{bmatrix} 7.5 & -0.05 & -0.5 \\ 0 & 0.35 & -0.2 \\ 3.5 & 0.2 & -1.2 \end{bmatrix} = \det A_1$$

$$= -2.203$$

$$\det \begin{bmatrix} 1.05 & 7.5 & -0.5 \\ -0.05 & 0 & -0.2 \\ 0.5 & 3.5 & -1.2 \end{bmatrix} = \det A_2$$

$$= -0.378$$

$$\det \begin{bmatrix} 1.05 & 0.05 & 7.5 \\ 0.05 & 0.35 & 0 \\ -0.5 & -0.2 & 3.5 \end{bmatrix} = \det A_3$$

$$= -0.11$$

Now we can use the solved determinants to arrive at solutions for Node voltages V_b, V_c and V_d as follows:

$$V_b = \frac{\det A_1}{\det A} = \frac{-2.203}{-0.3} = 7.378559 \text{ V}$$

$$V_c = \frac{\det A_2}{\det A} = \frac{-0.378}{-0.3} = 1.264657 \text{ V}$$

$$V_b = \frac{\det A_3}{\det A} = \frac{-0.11}{-0.3} = 0.368509 \text{ V}$$

Now we can apply Ohm's law to solve for the current through R_3 as follows:

$$I_{R3} = \frac{V_c}{R_3} = \frac{1.264657V}{10\Omega} = +0.126466 \text{ A}$$

As we have seen previously, the positive sign in the above current tells us that the effective current flowing through R_3 is in fact in the direction we chose when drawing up the circuit in figure shown above.

Example 2. *Consider Fig. 7.3 with the following parameters:*

$V_1 = 9\ V$

$R_1 = 200\ \Omega$

$R_2 = 20\ \Omega$

$R_3 = 10\ k\Omega$

$R_4 = 5\ k\Omega$
$R_5 = 15\ k\Omega$
$R_6 = 1\ k\Omega$

Find current through R_3 using Nodal Analysis method.

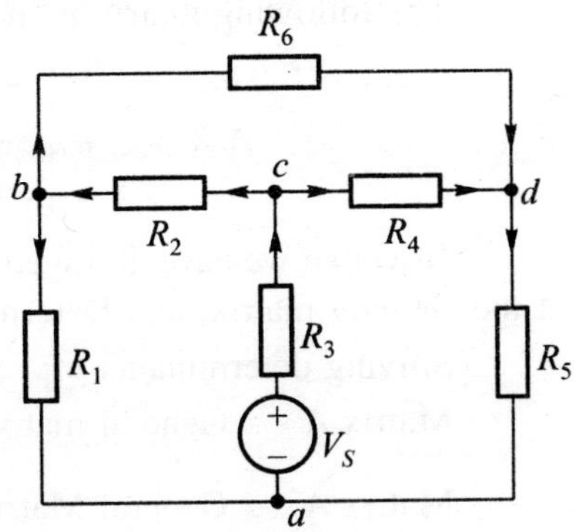

Fig. 2.19

KCL @ Node b:

$$i_2 = i_1 + i_6$$

Thus by applying Ohms law to above equation, we get

$$\frac{V_c - V_b}{R_2} - \frac{V_b}{R_1} - \frac{V_b - V_d}{R_6} = 0$$

Therefore,

$$-V_b\left(\frac{1}{R_1} + \frac{1}{R_2} + \frac{1}{R_6}\right) + V_c\left(\frac{1}{R_2}\right) + V_d\left(\frac{1}{R_6}\right) = 0 \quad \text{...(1)}$$

KCL @ Node c:

$$i_3 = i_2 + i_4$$

Thus by applying Ohms law to above equation, we get

$$\frac{V_b - V_c}{R_3} - \frac{V_c - V_b}{R_2} - \frac{V_c - V_d}{R_4} = 0$$

Therefore,

$$-V_b\left(\frac{1}{R_2}\right) - V_c\left(\left(\frac{1}{R_2} + \frac{1}{R_3} + \frac{1}{R_4}\right) + V_d\left(\frac{1}{R_4}\right)\right) = \frac{V_b}{R_3} \quad \text{...(2)}$$

KCL @ Node d:

$$i_5 = i_4 + i_6$$

Thus by applying Ohms law to above equation, we get

$$\frac{V_b - V_c}{R_3} - \frac{V_c - V_b}{R_2} - \frac{V_c - V_d}{R_4}$$

Therefore,

$$-V_b\left(\frac{1}{R_3}\right) - V_c\left(\frac{1}{R_4}\right) + V_d\left(\frac{1}{R_3} + \frac{1}{R_4} - \frac{1}{R_5}\right) = 0 \quad \text{...(3)}$$

$$G_1 = \frac{1}{R_1}$$

etc. thus equations (1), (2) and (3) will be re-written as follows:

$$V_b(-G_1 - G_2 - G_6) + V_c(G_2) + V_d(G_6) = 0 \quad \text{...(1)}$$

$$V_b(-G_2) + V_c(G_2 + G_3 + G_4) + V_d(-G_4) = (V_s \times G_3) \quad \text{...(2)}$$

$$V_b(-G_3) + V_c(-G_4) + V_d(G_3 + G_4 + G_5) = 0 \quad \text{...(3)}$$

Now we can create a matrix with the above equations as follows:

$$\begin{bmatrix} (-G_1 - G_2 - G_6) & (G_2) & (G_6) \\ (-G_2) & (G_2 + G_3 + G_4) & (-G_4) \\ (-G_4) & (G_4) & (G_3 + G_4 + G_5) \end{bmatrix} \cdot \begin{bmatrix} V_b \\ V_c \\ V_d \end{bmatrix} = \begin{bmatrix} 0 \\ V_s \times G_3 \\ 0 \end{bmatrix}$$

The following matrix is the above with values substituted:

$$A.\vec{X} = \vec{Y} \rightarrow \begin{bmatrix} -0.056 & 0.05 & 0.001 \\ -0.05 & 0.0503 & -0.0002 \\ -0.0001 & -0.0002 & 0.000367 \end{bmatrix} \cdot \begin{bmatrix} V_b \\ V_c \\ V_d \end{bmatrix} = \begin{bmatrix} 0 \\ 0.0009 \\ 0 \end{bmatrix}$$

Now that we have arranged equations (1), (2) and (3) into a matrix we need to get Determinants of the General matrix, and Determinants of alterations of the general matrix as follows:

Solving determinants of:

Matrix A : General matrix A from KCL equations

Matrix A_1 : General Matrix A with column 1 substituted by $\vec{Y}$.

Matrix A_2 : General Matrix A with column 2 substituted by $\vec{Y}$.

Matrix A_3 : General Matrix A with column 3 substituted by $\vec{Y}$.

As follows:

$$\det A = -9.8 \times 10^{-8}$$
$$\det A_1 = -1.7 \times 10^{-8}$$
$$\det A_2 = -1.8 \times 10^{-8}$$
$$\det A_3 = -1.5 \times 10^{-8}$$

Now we can use the solved determinants to arrive at solutions for Node voltages V_b; V_c and V_d as follows:

$$V_b = \frac{\det A_1}{\det A} = 0.17 \text{ V}$$

$$V_c = \frac{\det A_2}{\det A} = 0.19 \text{ V}$$

$$V_b = \frac{\det A_3}{\det A} = 0.15 \text{ V}$$

Now we can apply Ohm's law to solve for the current through R_3 as follows:

$$I_{R3} = \frac{V - V_c}{R_3} = \frac{9 \text{ V} - 0.19 \text{ V}}{10 \text{ k}\Omega} = 0.881 \text{ mA}$$

2. Mesh Analysis

The following is a general procedure for using Mesh or Loop Analysis method to solve electric circuit problems. The aim of this algorithm is to develop a matrix system from equations found by applying KVL around Loops or Meshes in an electric circuit. Carmer's rule is then used to solve the unknown Mesh Currents.

Once the Mesh Currents are solved, normal circuit analysis methods (Ohm's law; Voltage and Current Divider principles etc...) can then be used to find whatever circuit entity. Steps of Mesh/Loop Analysis

1. Choose a conventional current flow.
2. Identify and Number loops or meshes. (Usually 2 or 3 meshes/loops)
3. Apply KVL to identified meshes/loops and formulate circuit equations.
4. Create Matrix system from KVL equations obtained.
5. Solve Matrix for unknown Mesh Currents by using Cramer's rule (It is simpler although you can still use Gaussian method as well).
6. Used solved Mesh Currents to solve for the desired circuit entity.

The above algorithm is very basic and useful for 2 × 2 and 3 × 3 size matrices. Generally as the number of loops or meshes increase and the size of the matrix exceeds 3 × 3, numerical methods are employed with the aid of computers to solve such circuit networks.

Generally speaking, Mesh Analysis is shorter than Nodal Analysis although not always preferred.

Example 1. *Consider Fig. with the following Parameters:*

$$V_1 = 15\ V$$
$$V_2 = 7\ V$$
$$R_1 = 2\ \Omega$$
$$R_2 = 20\ \Omega$$
$$R_3 = 10\ \Omega$$
$$R_4 = 5\ \Omega$$
$$R_5 = 2\ \Omega$$
$$R_6 = 2\ \Omega$$

Find current through R_3 using Mesh Analysis method.

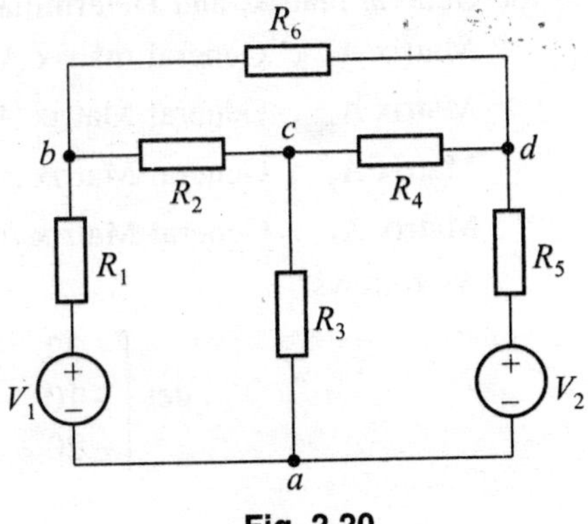

Fig. 2.20

Solution. We can see that there are three closed paths (loops) where we can apply KVL in, Loop 1, 2 and 3 as shown in figure above.

We can now apply KVL around the loops remembering passive convention when defining currents and voltages.

KVL around abca loop:

$$I_1 * R_1 + (I_1 - I_3) * R_2 + (I_1 - I_2) * R_3 = V_1$$

Therefore,

$$I_1(R_1 + R_2 + R_3) - I_2(R_3) - I_3(R_2) = V_1 \quad ...(1)$$

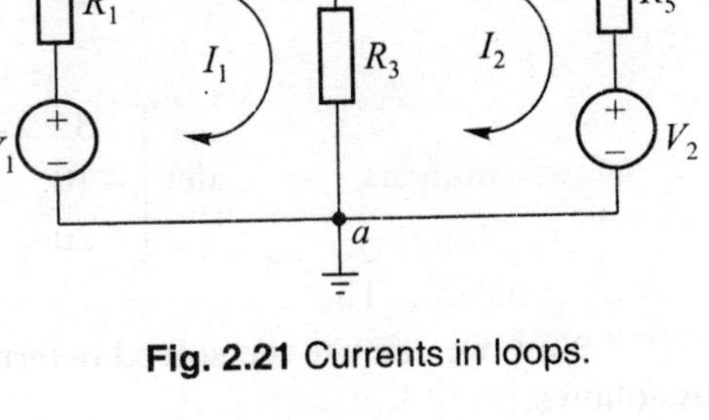

Fig. 2.21 Currents in loops.

KVL around abda loop:

$$(I_2 - I_1) * R_3 + (I_2 - I_3) * R_4 + I_2 * R_5 = V_2$$

Therefore,

$$-I_1(R_3) + I_2(R_3 + R_4 + R_5) - I_3(R_4) = V_2 \quad ...(2)$$

KVL around bcdb loop:

$$I_3 * R_6 + (I_3 - I_2) * R_4 + (I_3 - I_1) * R_2 = 0$$

Therefore,

$$-I_1(R_2) - I_2(R_4) + I_3(R_2 + R_4 + R_6) = 0 \quad ...(3)$$

Now we can create a matrix with the above equations as follows:

$$\begin{bmatrix} (R_1 + R_2 + R_3) & (-R_3) & (-R_2) \\ (-R_3) & (R_3 + R_4 + R_5) & (-R_4) \\ (-R_2) & (-R_4) & (R_2 + R_4 + R_6) \end{bmatrix} . \begin{bmatrix} I_1 \\ I_2 \\ I_3 \end{bmatrix} = \begin{bmatrix} V_1 \\ V_2 \\ 0 \end{bmatrix}$$

The following matrix is the above with values substituted:

$$A . \vec{X} = \vec{Y} \rightarrow \begin{bmatrix} 32 & -10 & -20 \\ -10 & 17 & -5 \\ -20 & -5 & 27 \end{bmatrix} . \begin{bmatrix} I_1 \\ I_2 \\ I_3 \end{bmatrix} = \begin{bmatrix} 15 \\ 7 \\ 0 \end{bmatrix}$$

Now that we have arranged equations (1), (2) and (3) into a matrix we need to get Determinants of the General matrix, and Determinants of alterations of the general matrix as follows:

Matrix A : General matrix A from KVL equations

Matrix A_1 : General Matrix A with column 1 substituted by $\vec{Y}$.

Matrix A_2 : General Matrix A with column 2 substituted by $\vec{Y}$.

Matrix A_3 : General Matrix A with column 3 substituted by $\vec{Y}$.

As follows:

$$\det \begin{bmatrix} 32 & -10 & -20 \\ -10 & 17 & -5 \\ -20 & -5 & 27 \end{bmatrix} = \det A$$

$$= 2388$$

$$\det \begin{bmatrix} 15 & -10 & -20 \\ 7 & 17 & -5 \\ 0 & -5 & 27 \end{bmatrix} = \det A_1$$

$$= 9100$$

$$\det \begin{bmatrix} 32 & 15 & -20 \\ -10 & 7 & -5 \\ -20 & 0 & 27 \end{bmatrix} = \det A_2$$

$$= 8798$$

$$\det \begin{bmatrix} 32 & -10 & 15 \\ -10 & 17 & 7 \\ -20 & -5 & 0 \end{bmatrix} = \det A_3$$

$$= 8370$$

Now we can use the solved determinants to arrive at solutions for Mesh Currents I_1, I_2 and I_3 as follows:

$$I_1 = \frac{\det A_1}{\det A} = 3.81 \text{ A}$$

$$I_2 = \frac{\det A_2}{\det A} = 3.68 \text{ A}$$

$$I_3 = \frac{\det A_3}{\det A} = 3.51 \text{ A}$$

Now we can solve for the current through R_3 as follows:

$$I_{R_3} = I_1 - I_2 = 0.126466 \text{ A}$$

The answer is as we expected it to be, flows in the direction of I_1.

Example 2. *Consider Figure shown below with the following parameters:*

$$V_1 = 9\ V$$
$$R_1 = 200\ k\Omega$$
$$R_2 = 20\ k\Omega$$
$$R_3 = 10\ k\Omega$$
$$R_4 = 5\ k\Omega$$
$$R_5 = 15\ k\Omega$$
$$R_6 = 1\ k\Omega$$

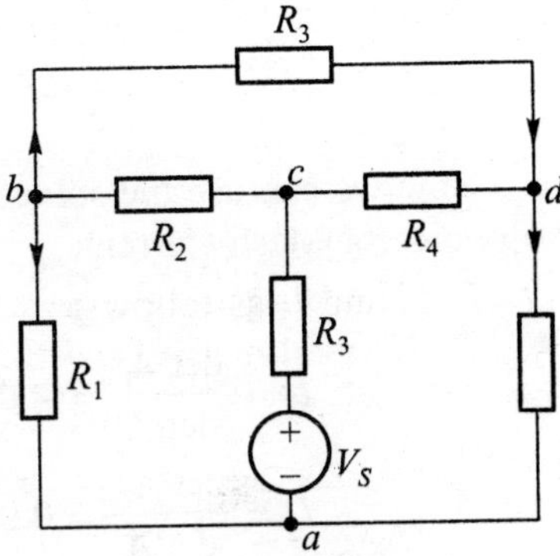

Fig. 2.22

Find current through R_3 using Mesh Analysis method.

KVL around abca loop:

$$I_1 * R_1 + (I_1 - I_3) * R_2 + (I_1 - I_2) * R_3 = -V_S$$

Therefore,

$$I_1 (R_1 + R_2 + R_3) - I_2 (R_3) - I_3 (R_2) = -V_S \qquad \text{...(1)}$$

KVL around abda loop:

$$(I_2 - I_1) * R_3 + (I_2 - I_3) * R_4 + I_2 * R_5 = V_S$$

Therefore,

$$-I_1 (R_3) + I_2 (R_3 + R_4 + R_5) - I_3 (R_4) = V_S \qquad \text{...(2)}$$

KVL around bcdb loop:

$$I_3 * R_6 + (I_3 - I_2) * R_4 + (I_3 - I_1) * R_2 = 0$$

Therefore,

$$-I_1 (R_2) - I_2 (R_4) + I_3 (R_2 + R_4 + R_6) = 0 \qquad \text{...(3)}$$

Now we can create a matrix with the above equations as follows:

$$\begin{bmatrix} (R_1 + R_2 + R_3) & (-R_3) & (-R_2) \\ (-R_3) & (R_3 + R_4 + R_5) & (-R_4) \\ (-R_2) & (-R_4) & (R_2 + R_4 + R_6) \end{bmatrix} . \begin{bmatrix} I_1 \\ I_2 \\ I_3 \end{bmatrix} = \begin{bmatrix} -V_s \\ V_s \\ 0 \end{bmatrix}$$

The following matrix is the above with values substituted:

$$A.\vec{X} = \vec{Y} \rightarrow \begin{bmatrix} 10220 & -10000 & -20 \\ -10000 & 30000 & -5000 \\ -20 & -5000 & 6020 \end{bmatrix} . \begin{bmatrix} I_1 \\ I_2 \\ I_3 \end{bmatrix} = \begin{bmatrix} -9 \\ 9 \\ 0 \end{bmatrix}$$

Now that we have arranged equations (1), (2) and (3) into a matrix we need to get Determinants of the General matrix, and Determinants of alterations of the general matrix as follows:

Solving determinants of:

Matrix A : General matrix A from KVL equations

Matrix A_1 : General Matrix A with column 1 substituted by $\vec{Y}$.

Matrix A_2 : General Matrix A with column 2 substituted by $\vec{Y}$.

Matrix A_3 : General Matrix A with column 3 substituted by $\vec{Y}$.

As follows:

$$\det A = 9.86 \times 10^{11}$$
$$\det A_1 = -857700000$$
$$\det A_2 = 11016000$$
$$\det A_3 = 6300000$$

Now we can use the solved determinants to arrive at solutions for Mesh Currents I_1; I_2 and I_3 as follows:

$$I_1 = \frac{\det A_1}{\det A} = -0.00086968 \text{ A}$$

$$I_2 = \frac{\det A_2}{\det A} = 0.00001117 \text{ A}$$

$$I_3 = \frac{\det A_3}{\det A} = 0.000006388 \text{ A}$$

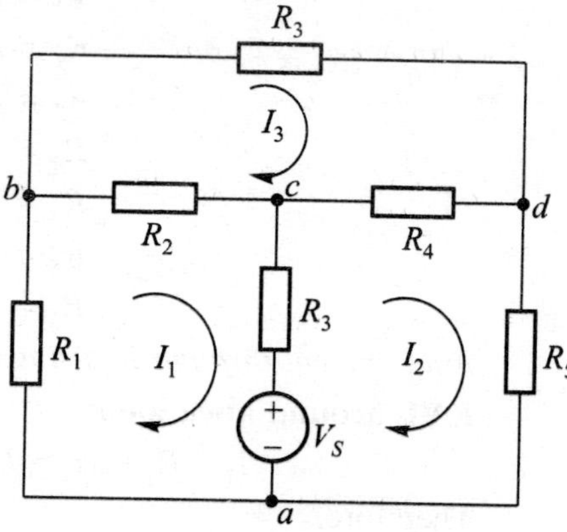

Fig. 2.23

Now we can solve for the current through R_3 as follows:

$$I_{R_3} = I_1 - I_2 = -0.881 \text{ mA}$$

The negative sign means that I_{R_3} is following in the direction of I_2.

Advantages of Nodal Analysis

Solves directly for node voltages

Current sources are easy

Voltage sources are either very easy or somewhat difficult

Works best for circuits with few nodes

Works for any circuit.

Advantages of Loop Analysis

Solves directly for some currents

Voltage sources are easy

Current sources are either very easy or somewhat difficult

Works best for circuits with few loops.

Disadvantages of Loop Analysis

Some currents must be computed from loop currents

Does not work with non-planar circuits

Choosing the super mesh may be difficult.

2.8 MORE ADVANCED APPLICATIONS

1. Voltage Dividers

The voltage divider is a very common and useful circuit configuration. Consider the circuit below, we add a current loop, and assume there is no current out at V_0,

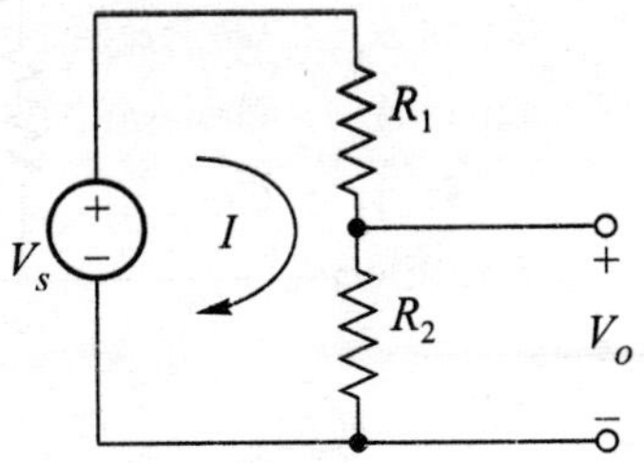

Fig. 2.24

First, sum the voltages about the loop,

$$\Sigma V = -V_5 + IR_1 + IR_2 = 0 \quad \therefore \quad I = \frac{V_5}{R_1 + R_2}$$

Next, find the output voltage, based on the current,

$$V_0 = IR_2 = \left(\frac{V_5}{R_1 + R_2}\right)R_2 = \boxed{V_5\left(\frac{R_2}{R_1 + R_2}\right)}$$

ASIDE: Variable resistors are often used as voltage dividers. As the wiper travels along the resistor the output voltage changes.

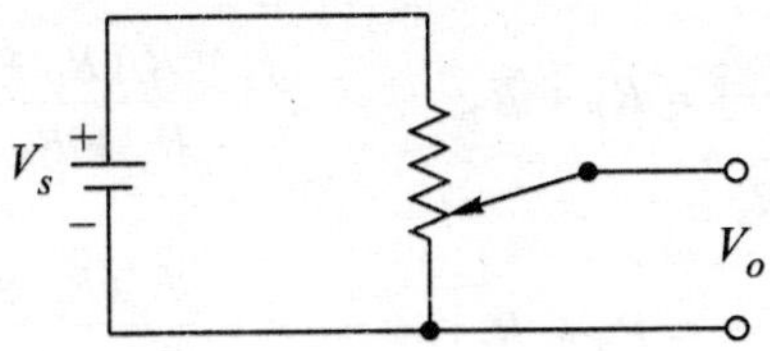

Fig. 2.25

2. The Wheatstone Bridge

The Wheatstone bridge is a very common engineering tool for magnifying and measuring signals. In this circuit a supply voltage V_s is used to power the circuit. Resistors R_1 and R_2 are generally equal, R_x is a resistance to be measured, and R_3 is a tuning resistor. An ammeter is shown in the center, and resistor R_3 is varied until the current in the center I_g is zero.

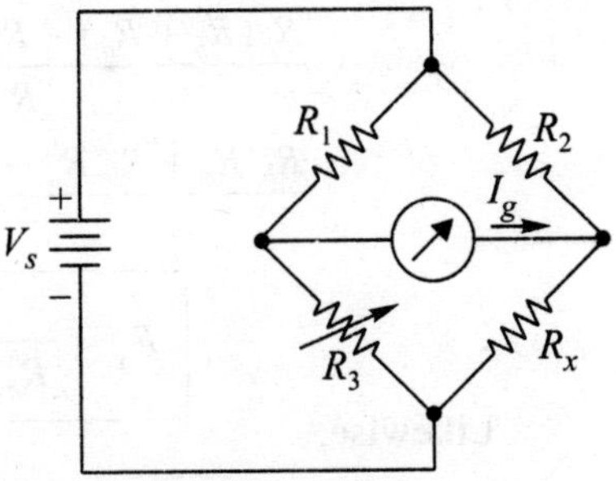

Fig. 2.26

For practice try to show the relationship below holds for a balanced bridge (*i.e.* $Ig = 0$)

$$R_x = \frac{R_2 \, R_3}{R_1}$$

3. Tee-to-Pi (Y to Delta) Conversion

It is fairly common to use a model of a circuit. This model can then be transformed or modified as required.

A very common model and conversion is the Tee to Pi conversion in electronics. A similar conversion is done for power circuits called delta to *y*.

We can find equivalent resistors considering that,

$$R_{ac} = \frac{1}{\dfrac{1}{R_b} + \dfrac{1}{R_c + R_a}} = R_1 + R_3 \qquad \therefore \qquad \frac{R_b(R_c + R_a)}{R_a + R_b + R_c} = R_1 + R_3 \qquad ...(1)$$

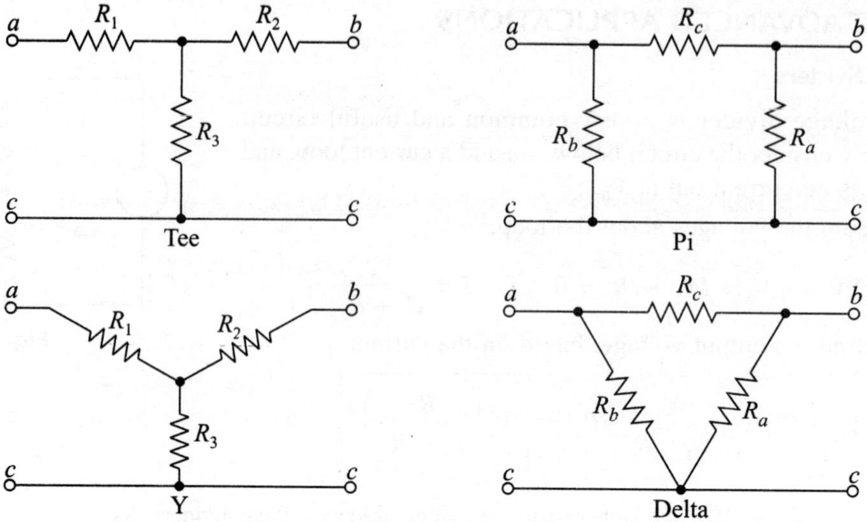

Fig. 2.27

$$R_{ab} = \frac{1}{\dfrac{1}{R_c} + \dfrac{1}{R_a + R_b}} = R_1 + R_2 \qquad \therefore \quad \frac{R_c(R_a + R_b)}{R_a + R_b + R_c} = R_2 + R_3 \qquad \text{...(2)}$$

$$R_{bc} = \frac{1}{\dfrac{1}{R_a} + \dfrac{1}{R_b + R_c}} = R_2 + R_3 \qquad \therefore \quad \frac{R_a(R_b + R_c)}{R_a + R_b + R_c} = R_2 + R_3 \qquad \text{...(3)}$$

To find R_1, (1) – (3) + (2)

$$\frac{R_b(R_c + R_a) - R_a(R_b + R_c) + R_c(R_a + R_b)}{R_a + R_b + R_c} = R_1 + R_3 - R_2 - R_3 + R_2 + R_3$$

$$\therefore \quad \frac{R_b\, R_c + R_b\, R_a - R_a\, R_b - R_a R_c + R_c\, R_a + R_c\, R_b}{R_a + R_b + R_c} = 2R_1$$

$$\therefore \quad \boxed{R_1 = \frac{R_b R_c}{R_a + R_b + R_c}}$$

Likewise,

$$\boxed{\begin{aligned} R_2 &= \frac{R_a R_c}{R_a + R_b + R_c} \\ R_3 &= \frac{R_a R_b}{R_a + R_b + R_c} \end{aligned}}$$

To find the equivalents the other way,

$$R_a + R_b + R_c = \frac{R_b R_c}{R_1} = \frac{R_a R_c}{R_2} = \frac{R_a R_b}{R_3}$$

$$\therefore \quad R_b = \frac{R_1}{R_2}\, R_a = \frac{R_3}{R_2}\, R_c$$

$$\therefore \quad R_a = \frac{R_2}{R_1} R_b = \frac{R_3}{R_1} R_c$$

$$\therefore \quad R_c = \frac{R_2}{R_3} R_b = \frac{R_1}{R_3} R_a$$

We can put these relationships into equations (1), to eliminate R_a and R_c

$$\frac{R_b\left(\frac{R_2}{R_3} R_b + \frac{R_2}{R_1} R_b\right)}{\frac{R_2}{R_1} R_b + R_b + \frac{R_2}{R_3} R_b} = R_1 + R_3$$

$$\frac{R_b\left(\frac{R_2\,R_1 + R_2\,R_3}{R_1\,R_3}\right)}{\frac{R_2 R_3 + R_1 R_3 + R_1 R_2}{R_1\,R_3}} = R_1 + R_3$$

$$\therefore \quad R_b = (R_1 + R_3)\left(\frac{R_2 R_3 + R_1 R_3 + R_1 R_2}{R_2\,R_1 + R_2 R_3}\right)$$

$$= \frac{R_2\,R_3 + R_1 R_3 + R_1 R_2}{R_2}$$

Likewise

$$R_a = \frac{R_2\,R_3 + R_1 R_3 + R_1 R_2}{R_1}$$

$$R_c = \frac{R_2 R_3 + R_1 R_3 + R_1 R_2}{R_3}.$$

SOLVED EXAMPLES

Example 1: *Find current I_4 which leaves the junction.*

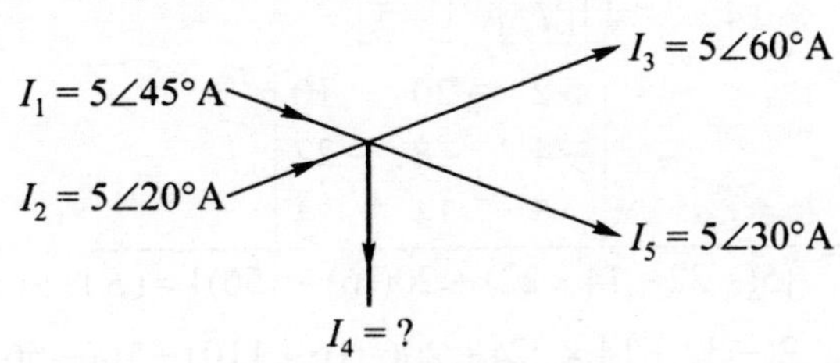

Fig. 2.28

Solution: Σ Currents enters junction = Σ currents leaves junction

$$I_1 + I_2 = I_3 + I_4 + I_5$$

$$5\angle 45° + 5\angle 20° = 5\angle 60° + I_4 + 5\angle 30°$$

$$\therefore \quad I_4 = 5\angle 45° + 5\angle 20° - 5\angle 60° - 5\angle 30°$$

$$I_4 = 5\,[\cos 45° + j \sin 45°] + 5[\cos 20° + j \sin 20°]$$
$$- 5[\cos 60° + j \sin 60°] - 5[\cos 30° + j \sin 30°]$$

$\therefore \quad I_4 = (3.54 + j3.54) + (4.7 + j1.71) - (2.5 + j4.33) - (4.37 + j2.5)$

$I_4 = 1.41 - j1.58 = 2.12\angle - 48.25°A.$

Example 2: *Calculate the branch currents in a bridge circuit shown below:*

$V_1 = 2\ V$
$R_1 = 5\ \Omega$
$R_3 = 20\ \Omega$
$V_2 = 4\ V$
$R_2 = 4\ \Omega$
$R_5 = 10\ \Omega$
$R_4 = 8\ \Omega$

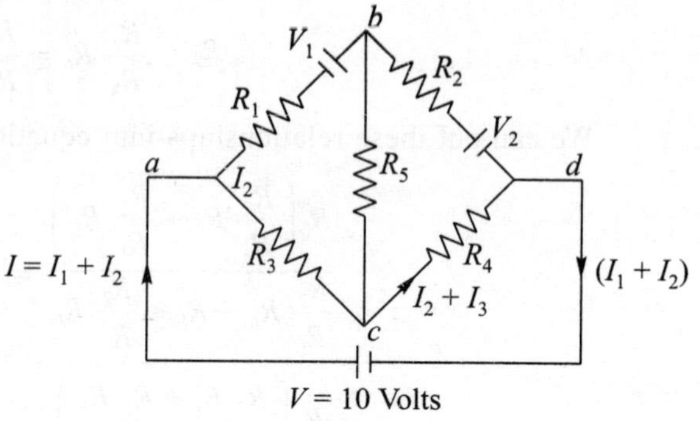

Fig. 2.29

Solution: Three variables *i.e.*, I_1, I_2, I_3 so, select any three loops then apply KVL each loop.

***(i)* Loop abcab**

Applied voltage in this loop = + 2 V

Apply KVL to this loop,

$$5I_1 - 20I_2 + 10I_3 = +2 \qquad ...(1)$$

***(ii)* Loop bdcb**

Applying KVL to this loop

$$(I_1 - I_3)R_2 - (I_2 - I_3)R_4 - I_3 R_5 = -4$$

$$4I_1 - 8I_2 - 22I_3 = -4 \qquad ...(2)$$

***(iii)* Applying KVL to this loop**

$$I_2R_3 + (I_2 + I_3)R_4$$

$$14I_2 + AI_3 = 5 \qquad ...(3)$$

Putting equations (1), (2), (3) in matrix form,

$$\begin{bmatrix} 5 & -20 & 10 \\ 4 & -8 & -22 \\ 0 & 14 & 4 \end{bmatrix} \begin{bmatrix} I_1 \\ I_2 \\ I_3 \end{bmatrix} = \begin{bmatrix} 2 \\ -4 \\ 5 \end{bmatrix}$$

$$\therefore \quad I_1 = \frac{\begin{bmatrix} 2 & -20 & 10 \\ -4 & -8 & -22 \\ 5 & 14 & 4 \end{bmatrix}}{\{5(-32 + 14 \times 22) + 20(16) + (56)\} = |R|}$$

$$I_1 = \frac{2(-32 + 14 \times 22) + 20(-16 + 110) + 10(-56 + 40)}{5(-32 + 308) + 320 + 560}$$

$$I_1 = 1.005 \text{ amperes}$$

$$I_2 = \frac{\begin{vmatrix} 5 & 2 & 16 \\ 4 & -4 & -22 \\ 0 & 5 & 4 \end{vmatrix}}{|R|} = \frac{5(-16 + 110) - 2(16) + 10(20)}{|R|}$$

$$I_2 = \frac{638}{2260} = 0.282$$

$$I_3 = \frac{\begin{bmatrix} 5 & -20 & 2 \\ 4 & -8 & -4 \\ 0 & 14 & 5 \end{bmatrix}}{|R|} = \frac{5(-40+56)+20(20)+2(56)}{2260}$$

$$I_3 = 0.262 \text{ amperes}$$

$$I = I_1 + I_2 = 1.005 + 0.282 = 1.287 \text{ amperes}$$

$$I_1 - I_3 = 0.743 \text{ amperes}$$

$$I_1 + I_3 = 0.544 \text{ amperes.}$$

Example 3: *Frame mesh equations for the circuit and determine the currents*

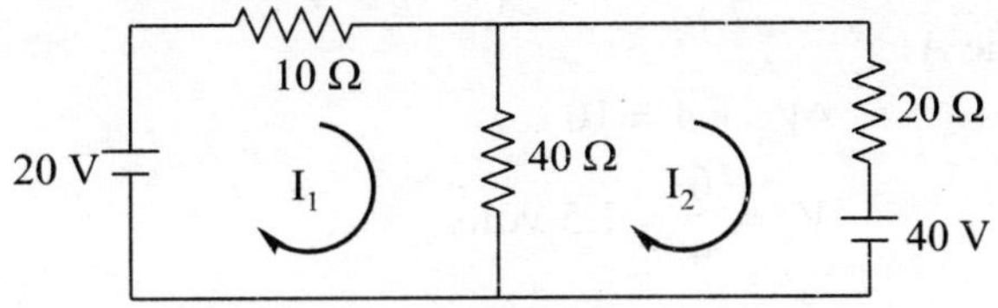

Fig. 2.30

Solution: For mesh 1: $20 - 10I_1 - 4(I_1 - I_2) = 0$

$$20 - 6I_1 + 4I_2 = 0$$

$$6I_1 - 4I_2 = 20$$

$$3I_1 - 2I_2 = 10 \quad \text{...(1)}$$

For mesh 2: $4(I_1 - I_2) - 20I_2 - 40 = 0$

$$4I_1 - 24I_2 - 40$$

$$2I_1 - 12I_2 - 20$$

$$I_1 - 6I_2 - 10 \quad \text{...(2)}$$

(1) × 3 ⇒ $9I_1 - 6I_2 = 30$

(2) ⇒ $I_1 - 6I_2 = 10$

(–) (+)

$$8I_1 = 20$$

$$I_1 = \frac{20}{8} = \frac{5}{2} = 2.5 \text{ Amp}$$

From (2) ⇒ $2.5 - 6I_2 = 10$

$$-6I_2 = 7.5$$

$$I_2 = \frac{-7.5}{6}$$

$$I_2 = -1.25 \text{ amp}$$

Negative sign here indicates that the actual current direction is opposite to the one assumed.

Example 4: *Determine the node voltage of the circuit by nodal analysis.*

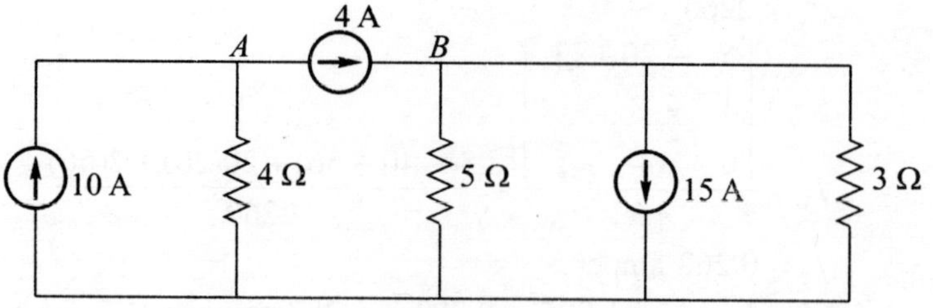

Fig. 2.31

Solution: Node A

$\therefore$ Σ currents entering node A = 10 Amp

and Σ currents leaving node $A = (4V_A + 4)$ amperes

Apply KCL at node A

$\therefore$
$$4V_A + 4 = 10$$
$$V_A = \frac{6}{4} = 1.5 \text{ volts}$$

Node B

Σ current entering node $B = 4$

and Σ currents leaving node B

$$= 5V_B + 15 + 3V_B = 8V_B + 15$$

Apply KCL at node B

$\therefore$
$$8V_B + 15 = 4$$
$$V_B = -1.4 \text{ volts}$$

Example 5: *Write the node voltage equations of the given circuit*

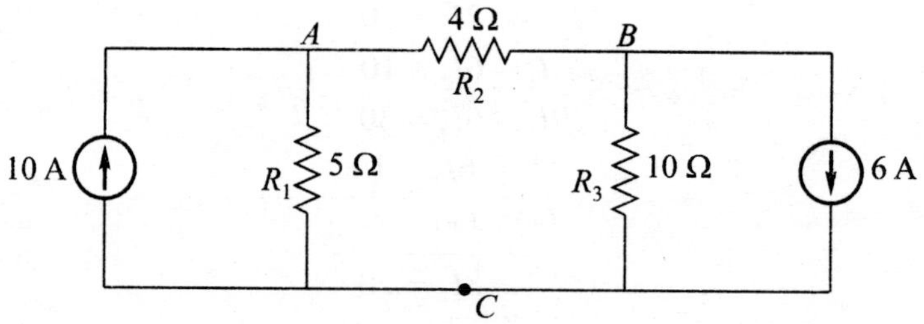

Fig. 2.32

Solution: Here $V_{AC} = V_A$, $V_{BC} = V_B$

Node A

$\therefore$ Σ currents entering node $A = \left[10 + \frac{V_B}{4}\right]$

$$= [10 + 0.25\ V_B] \text{ amp}$$

and Σ currents leaving node A

$$= \frac{V_A}{R_1} + \frac{V_A}{R_2} = (0.2 + 0.25)\ V_A$$
$$= 0.45\ V_A \text{ amperes}$$

Apply KCL at node A

$$\therefore \quad 0.45\ V_A = 10 + 0.25\ V_B$$

$$0.45\ V_A - 0.25\ V_B = 10 \quad ...(1)$$

Node B

$$\Sigma \text{ currents entering node } B = \frac{V_A}{R_2} = \frac{V_A}{R_4} = 0.25\ V_A \text{ amp}$$

and Σ currents leaving node B

$$= \left(\frac{V_B}{R_2} + \frac{V_B}{R_3} + 6\right) \text{ amp} = V_B\left[\frac{1}{4} + \frac{1}{10}\right] + 6$$

$$= V_B\,[0.25 + 0.1] + 6 = (0.35\ V_B + 6) \text{ amp}$$

Apply KCL at node A

$$0.35\ V_B + 6 = 0.25\ V_A$$

$$-0.25\ V_A + 0.35\ V_B = -6 \quad ...(2)$$

Put the equation in matrix form.

$$\begin{bmatrix} 0.45 & -0.25 \\ -0.25 & 0.35 \end{bmatrix}\begin{bmatrix} V_A \\ V_B \end{bmatrix} = \begin{bmatrix} 10 \\ -6 \end{bmatrix}$$

$$\therefore \quad V_A = \frac{\begin{vmatrix} 10 & -0.25 \\ -6 & 0.35 \end{vmatrix}}{\begin{vmatrix} 0.45 & -0.25 \\ -0.25 & 0.35 \end{vmatrix}} = \frac{3.5 - 15}{0.1575 - 0.0625}$$

$$V_A = \frac{2}{0.095} = 21.05 \text{ Volts}$$

$$V_B = \frac{\begin{vmatrix} 0.45 & 10 \\ -0.25 & -6 \end{vmatrix}}{\begin{vmatrix} 0.45 & -0.25 \\ -0.25 & 0.35 \end{vmatrix}} = \frac{-2.7 + 2.5}{0.095} = \frac{-0.2}{0.095}$$

$$V_B = -2.105 \text{ Volts}$$

$$\therefore \quad I_1 = \frac{V_A}{R_1} = \frac{21.05}{5} = 4.21 \text{ amp}$$

$$I_2 = \frac{V_A - V_B}{R_2} = \frac{21.05 + 2.105}{4} = \frac{23.155}{4} = 5.78 \text{ amp}$$

$$I_3 = \frac{V_B}{R_3} = \frac{-2.105}{10} = -0.2105 \text{ amp}$$

$I = 6 - 0.2105 = 5.78$ amp enter node B.

Example 6: *For the circuit given below, find current through R_2 by the mesh method of mesh current?*

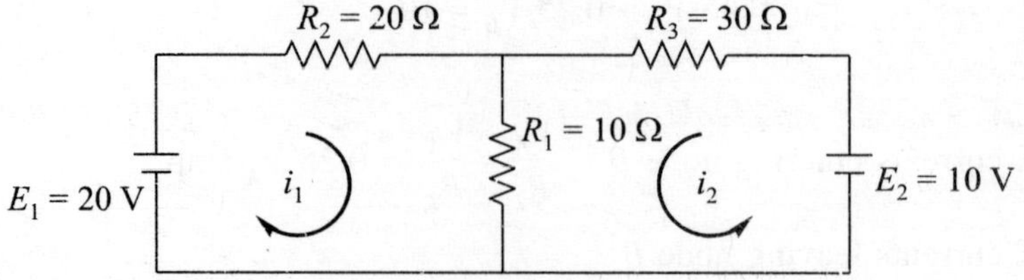

Fig. 2.33

Solution: For Mesh 1,

$$E_1 = I_1(20 + 10) + I_2(10)$$

$$20 = 30I_1 + 10I_2 \qquad ...(1)$$

For Mesh 2,

$$E_2 = 10I_1 + I_2(30 + 10)$$

$$10 = 10I_1 + 40I_2 \qquad ...(2)$$

(1) × ⇒ $\quad 80 = 120I_1 + 40I_2$

(2) ⇒ $\quad 10 = 10I_1 + 40I_2$

$\quad (-) \quad (-) \quad (-)$

$$70 = 110I_1$$

$$I_1 = \frac{70}{110} = \frac{7}{11} \text{ Amp.}$$

Example 7: *Find the voltage V_{ab} between points a and b in the following circuit by the method of node – voltage analysis.*

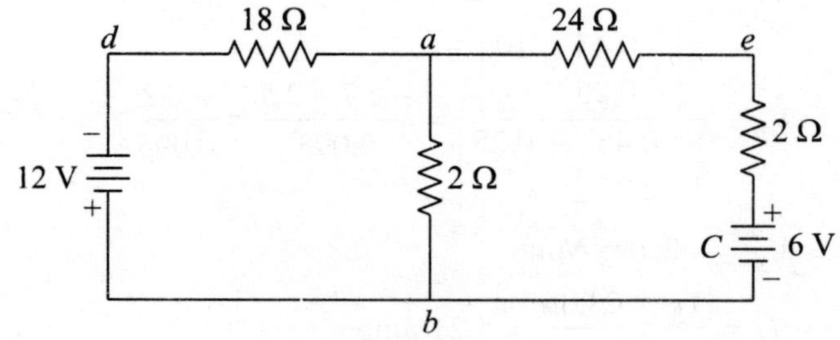

Fig. 2.34

Solution:

$$\frac{V_{ab} - (-12)}{18} + \frac{V_{ab}}{2} + \frac{V_{ab} - 6}{24 + 2} = 0$$

$$\frac{V_{ab} + 12}{18} + \frac{V_{ab}}{2} + \frac{V_{ab} - 6}{26} = 0$$

$$\frac{13V_{ab} + 156 + 117\,V_{ab} + 9V_{ab} - 54}{234} = 0$$

$$\frac{139\,V_{ab} + 102}{234} = 0$$

$$V_{ab} = \frac{1\Omega}{139} = 0.73\text{V.}$$

Example 8: *Determine the voltages of nodes 1 and 2 with respect to the reference of given figure below.*

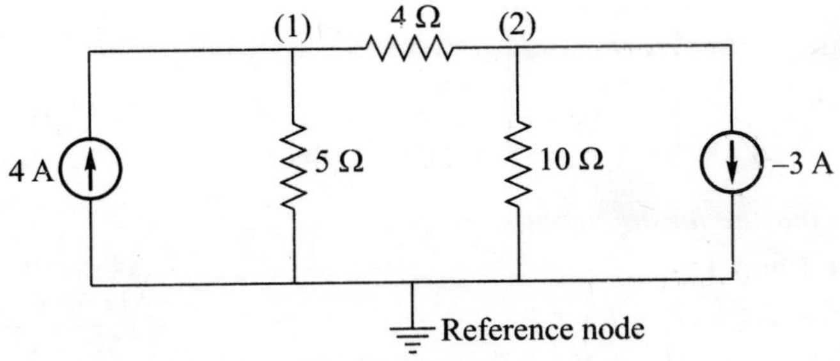

Fig. 2.35

Solution: Two principal nodes (1) and (2)

Node 1

$\therefore$ Σ currents enter node 1 $= \left[4 + \frac{V_2}{4}\right]$ amperes

$\therefore$ Σ currents leaving the node

$$= \frac{V_1}{5} + \frac{V_1}{4} = \frac{9\,V_1}{20} \text{ amperes}$$

Apply Kirchoff's current law at node 1

$$\therefore \qquad \frac{9}{20}\,V_1 = \frac{1}{4}\,V_2 + 4$$

$$\frac{9}{20}\,V_1 - \frac{1}{4}\,V_2 = 4 \qquad \text{...(1)}$$

Node 2

$\therefore$ Σ currents enter node $= \left[\frac{V_1}{4} + 3\right]$ A

$\therefore$ Σ currents leaving the node

$$\frac{V_2}{4} + \frac{V_2}{10} = V_2[0.25 + 0.1] = 0.35\,V_2 \text{ amperes}$$

Apply KCL at the node 2

i.e. Σ currents enter node = Σ currents leave node

$$\therefore \qquad 0.35\,V_2 = 0.25V_1 + 3$$

$$-\,0.25V_1 + 0.35V_2 = 3 \qquad \text{...(2)}$$

Arrange the equations in matrix form

$$\begin{bmatrix} \frac{9}{20} & -0.25 \\ -0.25 & 0.35 \end{bmatrix} \begin{bmatrix} V_1 \\ V_2 \end{bmatrix} = \begin{bmatrix} 4 \\ 3 \end{bmatrix}$$

Solving this, we get

$$V_1 = \frac{\begin{vmatrix} 4 & -0.25 \\ 3 & 0.35 \end{vmatrix}}{\frac{9 \times 0.35}{20} + (0.25)^2} = \frac{1.40 + 0.75}{0.1575 + 0.0625}$$

$$V_1 = \frac{2.15}{0.2200} = \frac{215}{22} = 9.77 \text{ volts}$$

$$V_2 = \frac{\begin{vmatrix} \frac{9}{20} & 4 \\ -0.25 & 3 \end{vmatrix}}{0.22} = 10.68 \text{ volts.}$$

Example 9: *Find the mesh currents.*

Solution: For Loop 1

$$= (I_1 - I_2) + (I_1 - I_3) \times 2$$
$$= (3I_1 - I_2 - 2I_3) \text{ volts}$$

Now, using KVL to this loop

$$3I_1 - I_2 - 2I_3 = 8 \qquad ...(1)$$

For Loop 2

Applying KVL to this loop

$$-I_1 + 9I_2 - 5I_3 = 0 \qquad ...(2)$$

For Loop 3

Applying KVL to this loop

$$-2I_1 - 5I_2 + 12I_3 = 4 \qquad ...(3)$$

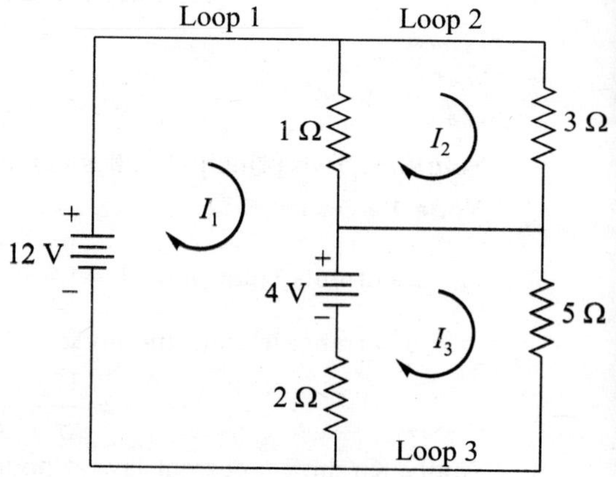

Fig. 2.36

Using (1), (2), (3)

$$\begin{bmatrix} 3 & -1 & -2 \\ -1 & 9 & -5 \\ -2 & -5 & 12 \end{bmatrix} \begin{bmatrix} I_1 \\ I_2 \\ I_3 \end{bmatrix} = \begin{bmatrix} 8 \\ 0 \\ 4 \end{bmatrix}$$

$$I_1 = \frac{\begin{bmatrix} 8 & -1 & -2 \\ 0 & 9 & -5 \\ 4 & -5 & 12 \end{bmatrix}}{|R|}, \qquad |R| = \begin{vmatrix} 3 & -1 & -2 \\ -1 & 9 & -5 \\ -2 & -5 & 12 \end{vmatrix}$$

$$I_1 = \frac{8(108 - 25) - 0 + 4(5 + 18)}{181}$$

$$= 3(108 - 25) + 1(-12 - 12) - (5 + 18) = 181$$

$$I_1 = \frac{756}{181} = 4.176 \text{ amp}$$

and

$$I_2 = \frac{\begin{vmatrix} 3 & 8 & -2 \\ -1 & 0 & -5 \\ -2 & 4 & 12 \end{vmatrix}}{|R|} = \frac{3(20) - 8(-12 - 10) - 2(-4)}{181}$$

$I_2 = 1.35$ amp

and $$I_3 = \frac{\begin{vmatrix} 3 & -1 & 8 \\ -1 & 9 & 0 \\ -2 & -5 & 4 \end{vmatrix}}{|R|} = \frac{3(36) + 1(-4) + 8(5+18)}{|8|}$$

$I_3 = 1.59$ amp.

Example 10: *Find the voltage at node N in the network.*

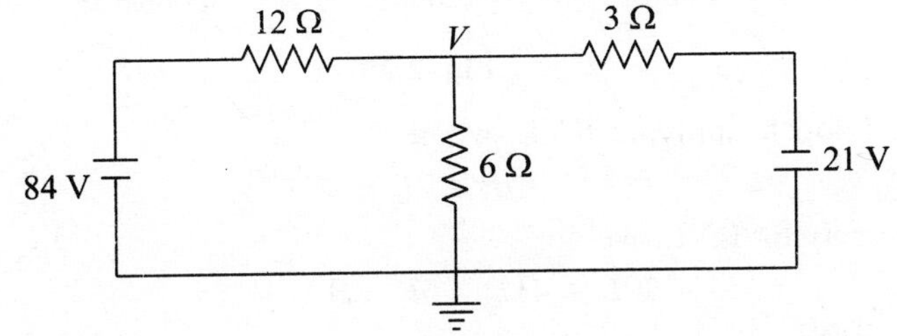

Fig. 2.37

Solution: The nodal equation is

$$\frac{V+74}{12} + \frac{V-0}{6} + \frac{V+21}{3} = 0$$

$$V\left[\frac{7}{12}\right] = -14$$

$$V = -24\text{V}$$

Negative sign indicates that R_3 is negative.

Example 11: *Find the potential difference V_{12} of the circuit shown below.*

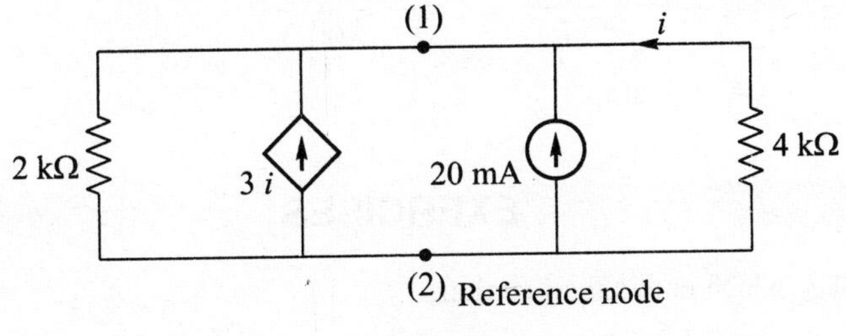

Fig. 2.38

Solution: Apply KCL at node (1)

$$\frac{V_{12}\left[1+\frac{1}{2}\right]}{2000} = 20 \times 10^{-3} + 3i$$

$$\frac{V_{12}}{2000}\left[1+\frac{1}{2}\right] = 20 \times 10^{-3} + 3\left[\frac{-V_{12}}{4000}\right]$$

$$V_{12} = 20 \times 10^{-3} \times \frac{4000}{6}$$

$$V_{12} = 13.3 \text{ volts.}$$

Example 12: *Find the loop current in the following circuit.*

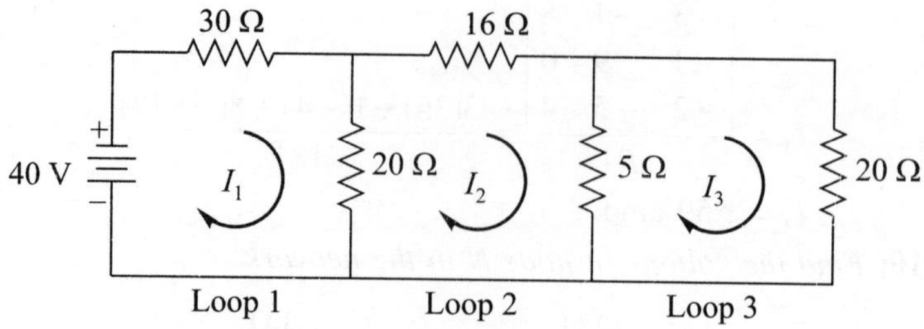

Fig. 2.39

Solution: For Loop 1, applying KVL, we get

$$5I_1 - 2I_2 = 4 \quad ...(1)$$

Foot Loop 2, applying KVL, we get

$$-20I_2 + 41I_2 - 5I_3 = 0 \quad ...(2)$$

For Loop 3, applying KVL, we get

$$-I_2 + 5I_3 = 0 \quad ...(3)$$

Using (1), (2), (3)

$$\begin{bmatrix} 5 & --2 & 0 \\ -20 & 41 & -5 \\ 0 & -1 & 5 \end{bmatrix} \begin{bmatrix} I_1 \\ I_2 \\ I_3 \end{bmatrix} = \begin{bmatrix} 4 \\ 0 \\ 0 \end{bmatrix}$$

$$|R| = 5[205 - 5] + 2[-100] = 800$$

$$I_1 = \frac{\begin{vmatrix} 4 & -2 & 0 \\ 0 & 41 & -5 \\ 0 & -1 & 5 \end{vmatrix}}{|R|} = \frac{4[205 - 5]}{800}$$

$$I_1 = 1 \text{ amp.}$$

EXERCISES

1. Find the current I_3 which enters the junction.

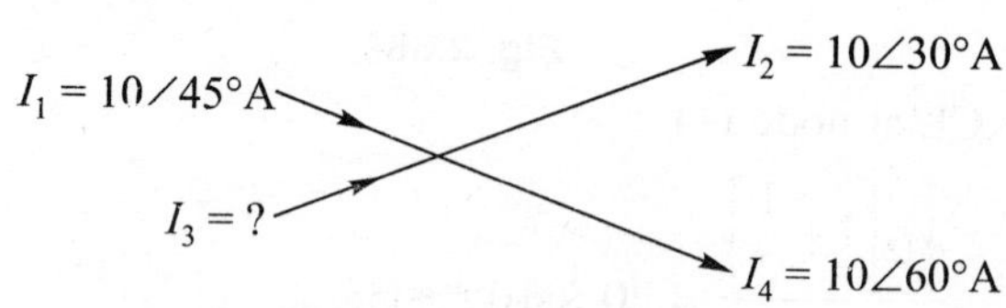

Fig. 2.40

2. Determine the branch currents and equivalent resistance of the bridge circuit at the terminals '*a*' and '*b*'.

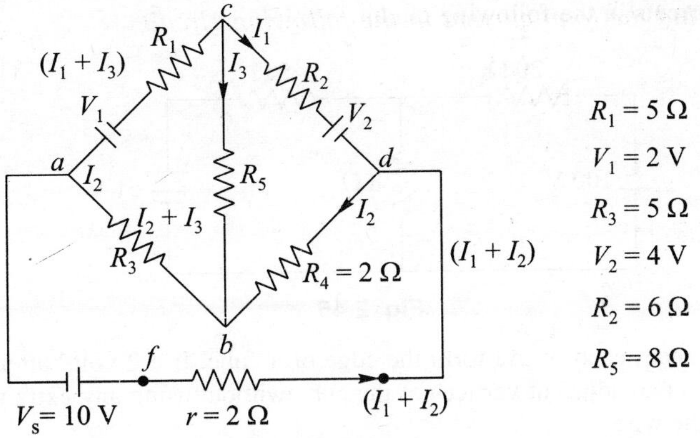

Fig. 2.41

3. Find the currents through the voltage source.

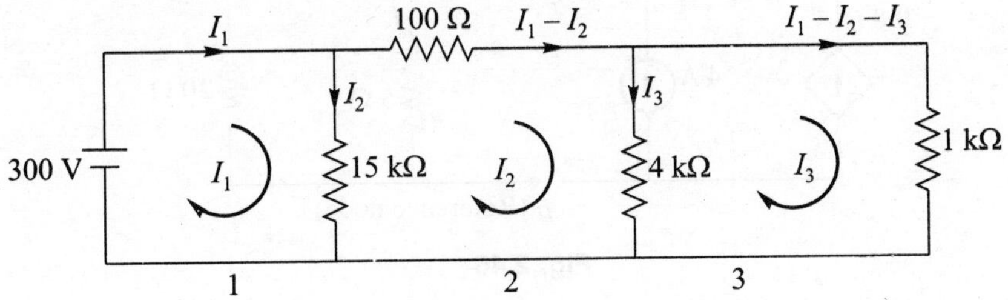

Fig. 2.42

4. Determine the node voltage of the circuit by nodal analysis.

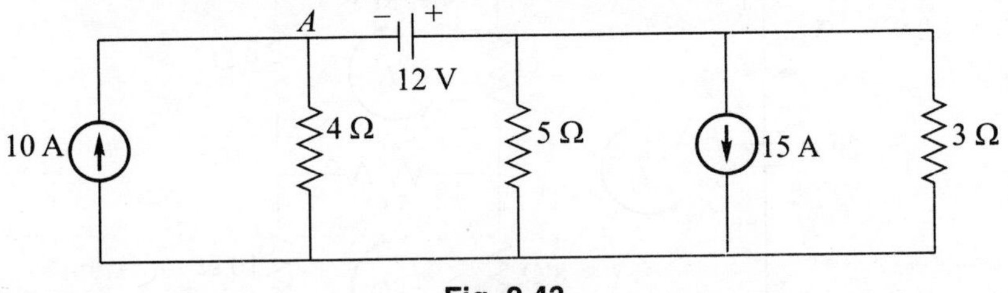

Fig. 2.43

5. Write node voltage equations of the given circuit.

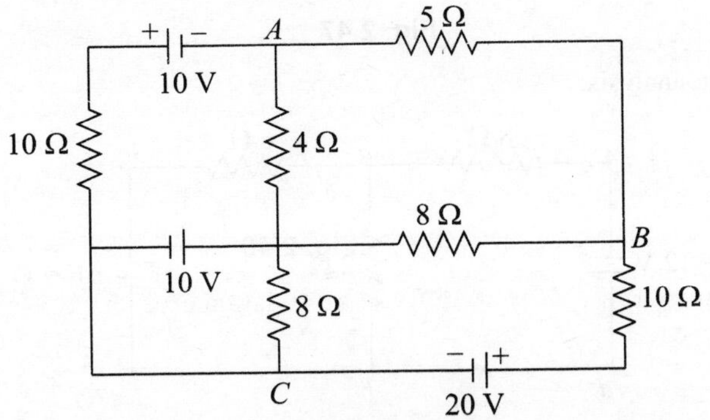

Fig. 2.44

6. Find the branch currents in the following circuit.

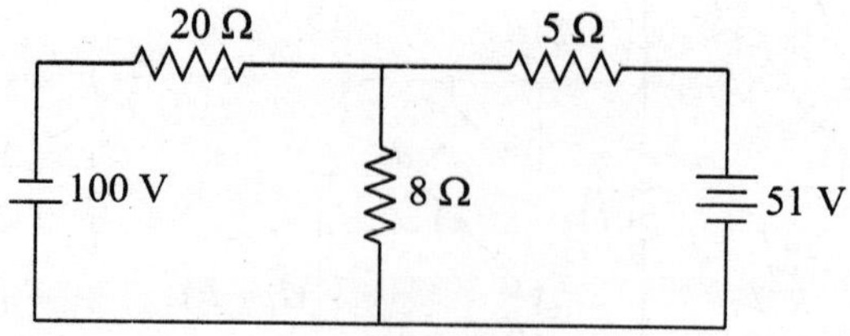

Fig. 2.45

7. Twelve equal resistances each of 2Ω form the edge of a cube. If a 2 volt battery of internal resistance 0.1Ω is connected to two adjacent vertices of the cube without using any extra resistance, calculate the current carried by the wires.

8. Find V_{ab} of the network shown below.

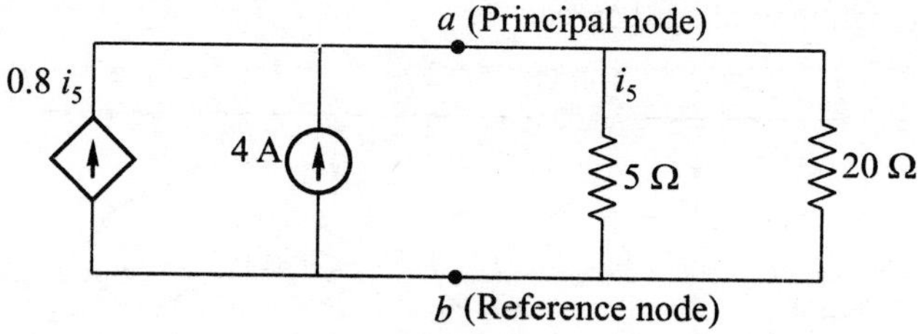

Fig. 2.46

9. Determine mesh currents.

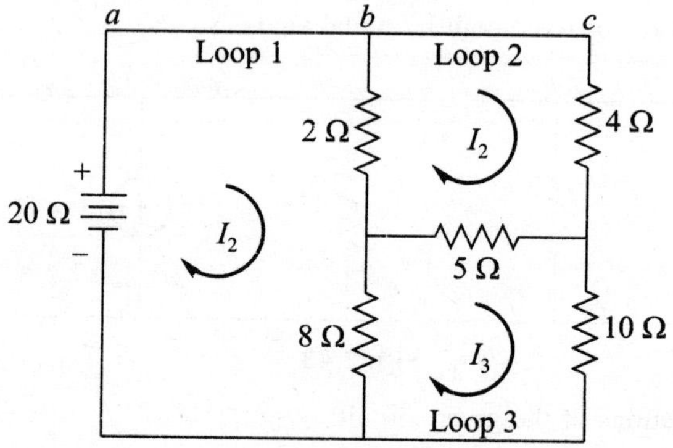

Fig. 2.47

10. Apply Node – voltage analysis.

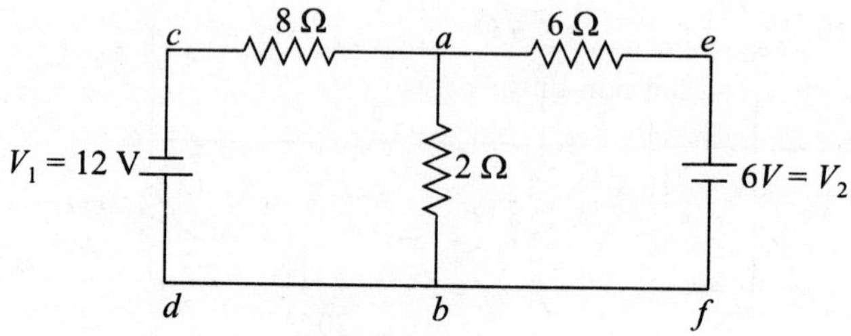

Fig. 2.48

11. Find the current through 20Ω resistance.

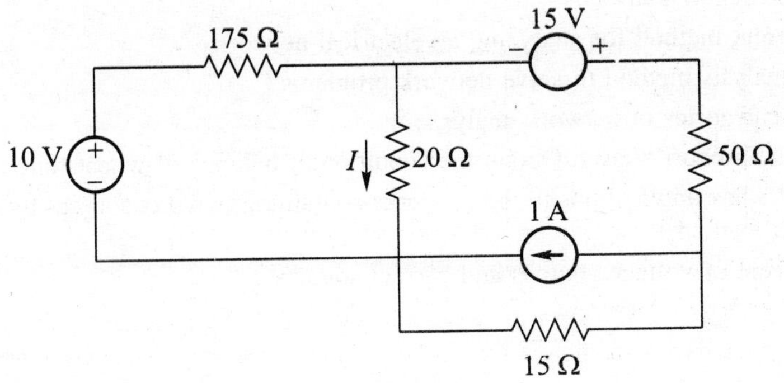

Fig. 2.49

12. Find mesh current for the following wheatstone bridge.

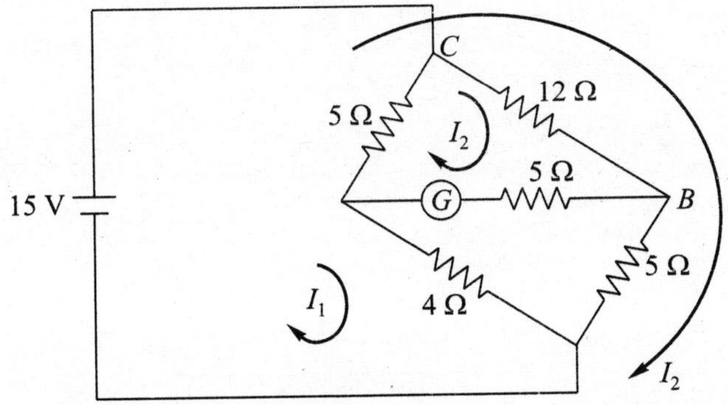

Fig. 2.50

QUESTIONS

A. Short answer type:

1. What is network? Define node, junction, branch, loop, mesh and tree.
2. Explain active and passive elements.
3. State Kirchoff's voltage law.
4. State Kirchoff's current law.
5. Distinguish between ideal and practical current sources.
6. State and Explain Kirchoff's laws.
7. What is an ideal voltage source? What are different types of voltage sources?
8. Explain lumped, distributed, linear and non-linear networks.
9. Write a short note on mesh analysis.
10. Write a short note on nodal analysis.

B. Long answer type:

1. State and prove Kirchoff's laws.
2. Explain loop current method for analyzing an electrical network.
3. Explain Nodal analysis method to solve network problems.
4. Explain various categories of network analysis.
5. State and explain Kirchoff's laws. Obtain the condition of balance of wheatstone bridge.
6. Explain Kirchoff's law applications to the networks containing two d.c. sources by loop current method and mesh current method.
7. Analyze the concept of voltage sources and current sources.

*C*HAPTER **3**

Network Theorems (A.C. and D.C.)

3.1 INTRODUCTION

Solving for currents and voltages in multi-loop electric circuits can be quite complicated, particularly for AC circuits. The voltage law and current law always apply, but using them may lead to long systems of equations. Some theorems provide the shortcuts and by applying them the complicated networks can be easily solved. These theorems are called the network theorems. These theorems are applicable for both A.C. and D.C. networks.

In AC networks, the impedances, voltages and currents are treated as phasors. Therefore, it is necessary to know the phase angle of each source and the amplitude of the voltage or current generated by the source.

In this section we shall discuss the following theorems:

1. Superposition Theorem
2. Thevenin's Theorem
3. Norton's Theorem
4. Maximum Power Transfer Theorem
5. Millman's Theorem
6. Reciprocity Theorem

3.2 SUPERPOSITION THEOREM

The superposition theorem for electrical circuits states that the total current in any branch of a bilateral linear circuit equals the algebraic sum of the currents produced by each source acting separately throughout the circuit.

Statement: In any linear network containing impedances and energy sources, the current flowing in any element is the vector sum of the currents that are separately caused to flow in that element by each energy source.

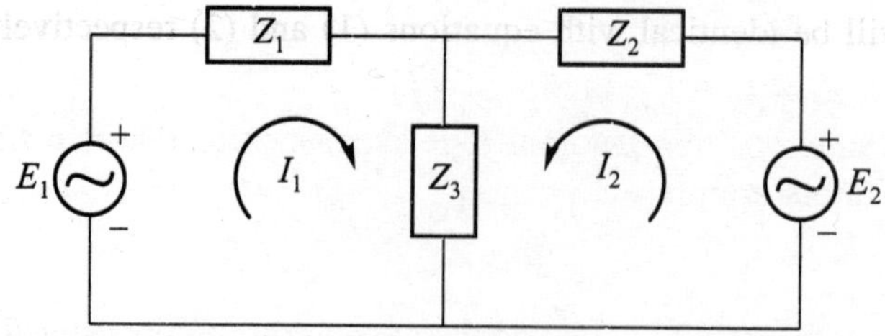

Fig. 3.1

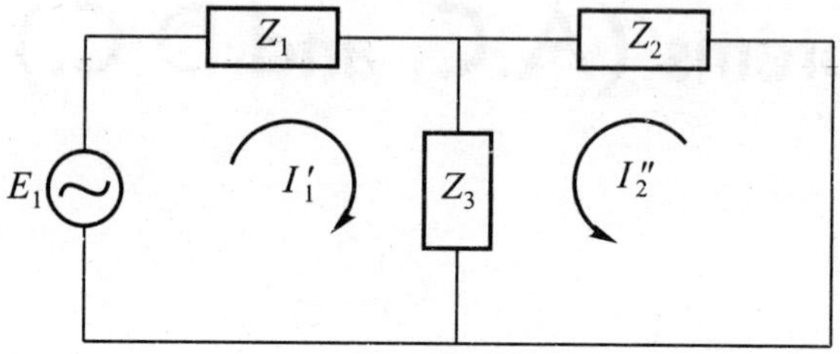

Fig. 3.2

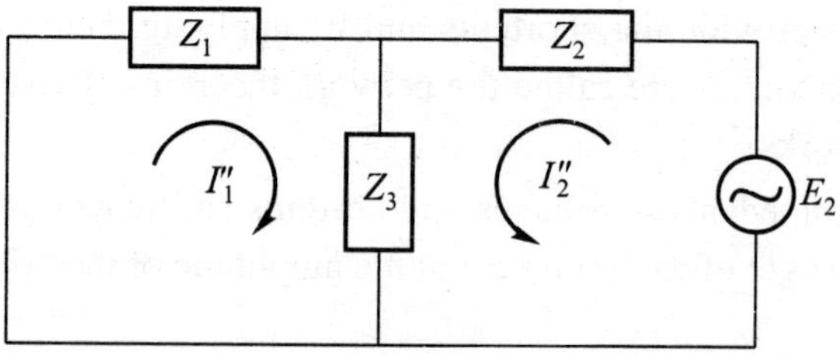

Fig. 3.3

Proof: Consider the two voltage sources E_1 and E_2 and three impedances Z_1, Z_2 and Z_3 as shown in Fig. (3.1). Let I_1 and I_2 be the two mesh currents then the mesh equations can be written as

$$E_1 = (Z_1 + Z_3)\, I_1 + Z_3 I_2 \quad \text{...(1)}$$

$$E_2 = (Z_3 I_1) + (Z_2 + Z_3)\, I_2 \quad \text{...(2)}$$

When E_1 is considered to actalone,

From Fig. (3.2), the mesh equations can be written as

$$E_1 = (Z_1 + Z_3) I'_1 + Z_3 I'_2 \quad \text{...(3)}$$

$$0 = Z_3 I'_1 + (Z_2 + Z_3) I'_2 \quad \text{...(4)}$$

When E_2 is considered alone,

From Fig. (3.3), the mesh equations can be written as

$$0 = (Z_1 + Z_3) I''_1 + Z_3 I''_2 \quad \text{...(5)}$$

$$E_2 = Z_3 I''_1 + (Z_2 + Z_3) I''_2 \quad \text{...(6)}$$

Adding equations (3) and (5), we get

$$E_1 = (Z_1 + Z_3)\, (I'_1 + I''_1) + Z_3 (I'_2 + I''_2) \quad \text{...(7)}$$

Similarly, adding equations (4) and (6), we get

$$E_2 = Z_3 (I'_1 + I''_1) + (Z_2 + Z_3)\, (I'_2 + I''_2) \quad \text{...(8)}$$

Equations (7) and (8) will be identical with equations (1) and (2) respectively, if

$$I_1 = I_1' + I_1'' \quad \text{and} \quad I_2 = I_2' + I_2''$$

This proves the truth of superposition theorem. The superposition theorem simplifies network calculations when several generators are present.

Example

To apply the superposition theorem to calculate the current through resistor R_1 in the two loop circuit shown, the individual current supplied by each battery is calculated with the other battery replaced by a short circuit.

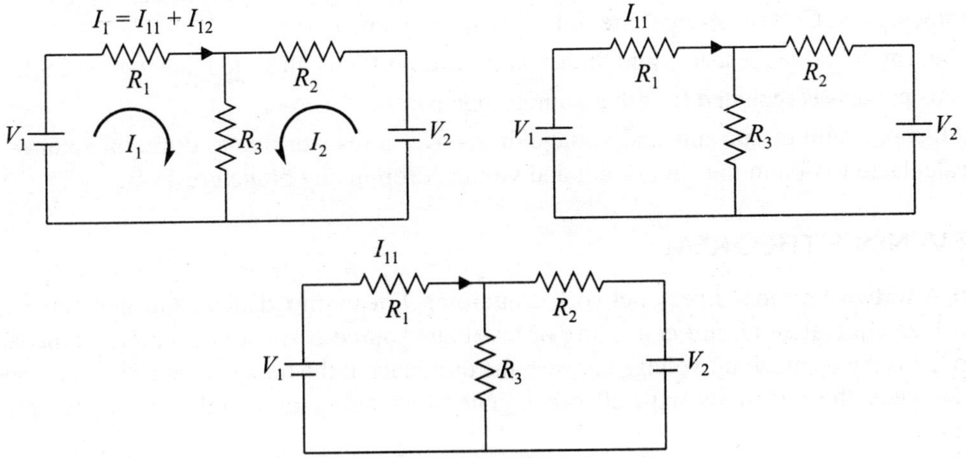

Fig. 3.4

$$I_1 = I_{11} + I_{12}$$

$$I_1 = \frac{V_1}{R_1 + R_2 \parallel R_3} \qquad I_{12} = \frac{-V_d}{R_1}$$

$$V_d = \frac{R_1 \parallel R_3 V_2}{R_2 + R_1 \parallel R_3}$$

$R_1 \parallel R_3$ means the parallel resistance of R_1 and R_3. The calculated currents are

$$I_{11} = \frac{V_1}{R_1 + R_2 \parallel R_3}$$

$$I_1 = -[V_d / R_1]$$

with a resultant current in R_1 of $I_1 = I_{11} + I_{12}$

Hence the superposition theorem is proved.

Advantages

1. This theorem permits the solution of the networks without setting up a large number of simultaneous equations because at a time one generator is used.
2. If the new generators are introduced into the system, it is not necessary to solve the network from beginning provided the internal impedances of the generators are zero.
3. If voltages of different frequencies are introduced, this theorem permits a solution to be obtained for each individual frequency.

In case of A.C. network, according to superposition theorem, "In any network containing more than one voltage or current source, the current through any branch is the phasor sum of the currents due to each sources acting independently."

Thus, while analyzing A.C. networking by superposition theorem, one source is considered at a time and amplitude and phase angle of current due to this single source is calculated through the desired branch. This process is repeated for all other sources one by one and then finally phasor sum of the currents through the desired branch is found to get the actual current.

Application Procedure

1. Consider only one source at a time. Replace all other voltage sources by their internal impedances. Current sources are to be replaced by open circuits.
2. Current in various resistors and their voltage drop is then calculated due to this single source.
3. This process is repeated for other sources one by one.
4. Algebraic sum of currents and voltage drops over a resistor due to different sources is then calculated to obtain the net current and voltage drop in any branch/resistor.

3.3 THEVENIN'S THEOREM

Statement: Any two terminal linear network containing linear impedances and generators can be replaced with an equivalent circuit consisting of a voltage source E^1 in series with an impedance Z^1. The value of E^1 is the open circuit voltage between the terminals of the network, and Z^1 is the impedance measured between the terminals with all other generators being removed (but not their internal impedances).

Proof:

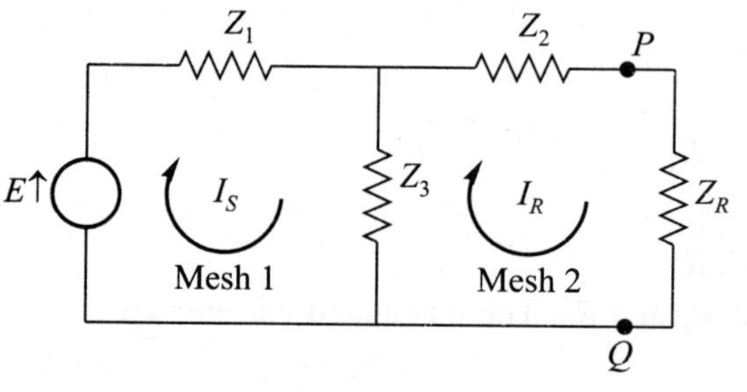

Fig. 3.5

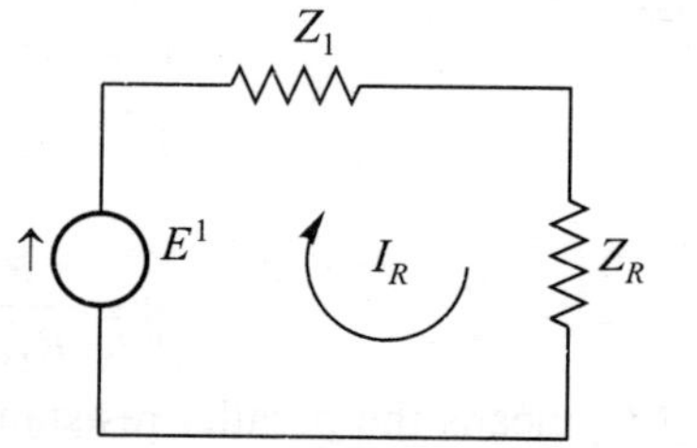

Fig. 3.6

Consider a two terminal network, containing one active mesh and other passive mesh and also assume that load impedance Z_R is appearing between two terminals.

In Fig. 3.5, I_S and I_R represent the loop currents flowing in active and passive network respectively. Our aim is to calculate E^1 and Z^1 from Fig. 3.5 and then to show that it is equivalent to a circuit shown in Fig. 3.6.

Applying the voltage law equations in mesh 1 and mesh 2 respectively, we have

$$Z_1 I_S + Z_3 I_S - Z_3 I_R = E \quad ...(1)$$

and

$$(Z_2 + Z_3 + Z_R)\, I_R - Z_3 I_S = 0 \quad ...(2)$$

From equation (2), we have

$$I_S = \frac{[Z_2 + Z_3 + Z_R]I_R}{Z_3}$$

Substituting this value of I_1 in eqn. (1), we get

$$E = \frac{[Z_2 + Z_3 + Z_R]I_R}{Z_3}(Z_1 + Z_3) - Z_3 I_R$$

$$I_R = \frac{E}{\frac{[Z_2 + Z_3 + Z_R][Z_1 + Z_3]}{Z_3} - Z_3}$$

$$I_R = \frac{EZ_3}{Z_2[Z_1 + Z_3] + Z_1Z_3 + Z_3^2 + Z_R(Z_1 + Z_3) - Z_3^2}$$

$$I_R = \frac{EZ_3}{Z_2[Z_1 + Z_3] + Z_1Z_3 + Z_R(Z_1 + Z_3)}$$

$$I_R = \frac{\frac{EZ_3}{Z_1 + Z_3}}{Z_2 + \frac{Z_1Z_3}{Z_1 + Z_3} + Z_2} \quad \text{...(3)}$$

Again from Fig. 3.5, the open circuit voltage at terminals [*P, Q*] *i.e.*, when load is disconnected, is given by

$$E^1 = \frac{EZ_3}{Z_1 + Z_3} = E\left[\frac{Z_3}{Z_1 + Z_3}\right] \quad \text{...(4)}$$

The impedance between terminals *P*, *Q* after disconnecting the load Z_R in Fig. 3.5 is given by

$$Z^1 = Z_2 + \frac{Z_1Z_3}{Z_1 + Z_3} \quad \text{...(5)}$$

Substituting the values of E^1 and Z^1 from (4), (5) in equation (3), we have

$$I_R = \frac{E^1}{Z^1 + Z_R} \text{ (shown in Fig. 3.6)} \quad \text{...(6)}$$

Which is the current expression for network in Fig. 3.6. The truth of Thevenin's theorem is thus, established.

Thevenin's theorem can be used as a circuit analysis method and is particularly useful if the load is to take a series of different values. It is not as powerful as mesh or nodal analysis is large networks because the use of Mesh or Nodal analysis is usually necessary in any Thevenin exercise, so it might as well be used from the start. However, Thevenin's equivalent circuits of Transistors, Voltage Sources such as batteries etc., are very useful in circuit design.

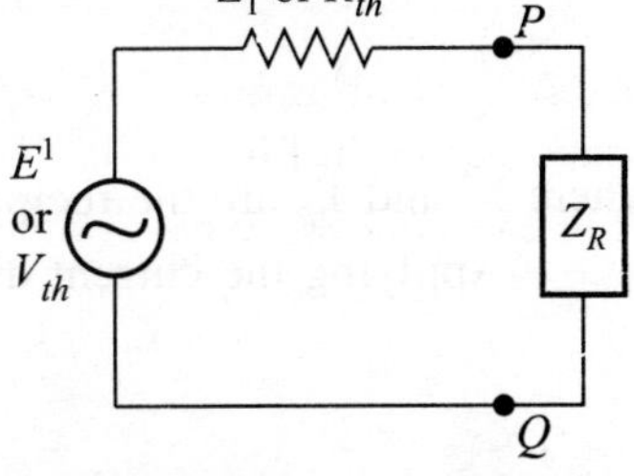

Fig. 3.7. Thevenin's equivalent

Thus Thevenin's theorem states that "Any linear circuit containing several voltages and resistors can be replaced by just a Single Voltage in series with a Single Resistor". In other words, it is possible to simplify any "Linear" circuit, no matter how complex, to an equivalent circuit with just a single voltage source in series with a resistance connected to a load as shown figure. Thevenin's Theorem is especially useful in analyzing power or battery systems and other interconnected circuits where it will have an effect on the adjoining part of the circuit.

Application Procedure

The basic procedure for solving **Thevenin's** Analysis equations is as follows:

1. Remove the load resistor Z_R or component concerned.
2. Find Rth by shorting all voltage sources or by open circuiting all the current sources.
3. With load terminals P and Q open, the open circuit voltage (V_{oc}) is calculated. This is called Thevenin's voltage V_{th}.
4. Find the current flowing through the load resistor Z_R.

3.4 NORTON'S THEOREM

Statement: Any two terminal network consisting of linear impedances and generators can be replaced by an equivalent circuit containing a current source I^1 in parallel with an admittance Y^1. The value of I^1 is the short circuited current between the terminals of the network, and Y^1 is the admittance measured between the terminals with all generators removed (but not their admittances).

Proof:

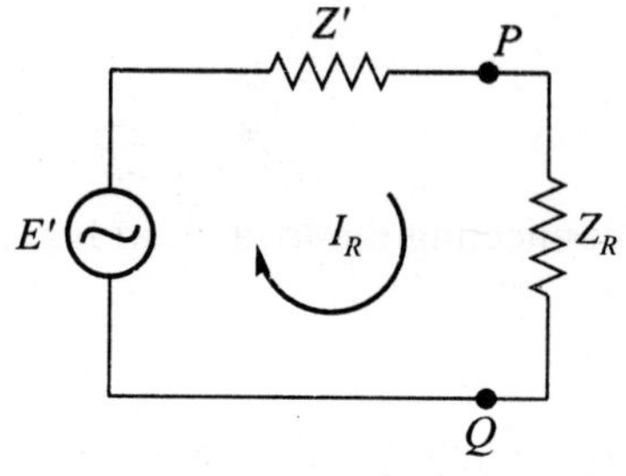

Fig. 3.8

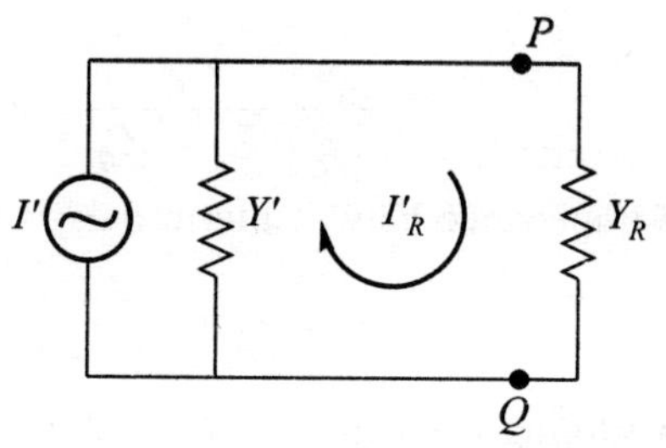

Fig. 3.9

In the Thevenin's equivalent network as shown in Fig. 3.8, the load impedance Z_R is appearing between two terminals P and Q. Now this Thevenin's equivalent circuit may be easily converted into a circuit containing current source I^1 in parallel with Y^1 and Y_R appearing between two terminals P and Q as shown in Fig. (3.9).

The value of I_R from Fig. (3.8) is given by

$$I_R = \frac{E'}{Z' + Z_R} = \frac{E'}{\frac{1}{Y'} + \frac{1}{Y_R}} = \frac{EY' Y_R}{Y' + Y_R} \quad ...(1)$$

where Y' and Y_R are the reciprocal of Z' and Z_R respectively, known as admittances.

Applying the current division law in Fig. (3.9), we have

$$I'_R = \frac{I' Y_R}{Y' + Y'_R} \quad ...(2)$$

The load current I'_R can be made equal to I_R. Then the comparison of equations (1) and (2) gives

$$I' = E'Y' = \frac{E'}{Z'} \qquad ...(3)$$

Equation (3) clearly indicates that circuit (1) and (2) are the same.

Thus, we see that interchange of voltage and current sources with the help of Thevenin's and Norton's theorems give a method of circuit analysis.

Voltage source is removed from a circuit by short circuiting it's e.m.f. whereas a current source is removed by opening it's circuit.

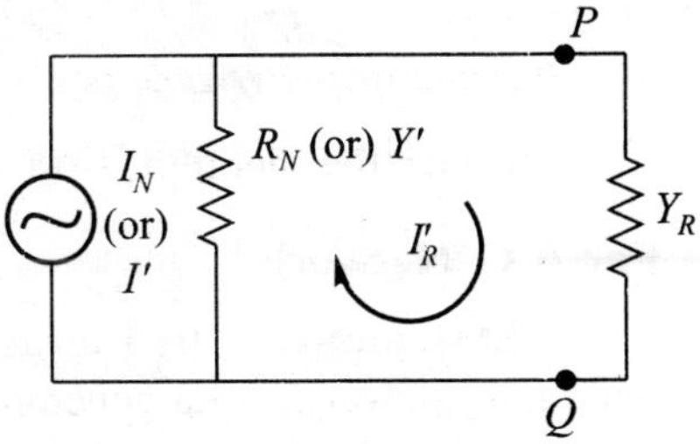

Fig. 3.10. Norton's equivalent circuit.

Application Procedure

The following procedure is followed to Nortonise a circuit:

1. Remove the load resistor Z_R or component concerned.
2. The short circuited current flows between the terminals, when the load is removed, is found. This current is called Norton's current I_N.
3. The Norton's resistance R_N of the circuit as viewed back or looked into from open terminals (when the load is removed) is found.
4. Circuit is replaced by Norton current sources I_N in parallel with Norton resistance R_N connected to load Z_R.

Norton's theorem is preferable when network impedance is high compared with load impedance, such as in the equivalent circuit for a pentode tube amplifier.

3.5 MAXIMUM POWER TRANSFER THEOREM

From D.C. Network

Statement: The power delivered by an active network to a load across it's terminals is maximum, when the load resistance is equal to the internal resistance of the d.c. source.

Proof: Let us consider a two terminal active linear network connected to a two terminal passive linear network. In Fig. 3.11, Z_R represents the equivalent impedance of a psssive linear network.

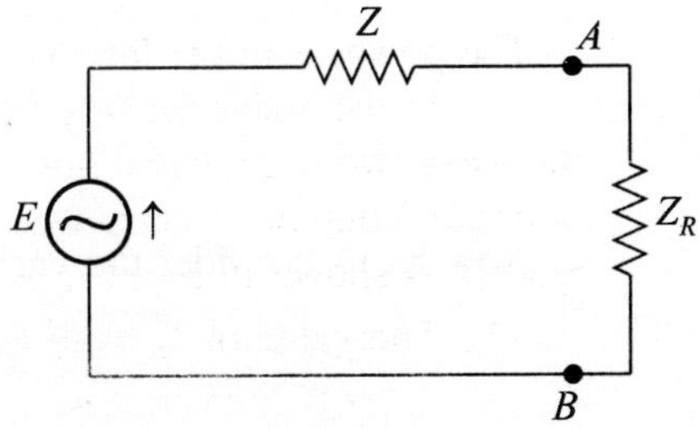

Fig. 3.11

The power delivered to the load is

$$P_L = I^2 Z_R = \frac{E^2 Z_R}{(Z + Z_R)^2}$$

To arrive at the condition for maximum value of P_L, evaluate $\frac{dP_L}{dZ_R}$ and equate it to zero.

$$\frac{dP_L}{dZ_R} = \frac{[Z+Z_R]^2\, E^2 - 2E^2\, Z_R[Z+Z_R]}{(Z+Z_R)^4} = 0$$

$$\frac{dP_L}{dZ_R} = [Z + Z_R] - 2Z_R = 0$$

$$Z = Z_R$$

Hence the theorem is proved.

For A.C. Network

Statement: A two terminal network will absorb maximum power from a generator if the load impedance is the complex conjugate of the internal impedance of the generator.

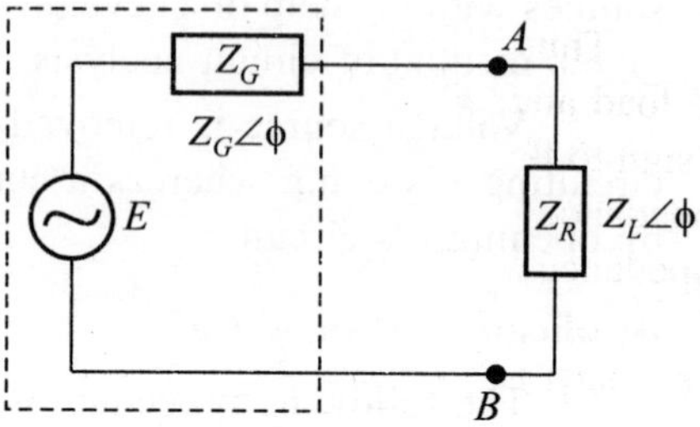

Fig. 3.12

Let us consider a two terminal active linear network connected to a two terminal passive linear network. In Fig. 3.12, Z_R represents the equivalent impedance of a passive linear network.

Here, the generator impedance

$$Z_G\,\angle\phi = R_G + jX_G$$

Where phase angle is given by $\phi = \dfrac{X_G}{R_G}$

And, the load impedance

$$Z_R\,\angle\phi = R_R + jX_R$$

Where $\quad \tan\phi = \dfrac{X_R}{R_R}$

The current I flowing in the circuit is given by

$$I = \frac{E}{[R_R + R_G] + j[X_R + X_G]} \qquad \text{...(1)}$$

The power P in the load is

$$P = I^2 R_R = \frac{E^2 R_R}{[R_R + R_G]^2 + [X_R + X_G]^2} \qquad \text{...(2)}$$

Now let us consider the variation of P with X_R. The power P will be maximum. When

$$\frac{dP}{dX_R} = \frac{-2E^2 R_R\,[X_R + X_G]}{[R_R + R_G]^2 + [X_R + X_G]^2} = 0$$

(or), the condition of maximum power is

$$X_R = -X_G \qquad \text{...(3)}$$

Then, the power in the load becomes

$$P_{\max} = \frac{E^2 R_R}{[R_R + R_G]^2} \qquad \text{...(4)}$$

Againing, considering the variation of P with R_R. For it, differentiating eqn. (2) with respect to R_R and equating derivative to zero. Hence power will be maximum when,

$$\frac{dP}{dR_R} = \frac{E^2(R_R + R_G)^2 - 2E^2 R_R (R_R + R_G)}{[R_R + R_G]^4} = 0$$

$$[R_R + R_G]^2 - 2R_R^2 - R_R R_G = 0$$

$$R_R = R_G$$

Now, $$P_{max} = \frac{E^2 R_R}{(ZR_R)^2} = \frac{E^2}{4R_R}$$

Thus the power absorbed by the load will be maximum when the resistive components of both the load and generator impedances are equal and also the reactance of the load is equal but opposite in sign to the reactance of the generator, *i.e.*, when the load impedance is the conjugate of the internal impedance of the generator. Hence, if the generator is in the form $R_G + jX_G$, then the value of load impedance for maximum power is $R_G - jX_G$.

3.6 MILLMAN'S THEOREM

(1) When a number of constant voltage generators of emf's $E_1, E_2, E_3, \ldots E_n$ and internal admittances $Y_1, Y_2, Y_3, \ldots Y_n$ respectively are connected in parallel, then resultant potential difference E across output is given by

$$E = \frac{E_1Y_1 + E_2Y_2 + E_3Y_3 + \ldots + E_nY_n}{Y_1 + Y_2 + Y_3 + \ldots + Y_n} = \frac{\sum_{r=1}^{n} E_rY_r}{\sum_{r=1}^{n} Y_r}$$

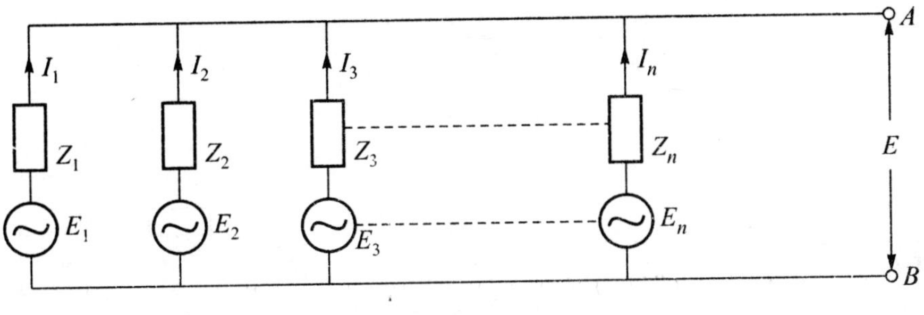

Fig. 3.13

Proof: Let $I_1, I_2, I_3, \ldots I_n$ be the currents through generators respectively. Now, the potential difference across output terminal A and B is given by

$$E = E_1 - I_1Z_1 \quad \text{(or)} \quad \frac{E}{Z_1} = \frac{E_1}{Z_1} - I_1 \qquad \ldots(1)$$

$$E = E_2 - I_2Z_2 \quad \text{(or)} \quad \frac{E}{Z_2} = \frac{E_2}{Z_2} - I_2 \qquad \ldots(2)$$

$$E = E_3 - I_3Z_3 \quad \text{(or)} \quad \frac{E}{Z_3} = \frac{E_3}{Z_3} - I_3 \qquad \ldots(3)$$

....

....

$$E = E_n - I_n Z_n \quad \text{(or)} \quad \frac{E}{Z_n} = \frac{E_n}{Z_n} - I_n \qquad \text{...}(n)$$

Adding all these equations, we get

$$E\left[\frac{1}{Z_1} + \frac{1}{Z_2} + \frac{1}{Z_3} + \ldots + \frac{1}{Z_n}\right] = \left[\frac{E_1}{Z_1} + \frac{E_2}{Z_2} + \ldots + \frac{E_n}{Z_n}\right] - [I_1 + I_2 + I_3 + \ldots + I_n]$$

$$E\left[\frac{1}{Z_1} + \frac{1}{Z_2} + \frac{1}{Z_3} + \ldots + \frac{1}{Z_n}\right] = \left[\frac{E_1}{Z_1} + \frac{E_2}{Z_2} + \ldots + \frac{E_n}{Z_n}\right]$$

$$[\therefore \; I_1 + I_2 + I_3 + \ldots + I_n = \Sigma I = 0 \text{ (Kirchoff's law)}]$$

$$E = \frac{E_1Y_1 + E_2Y_2 + E_3Y_3 + \ldots + E_nY_n}{Y_1 + Y_2 + Y_3 + \ldots + Y_n} \left[\because \; Y = \frac{1}{Z}\right]$$

$$E = \frac{\sum_{r=1}^{n} E_r Y_r}{\sum_{r=1}^{n} Y_r}$$

(2) When a number of constant current generators of currents I_1, I_2, I_3, ... I_n and parallel impedances Z_1, Z_2, Z_3, ..., Z_n respectively be connected in series as to form a closed loop, then the resultant curve through the loop is given by

$$I = \frac{I_1Z_1 + I_2Z_2 + \ldots + I_nZ_n}{Z_1 + Z_2 + \ldots + Z_n} = \frac{\sum_{r=1}^{n} I_r Z_r}{\sum_{r=1}^{n} Z_r}$$

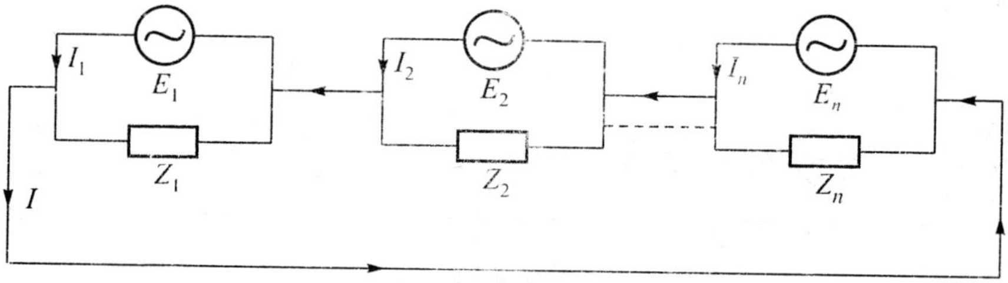

Fig. 3.14

Proof: The current I through series circuit is given by

$$I = I_1 - \frac{E_1}{Z_1} \quad \text{(or)} \quad IZ_1 = I_1Z_1 - E_1 \qquad \text{...(1)}$$

$$I = I_2 - \frac{E_2}{Z_2} \quad \text{(or)} \quad IZ_2 = I_2Z_2 - E_2 \qquad \text{...(2)}$$

....

....

$$I = I_n - \frac{E_n}{Z_n} \qquad IZ_n = I_2Z - E_n \qquad \text{...(3)}$$

Adding all these equations, we get

$$I\,[Z_1 + Z_2 + \ldots + Z_n] = [I_1Z_1 + I_2Z_2 + \ldots + I_nZ_n] - [E_1 + E_2 + \ldots + E_n]$$

$[\therefore$ But $E_1 + E_2 + \ldots + E_n = \Sigma E = 0$, Kirchoff's law$]$

$$I\,[Z_1 + Z_2 + Z_3 + \ldots + Z_n] = I_1Z_1 + I_2Z_2 + \ldots + I_nZ_n$$

$$I = \frac{I_1Z_1 + I_2Z_2 + I_3Z_3 + \ldots + I_nZ_n}{Z_1 + Z_2 + Z_3 + \ldots + Z_n}$$

$$I = \frac{\sum_{r=1}^{n} I_rZ_r}{\sum_{r=1}^{n} Z_r}$$

Hence the Millmann's theorem is proved.

3.7 RECIPROCITY THEOREM

Statement: "If an e.m.f. applied in one mesh of network of linear impedance produces a certain current in the second mesh, then the same e.m.f. acting in the second mesh will give an identical current in the first mesh".

Proof: Consider the circuit as shown in Fig. (3.15). Let us calculate the value of $\frac{E}{I_2}$ with the help of Kirchoff's law

Applying this law, we have

$$E = [Z_1 + Z_3]\,I_1 - I_2Z_3 \qquad \ldots(1)$$

$$0 = -\,Z_3I_1 + [Z_2 + Z_3]I_2 \qquad \ldots(2)$$

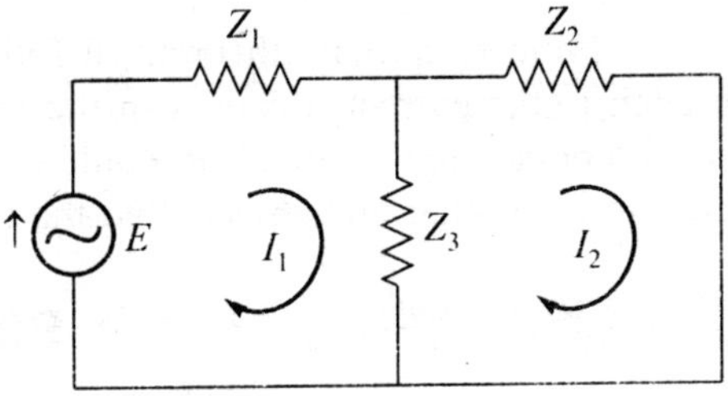

Fig. 3.15

Solving the equations (1) and (2), we get

From (2) $\Rightarrow I_1Z_3 = [Z_2 + Z_3]I_2$

$$I_1 = \frac{[Z_2 + Z_3]I_2}{Z_3}$$

Substituting the value of I_1 in eqn. (1), we get

$$E = \frac{(Z_1 + Z_3)(Z_2 + Z_3)}{Z_3}\,I_2 - I_2\,Z_3$$

$$E = I_2\left[\frac{(Z_1 + Z_3)(Z_2 + Z_3)}{Z_3} - Z_3\right]$$

$$\frac{E}{I_2} = \frac{Z_1Z_2 + Z_1Z_3 + Z_3Z_2 + Z_3^2 - Z_3^2}{Z_3}$$

$$\frac{E}{I_2} = \frac{Z_1Z_2 + Z_1Z_3 + Z_2Z_3}{Z_3} \qquad \ldots(3)$$

Now, let the position of the battery be changed and the circuit be as shown in Fig. 3.16. Applying Kirchoff's law, we have

$$E = [Z_2 + Z_3]I'_2 - Z_3I'_1 \qquad \ldots(4)$$

$$0 = -Z_3 I'_2 + (Z_1 + Z_3) I'_1 \quad \text{...(5)}$$

From (5), $Z_3 I'_2 = (Z_1 + Z_3) I'_1$

$$I'_2 = \left[\frac{Z_1 + Z_3}{Z_3}\right] I'_1$$

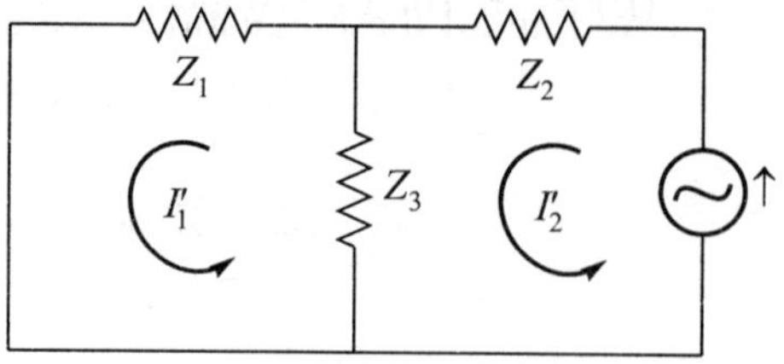

Fig. 3.16

Substituting the value of I'_2 in eqn. (1), we get

$$E = [Z_2 + Z_3]\left[\frac{Z_1 + Z_3}{Z_3}\right] I'_1 - Z_3 I'_1$$

$$E = I'_1\left[\frac{(Z_1 + Z_3)(Z_2 + Z_3)}{Z_3} - Z_3\right]$$

$$E = I'_1\left[\frac{Z_1Z_2 + Z_1Z_3 + Z_2Z_3 + Z_3^2 - Z_3^2}{Z_3}\right]$$

$$\frac{E}{I'_1} = \frac{Z_1Z_2 + Z_1Z_3 + Z_2Z_3}{Z_3} \quad \text{...(6)}$$

Now, it is clear that eqn. (3) and eqn. (6) are equal.

So,
$$\frac{E}{I_2} = \frac{E}{I'_1}$$
$$I_2 = I'_1$$

Which proves the reciprocity theorem.

From reciprocity theorem, it follows that in any linear network, the ratio of the e.m.f. in one branch to the current it causes in the second branch is the same as the ratio of e.m.f. placed in the second branch to the current it would cause in the first branch. The ratio of e.m.f. in one branch to the current in another branch is called the transfer impedance.

SOLVED EXAMPLES

Example 1: *Find the value of R_L corresponding to which where is a maximum power transfer to the load resistor and also find this maximum power. The circuit is shown below.*

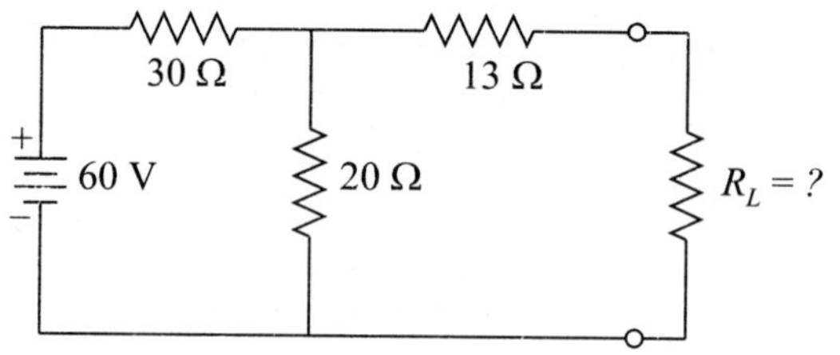

Fig. 3.17

Solution: The load resistor is connected across terminals *A* and *B*

$= V_{th} = V_{OC}$ when terminals *A* and *B* are kept open.

(*i*) ∴ $$V_{th} = \left[\frac{60}{30 + 20}\right] \times 20\ \Omega = \frac{60}{50} \times 20\Omega = 1.2 \times 20 = 24 \text{ volts}$$

(*ii*) R_{eq} = Total resistance viewed from terminals, *A* and *B* kept open and voltage source to be made dead.

$$\therefore \quad R_{eq} = \frac{30 \times 20}{30 + 20} + 13 = \frac{600}{50} + 13 = 12 + 13 = 25\ \Omega$$

∴ Thevenin's equivalent circuit is shown in figure.

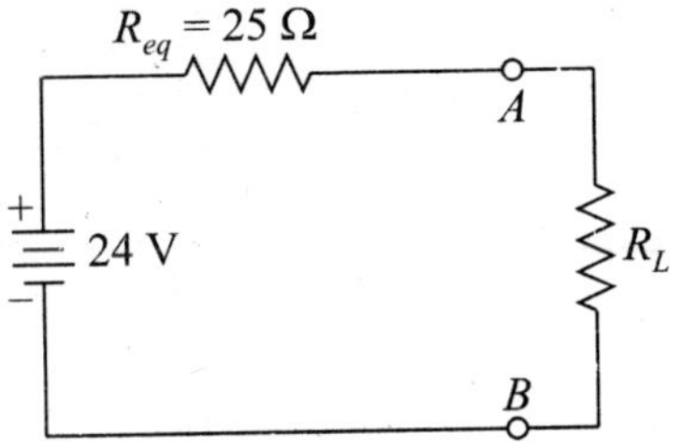

Fig. 3.18

As per maximum power transfer theorem

$$\therefore \quad R_L = R_{eq} \quad \text{or} \quad R_L = 25\ \Omega$$

$$\therefore \quad P_L = I_L^2 R_L = \left[\frac{24}{25 + R_L}\right]^2 R_L = \frac{24^2 R_L}{(25 + R_L)^2}$$

$$\therefore \quad \frac{dP_L}{dR_L} = (24)^2 \frac{\{(25 + R_L)^2 \times 1 - R_L \times 2(25 + R_L)\}}{(25 + R_L)^2}$$

$$(25 + R_L)^2 = 2R_L\,(25 + R_L)$$

$$25 + R_L = 2R_L$$

$$R_L = 25\ \Omega$$

$$\therefore \quad P_L\,(\text{Max}) = \frac{(24)^2}{(24 + 25)^2} \times 25 = 5.5 \text{ watts}$$

Example 2: *Use superposition theorem to find i_p in the circuit shown below?*

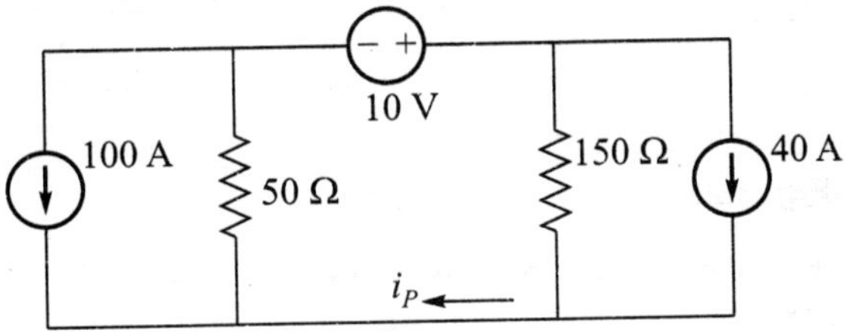

Fig. 3.19

Solution: (*i*) Let only 100 A source exists and other sources are set to zero.

$$i_{P_1} = -100\left(\frac{50}{50 + 150}\right) = -25 \text{ Amp}$$

(*ii*) Now 10 V source exists and current sources set to zero or open circuit the current source terminals.

$$\therefore \quad i_P = \frac{10 \text{ volts}}{150 + 50} = \frac{10}{200} = \frac{1}{20} \text{ amperes}$$

(*iii*) Now 40 A source exists and other sources dead.

$$\therefore \quad i_{P_3} = 40 \times \frac{150}{50 + 150} = 30 \text{ Amps}$$

As per superposition theorem

$$i_P = i_{P_1} + i_{P_2} + i_{P_3}$$

$$i_P = -30 + \frac{1}{20} + 30 = \frac{1}{20} = -0.05\text{A} = 50 \text{ mA}$$

Example 3: *Find (a) Thevenin's equivalent circuit; (b) Norton's circuit for the active network shown in Fig. 3.20.*

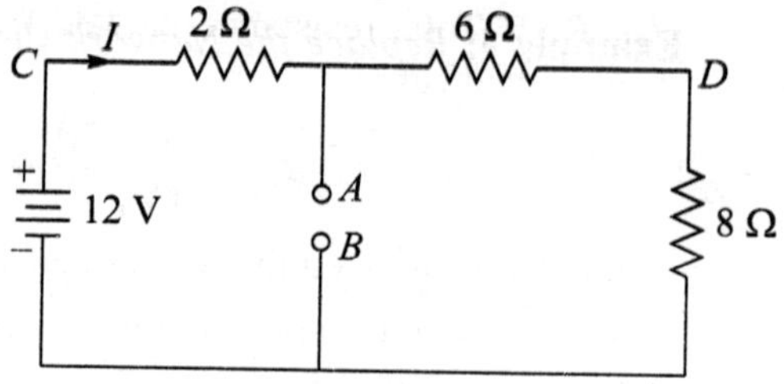

Fig. 3.20

Solution: (*i*) Thevenin equivalent circuit

$\therefore \quad V_{th} = V_{AB}$

Loop *ACBA*:

$\therefore \quad V_{AB} - 12 = -I \times 2$ (apply KVL)

Where $\quad I = \dfrac{12}{2+6+8} = \dfrac{12}{16} = 0.75$ Amp

Where $V_{AB} - 12 = -I \times 2$

$$V_{AB} = 12 - 0.75 \times 2 = 10.5 \text{ volts}$$

$$\therefore \quad R_{eq} = \frac{2 \times 10.5}{2 + 10.5} = \frac{21}{12.5}$$

$$R_{eq} = 1.68\ \Omega$$

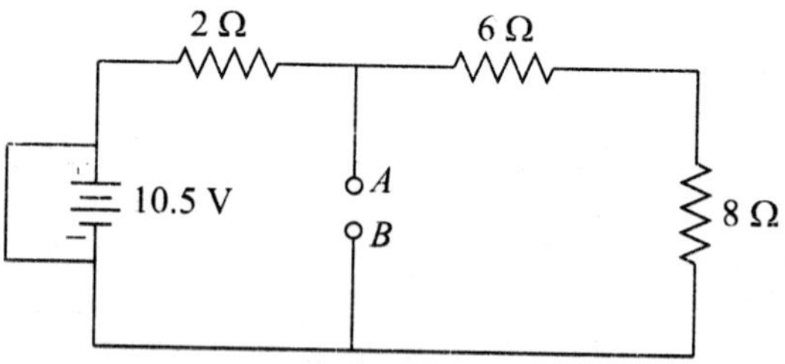

Fig. 3.21

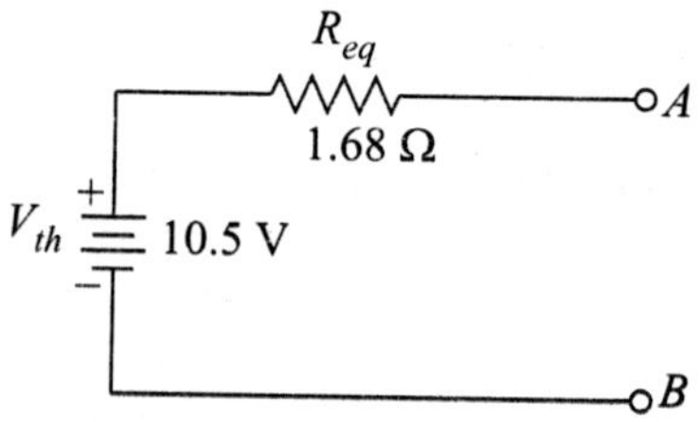

Fig. 3.22

Now, Thevenin's equivalent circuit is shown below.

(*ii*) Norton's equivalent circuit

$$I_N = \frac{12}{2} = 6 \text{ Amp}$$

$$R_{eq} = 1.68\ \Omega$$

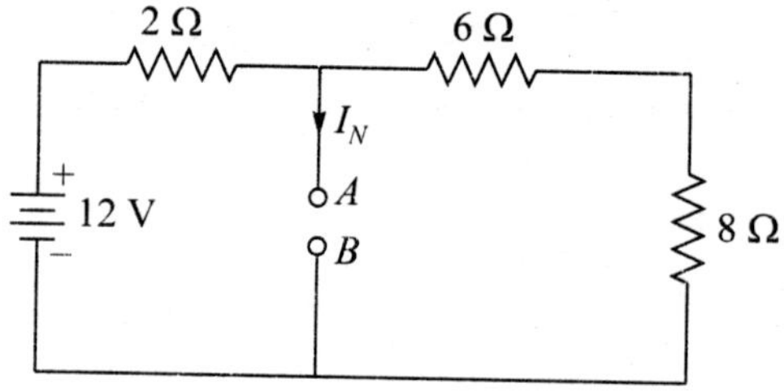

Fig. 3.23

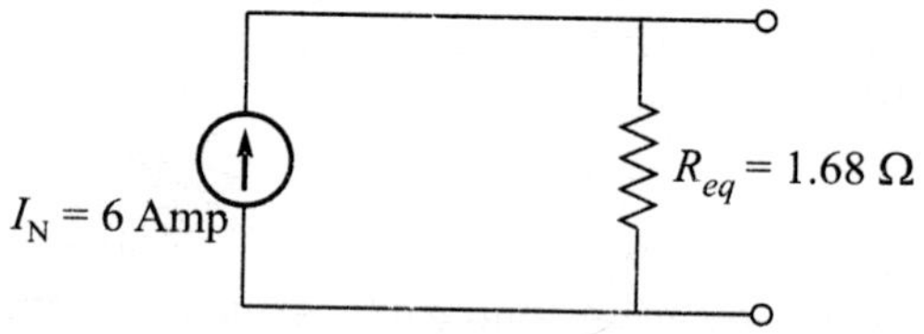

Fig. 3.24

Example 4: *Replace the Network shown below using Thevenin's Network.*

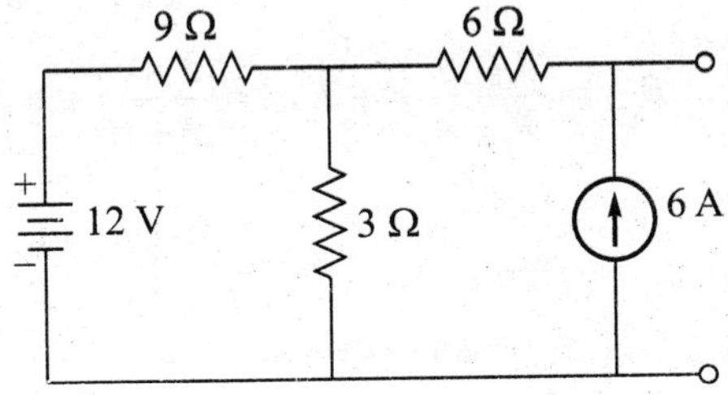

Fig. 3.25

Solution: $V'_{th} = \frac{12}{9+3} \times 3 = 3$ volts

$\therefore \quad R_P = \frac{9 \times 3}{9+3} = \frac{27}{12} = \frac{9}{4} = 2.25\ \Omega$

Now 2.25 Ω in series with 6 Ω resistor

$\therefore \quad R_t = 2.25 + 6 = 8.25\ \Omega$

$\therefore \quad V''_{AB} = 6R_t = 6 \times 8.25 = 49.5$ volts

$V''_{th} = V''_{AB} = 49.5$ volts

$\therefore \quad V_{th} = V'_{th} + V''_{th} = 3 + 49.5 = 52.5$ volts

$R_{th} = \frac{9 \times 3}{9+3} + 6 = 8.25\ \Omega.$

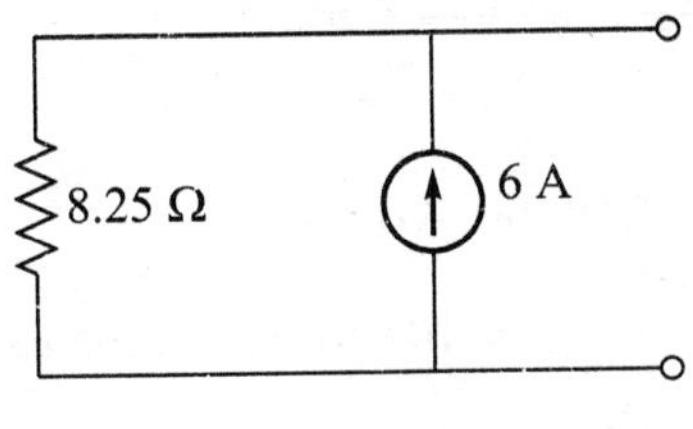

Fig. 3.26

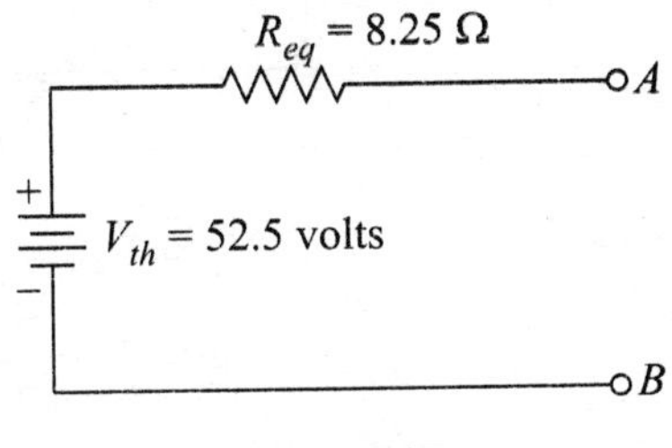

Fig. 3.27

Example 5: *Apply Thevenin's Theorem to the following network.*

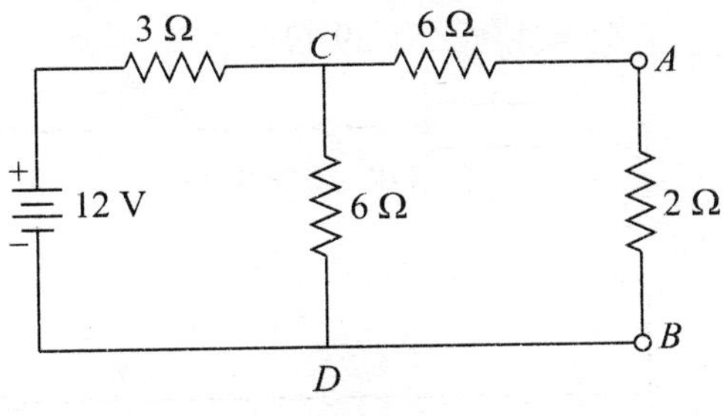

Fig. 3.28

Solution:

$$V_{th} = V_{CD} = 6I_{CD} = 6 \times \frac{12}{3+6} = \frac{72}{9} = 8 \text{ volts}$$

$$R_{eq} = \frac{R_{CD} \times 3}{R_{CI} + 3} + R_{AC} = \frac{3 \times 6}{3+6} + 6 = 2 + 6 = 8\ \Omega$$

$$I_L = \frac{V_{th}}{R_{eq} + R_L} = \frac{8}{8+2} = 0.8 \text{ Amps.}$$

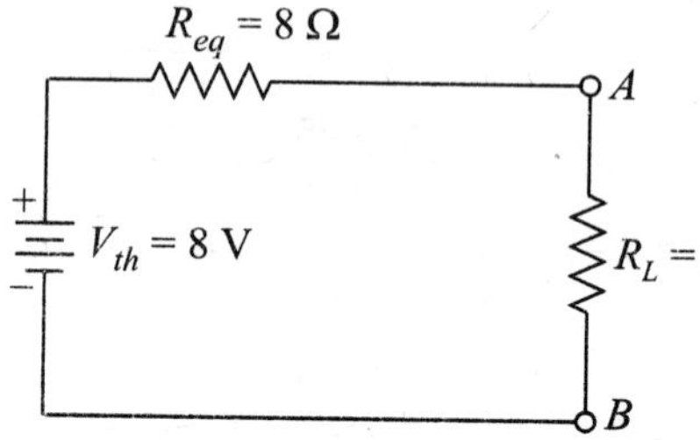

Fig. 3.29

Example 6: *Replace the following network by Thevenin's Network.*

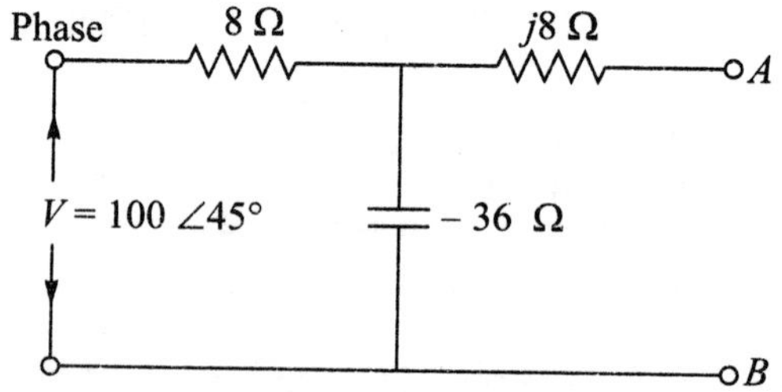

Fig. 3.30

Solution:

$$V_{th} = V_{AB} = I(-j6) = 6I\angle -90° \text{ volts}$$

Now

$$I = \frac{V}{8 - j6} = \frac{100\angle 45°}{10\angle -36.86°} = 10\angle 81.86° \text{ amperes}$$

$$V_{th} = 6\angle -90° \times 10\angle 81.86° = 60\angle -8.4° \text{ volts}$$

$$Z_{eq} = \frac{-j48}{8 - j6} + j8 = \frac{-j48 + j64 - j^2 48}{8 - 6j}$$

$$Z_{eq} = 5.06\angle 55.29° \Omega.$$

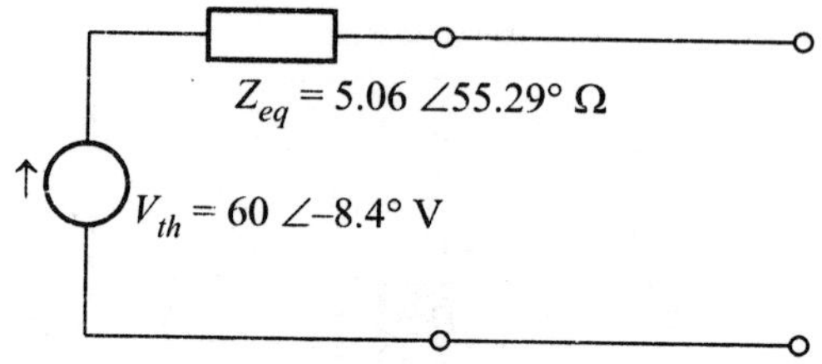

Fig. 3.31

Example 7: *Verify Norton's Theorem.*

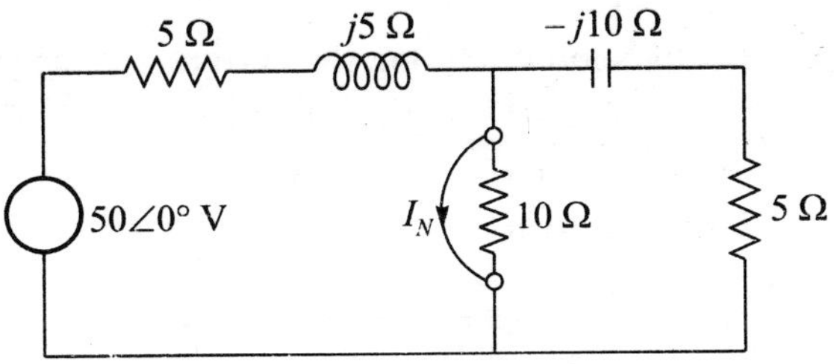

Fig. 3.32

Solution:

$$I_N = I_{S.C.} = \frac{50\angle 0^\circ}{5 + j5} = \frac{50\angle 0^\circ}{7.07\angle 45^\circ} = 7.07\angle -45^\circ \text{ amp}$$

Now

$$Z_{eq} = \frac{(5 + j5)(5 - j10)}{(5 + j5) - j10 + 5} = \frac{7.07\angle 45^\circ \times 11.18\angle -63.43^\circ}{10 - j5}$$

$$Z_{eq} = \frac{79\angle -18.43^\circ}{11.18\angle -26.57^\circ} - 7.07\angle 8.14^\circ \Omega$$

Norton's equivalent circuit with load resistor of 10Ω

∴

$$I_{10} = I_N \times \frac{Z_{eq}}{Z_{eq} + 10}$$

$$I_{10} = 7.07\angle -45^\circ \times \frac{7.07\angle 8.14^\circ}{7.07\angle 8.14^\circ + 10}$$

$$I_{10} = 7.07\angle -45^\circ \times \frac{7.07\angle 8.14^\circ}{17\angle 3.37^\circ} = \frac{50\angle -36.87^\circ}{17\angle 3.37^\circ}$$

$$I_{10} = 2.94\angle -40.24^\circ \text{ Amp}$$

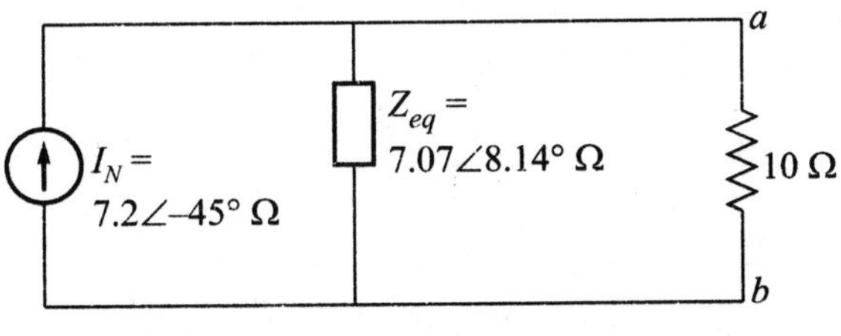

Fig. 3.33

Example 8: *Use Superposition theorem to find i_p in the following circuit.*

Solution: We shall calculate the current i_p by short circuiting 6V source *i.e.*

$$R_{T1} = 2 + \frac{1 \times 3}{1 + 3} = 2 + \frac{3}{4} = \frac{11}{4}\Omega$$

So, the current generated by 12V source is

$$I_1 = \frac{12}{11/4} = \frac{48}{11} \text{ amp}$$

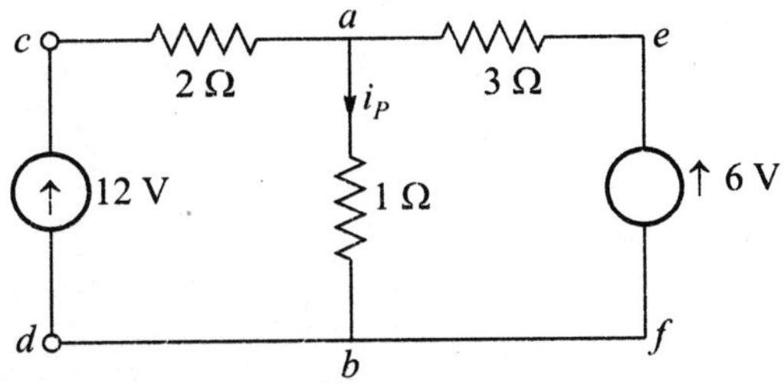

Fig. 3.34

$$I_{P1} = \frac{48}{11} \times \frac{3}{4} = \frac{46}{11} \text{ amp}$$

Now, calculate the current by short circuiting the current 12V source.

$$R_{T2} = 3 + \frac{2 \times 1}{2+1} = 3 + \frac{2}{3} = \frac{11}{3}$$

So, current generated by 6V source is

$$I_2 = \frac{6}{11/3} = \frac{18}{11}$$

Hence $$= \frac{18}{11} \times \frac{2}{3} = \frac{12}{11} = i_{P2}$$

$\therefore$ $$i_P = i_{P1} + i_{P2} = \frac{36}{11} + \frac{12}{11} = \frac{48}{11} = 4.36\Omega$$

Example 9: *Apply Reciprocity Theorem for the following active network.*

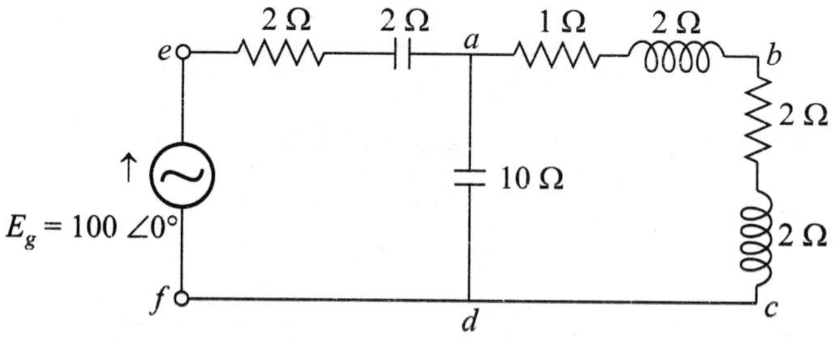

Fig. 3.35

Solution: Let us calculate the current in the branch bc due to $E_g = 100\ \angle 0$ inserted in ef branch.

$$Z_{ac} = \frac{Z_{abc} Z_{ad}}{Z_{abc} + Z_{ad}} = \frac{(3+j4)(-j10)}{3+j4-j10} = 6.67 + j3.33\Omega$$

$$Z_{cc} = Z_{ea} + Z_{ac} = 2 - j2 + 6.67 + j3.33 = 8.67 + j1.33\Omega$$

$$I_{ec} = \frac{V_{fe}}{Z_{ec}} = \frac{100 + j0}{8.67 + j1.33} = 11.27 - j1.732 \text{ amp}$$

$$V_{ac} = I_{ec} Z_{ac} = (11.27 - j1.732)\ (6.67 + j3.33) = 81 + j26 \text{ volt}$$

$$I_{bc} = \frac{V_{ac}}{Z_{abc}} = \frac{8 + j26}{3 + j4} = 13.88 - j9.84 \text{ amp}$$

Now, $$Z_{af} = \frac{(2-j2)(-j10)}{(2-j2-j10)} = 1.352 - j18.92\Omega$$

$$Z_{caf} = 3 + j4 + 1.352 + j1.8692 = 4.352 + j2.108\Omega$$

$$I_{caf} = \frac{100 + j0}{4.352 + j2.108} = 18.6 - j9.02 \text{ amp}$$

$$V_{ef} = (18.6 - j9.02)\ (1.352 - j1.892) = 8.07 - j47.4 \text{ volt}$$

$$I_{ef} = \frac{8.01 - j47.4}{2 - j2} = 13.88 - j9.84 \text{ amp}$$

$$I_{bc} = I_{ef}$$

Example 10: *Find the open circuit voltage and Thevenin resistance of the two terminal network.*

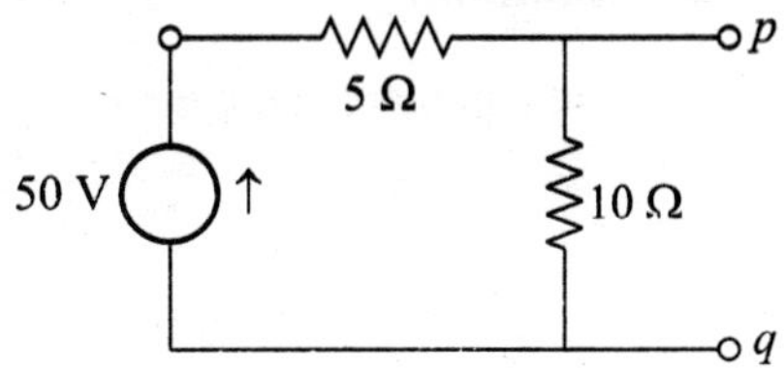

Fig. 3.36

Solution: Given that

$$E = 50\text{V},\ Z_1 = 5\Omega,\ Z_2 = 0\Omega$$
$$Z_3 = 10\Omega$$

Then by Thevenin Components we have open circuited voltage as

$$E' = \frac{EZ_3}{Z_1 + Z_3} = \frac{50 \times 10}{5 + 10} = \frac{500}{15} = 33.33\text{V}$$

and
$$Z' = Z_2 + \frac{Z_1 Z_3}{Z_1 + Z_3} = 0 + \frac{5 \times 10}{5 + 10}$$
$$Z' = \frac{50}{15} = 3.33\Omega$$

Example 11: *Calculate the current through load impedance $Z_{R'}$ using Thevenin's Theorem.*

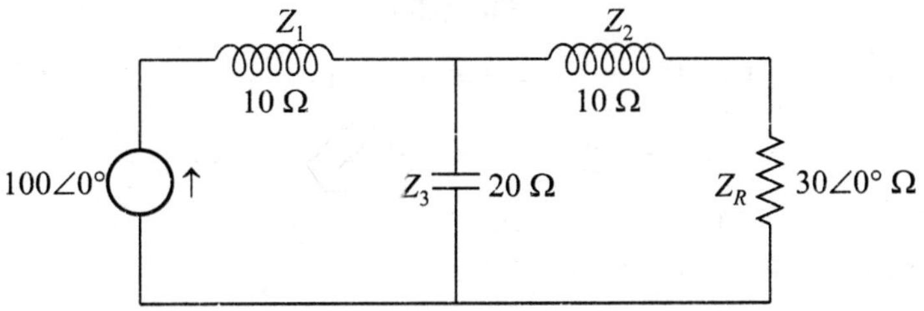

Fig. 3.37

Solution:
$$I_R = \frac{E'}{Z' + Z_R}$$

where
$$E' = \frac{EZ_3}{Z_1 + Z_3} = \frac{(100 + j0)(-20j)}{j10 - j20} = (200 + j0) \text{ volt}$$
$$Z' = Z_2 + \frac{Z_1 Z_3}{Z_1 + Z_3} = j10 + \frac{(j10)(-j20)}{j10 - j20} = j10 + j20 = j30\Omega$$
$$I_R = \frac{E'}{Z' + Z_R} = \frac{200 + j0}{j30 + 30 + j0} = \frac{200}{30 + j30} = (3.33 - j3.33) \text{ amp}$$
$$I_R = 4.72 \angle -5° \text{ amp}$$

Example 12: *Calculate the current I shown in figure below using Millman's Theorem.*

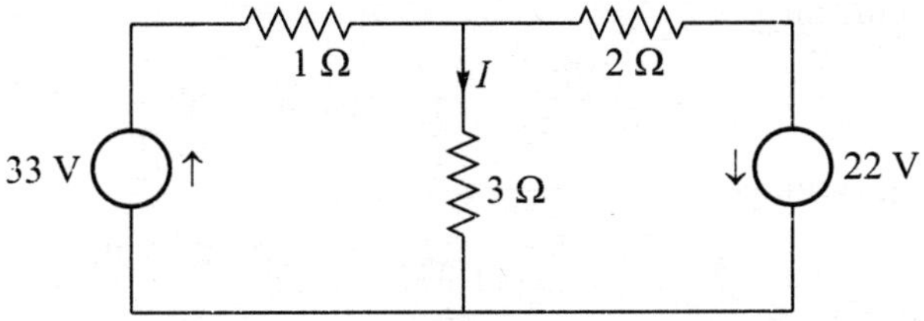

Fig. 3.38

Solution:

$$E' = \frac{E_1Y_1 + E_2Y_2}{Y_1 + Y_2} = \frac{33 \times 1 - (22/2)}{1 + (1/2)} = \frac{44}{3} \text{ volt}$$

$$Z' = \frac{1}{Y_1 + Y_2} = \frac{1}{1 + (1/2)} = \frac{2}{3}\Omega$$

$\therefore$

$$I = \frac{44/3}{3 + (2/3)} = 4 \text{ amp}$$

Example 13: *Find the current I in the circuit given below using Superposition Theorem.*

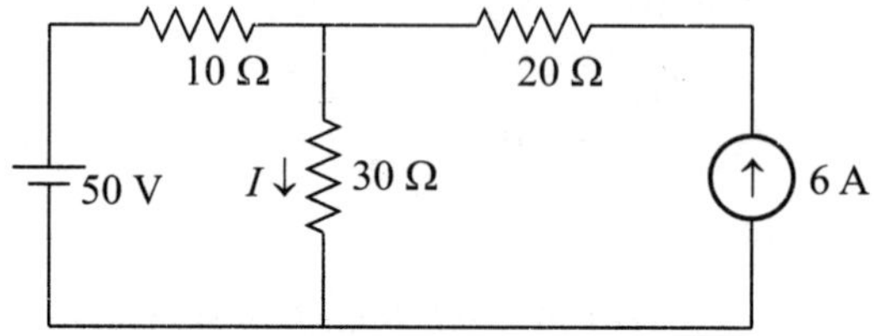

Fig. 3.39

Solution: Considering first the voltage source alone,

$$I_1 = \frac{50}{40} = 1.25 \text{ amp}$$

Now, considering the current source only.

The current flowing through 10Ω resistor

$$I_2 = 2 \times \frac{10}{10 + 30} = \frac{20}{40} = 0.5 \text{ amp}$$

Applying superposition theorem,

$$I = I_1 + I_2 = 1.25 + 0.5 = 1.75 \text{ amp.}$$

Example 14: *Thevinise the following network*

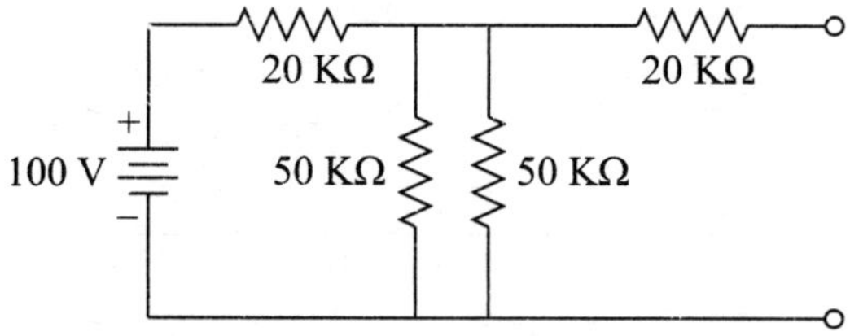

Fig. 3.40

Solution: Source current

$$I = \frac{100}{20 + 25} = \frac{100}{45} = 2.22 \text{ amp}$$

Therefore, Thevenin voltage

$$V_{th} = I \times 25 = 2.22 \times 25 = 55.55 \text{ volt}$$

Now, Thevenin resistance

$$R_{th} = 20 \text{ K}\Omega + \frac{25 \times 20}{25 + 20} \text{ K}\Omega$$

$$= 20 + \frac{500}{45} = 20 + 11.11 = 31.11 \text{ K}\Omega$$

Now, Thevenin's equivalent Network is shown as

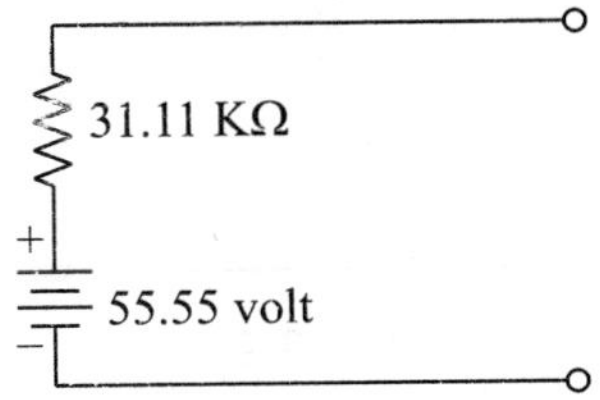

Fig. 3.41

Example 15: *Apply Norton's Theorem to the following network.*

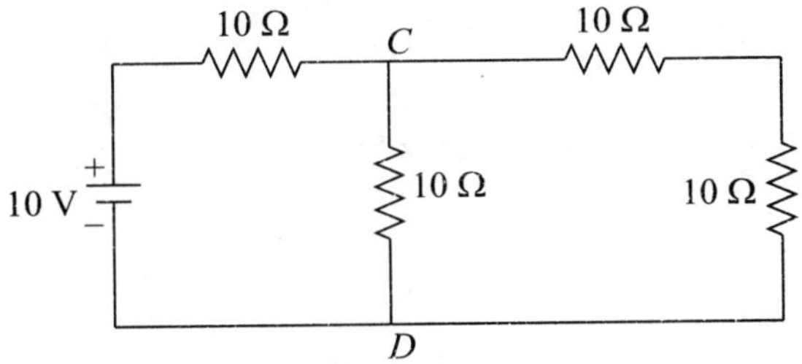

Fig. 3.42

Solution: $$I = \frac{10}{10 + (10 \,||\, 10)} = \frac{10}{15} = 0.66 \text{ amp}$$

$$I_N = \frac{0.66}{2} = 0.33 \text{ amp } (I_N \text{ is equally divided at } C)$$

Now, $$R_N = (10 \,||\, 10) + 10 = 15\ \Omega$$

Now, Norton's equivalent Network is

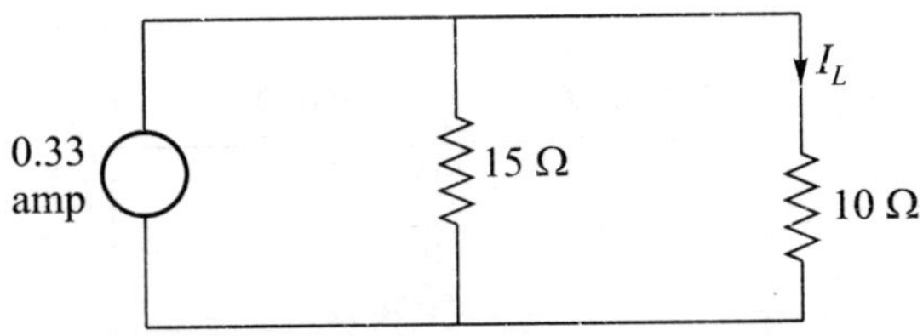

Fig. 3.43

Example 16: *Prove Reciprocity Theorem for 1Ω branch.*

Solution: Total current in the circuit

$$I = \frac{36}{[(3+1)//12]+2+4} = \frac{36}{3+2+4} = 4 \text{ amp}$$

$$I_L = 4 \times \frac{12}{12+3+1} = 3 \text{ amp}$$

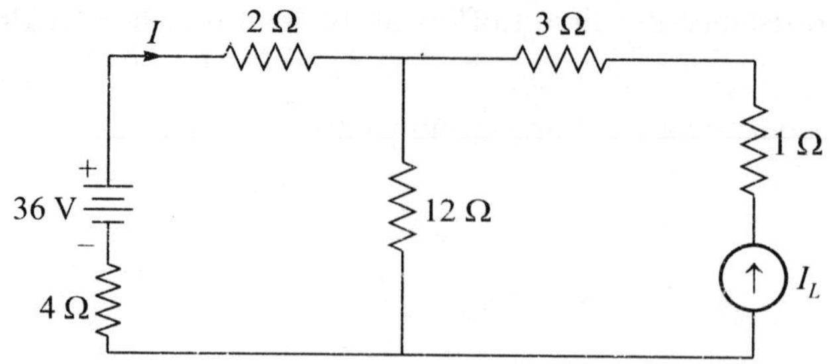

Fig. 3.44

According to Reciprocity Theorem, replacing voltaging source in 1Ω branch,

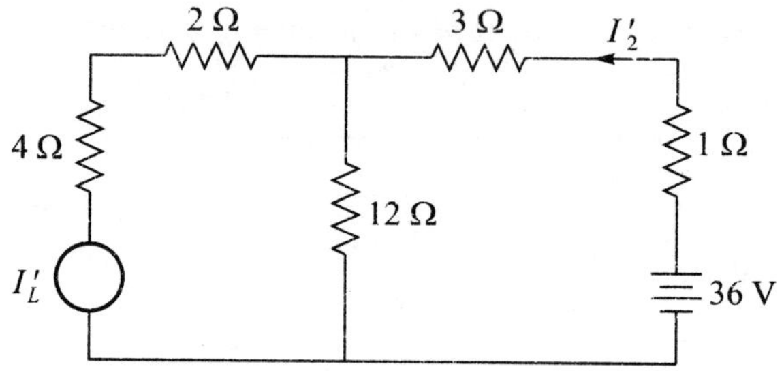

Fig. 3.45

$$I' = \frac{36}{[(4+2)//12]\,3+1} = \frac{36}{4+3+1} = 4.5 \text{ amp}$$

$$I'_L = 4.5 \times \frac{12}{12+4+2} = 3 \text{ Amp}$$

$$I_L = I'_L$$

Hence, Reciprocity Theorem is proved.

Example 17: *Find the value of R_L for which the power transferred to the load is maximum. Also find power.*

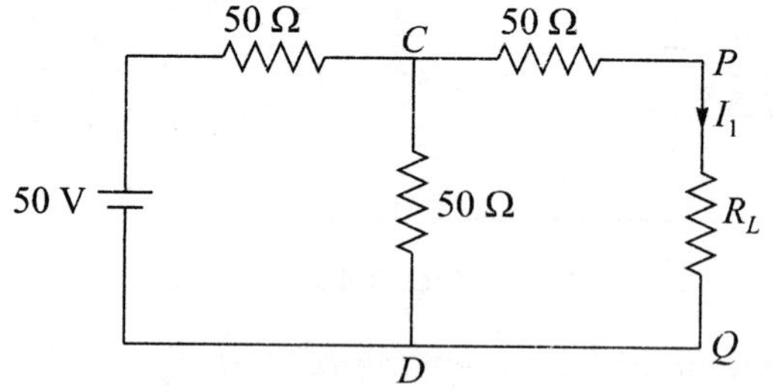

Fig. 3.46

Solution: The resistance the circuit offers at the load is (when looking back)

$$50 + 50//50 = 75\ \Omega$$

According to Maximum Power transfer theorem, it is maximum when $R_L = 75\ \Omega$

When $R_L = 75\ \Omega$, the total resistance of the circuit is

$$50 + 50//75 = 80\ \Omega$$

$$I = \frac{50\text{ V}}{80} = 0.625\text{ amp}$$

$$I_L = 0.625 \times \frac{50}{50+75} = 0.25\text{ amp}$$

$$P_{max} = I_L^2 R_L = (0.25)^2 \times 75 = 4.6875\text{ watt}$$

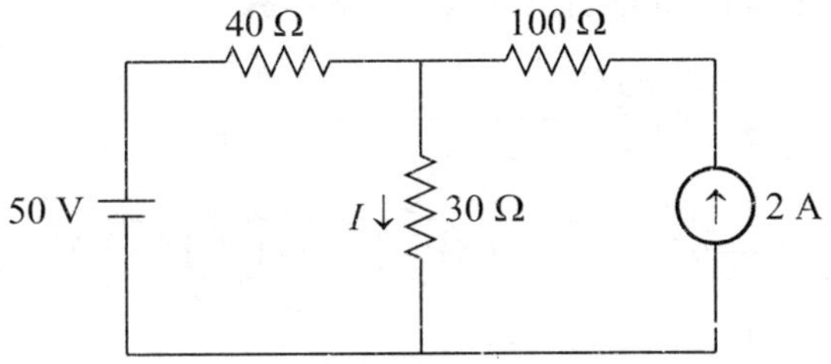

Fig. 3.47

Example 18: *Find the current I in the circuit given in Fig. 3.48 using superposition theorem.*

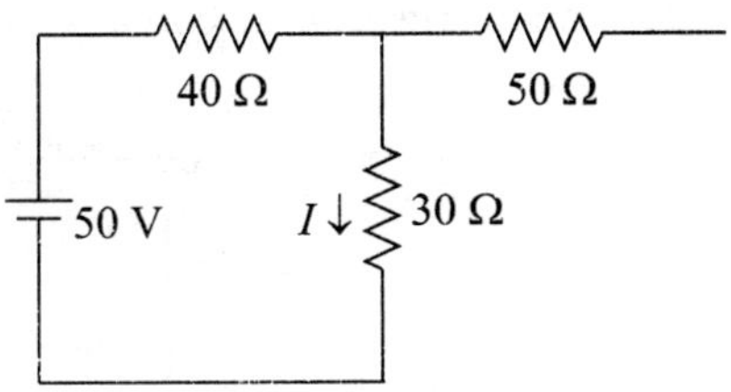

Fig. 3.48

Solution: To do this, first consider only voltage source

The $$I_1 = \frac{50}{70} = 0.71\text{ amp}$$

Now consider only current source.

The current through 30 Ω resistor is given by

$$I_2 = \frac{2 \times 40}{40+30} = \frac{80}{70} = 1.14\text{ amp}$$

Now, the total current through 30 Ω resistor is given by

$$I = I_1 + I_2 = 0.71 + 1.14 = 1.85\text{ amp.}$$

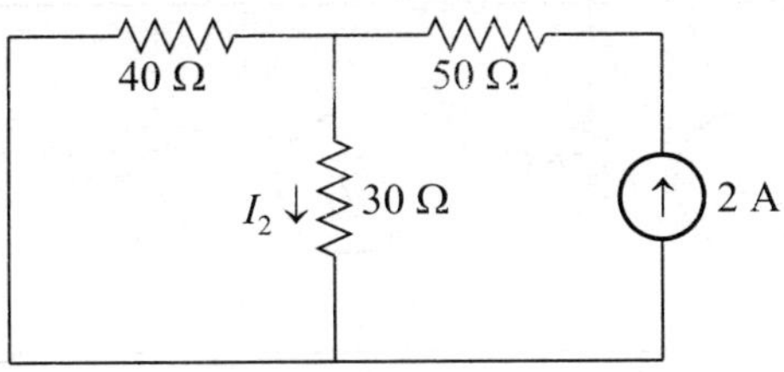

Fig. 3.49

EXERCISE

1. Use Superposition Principle to find the current I_1 in the circuit shown below.

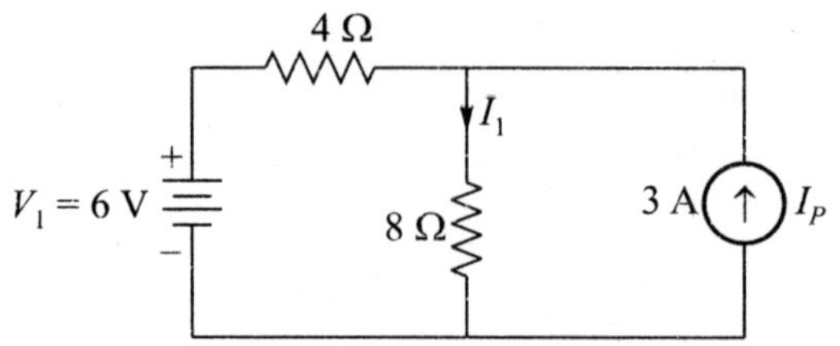

Fig. 3.50

2. Apply the Superposition theorem in the following network.

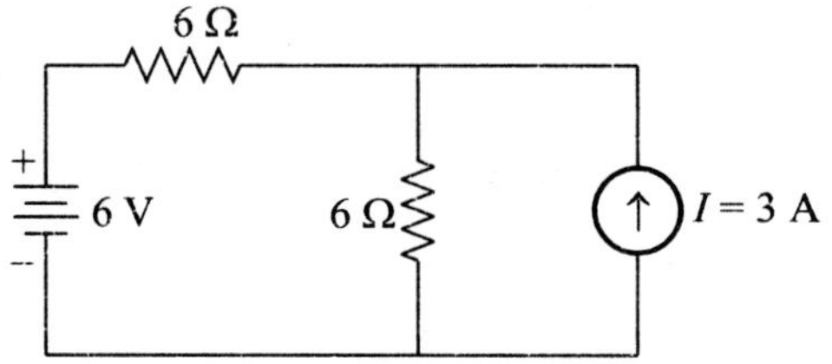

Fig. 3.51

3. Apply Thevenin's, Norton's theorem's for the networks shown below.

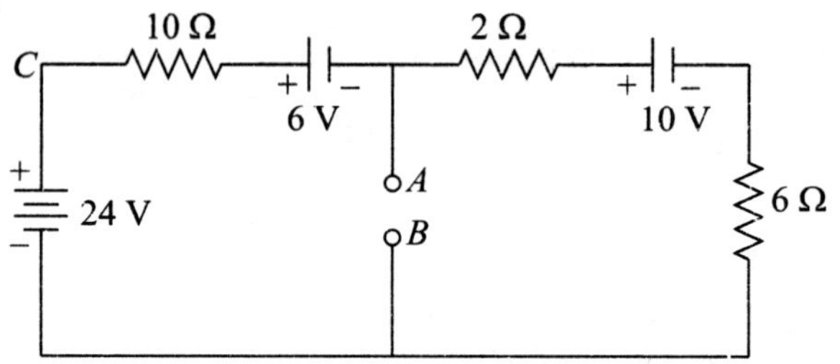

Fig. 3.52

4. Replace the active network shown below with Thevenin's and Norton's networks.

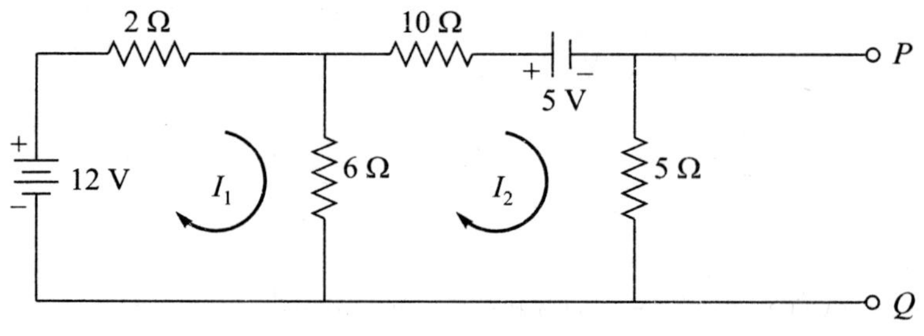

Fig. 3.53

5. Apply Thevenin's theorem to the following network.

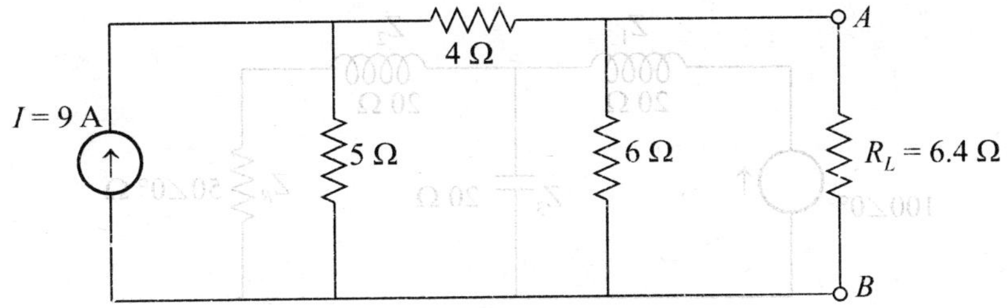

Fig. 3.54

6. Replace the following network by Thevenin's equivalent network.

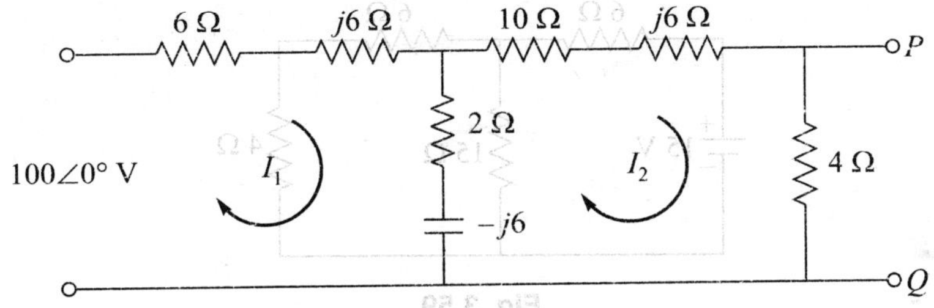

Fig. 3.55

7. Use Superposition theorem to find the current i_p in branch *ce* in the circuit.

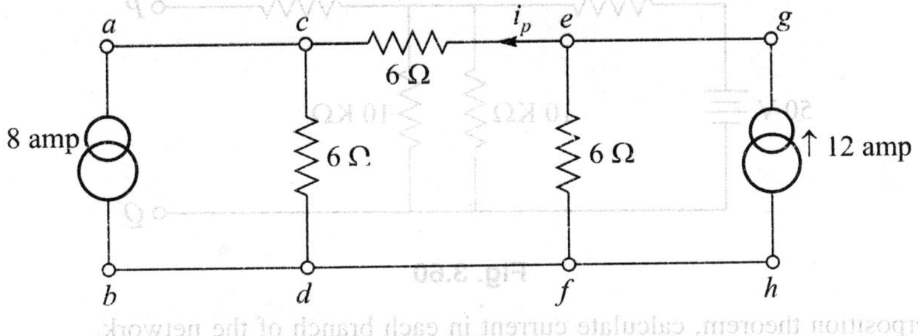

Fig. 3.56

8. Find the short circuit current and Thevenin resistance of the two terminal network.

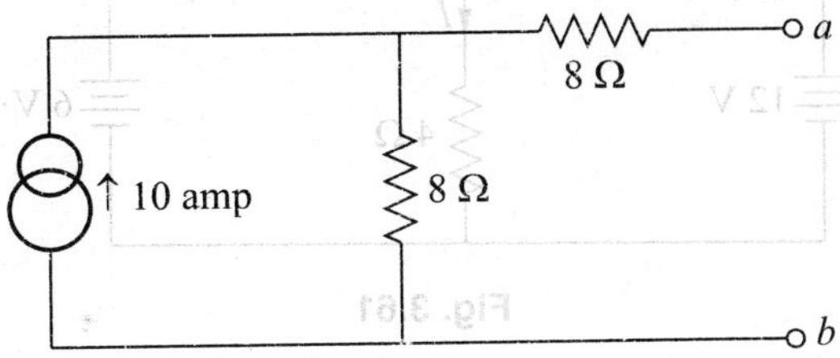

Fig. 3.57

9. Calculate current through Z_R using Norton's theorem.

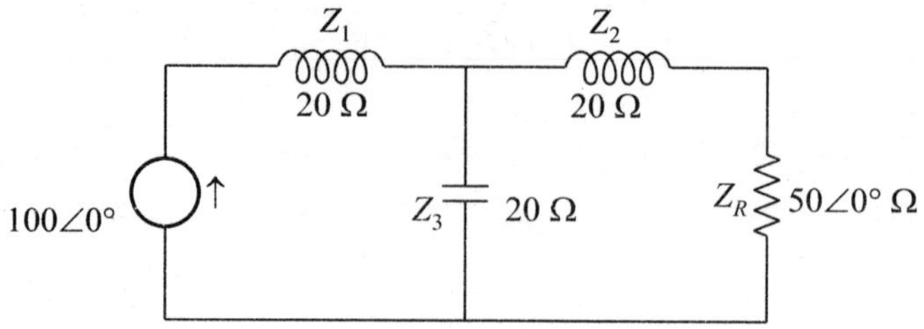

Fig. 3.58

10. Prove reciprocity theorem for the 15 Ω branch in the circuit.

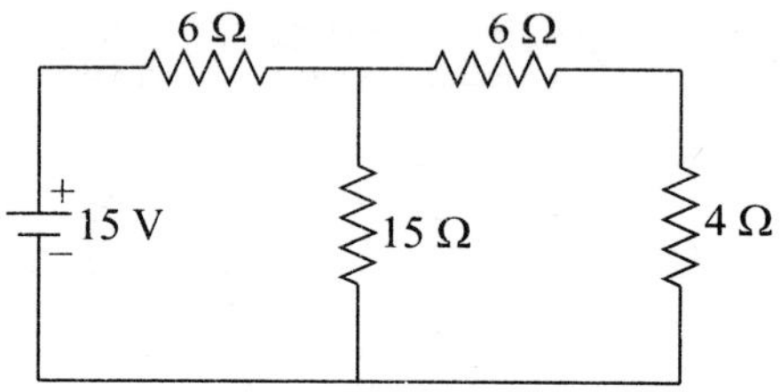

Fig. 3.59

11. Show the Thevenin's equivalent network for the following network.

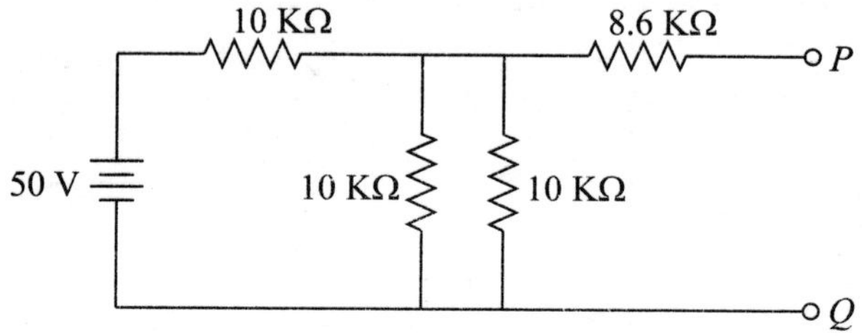

Fig. 3.60

12. Using superposition theorem, calculate current in each branch of the network.

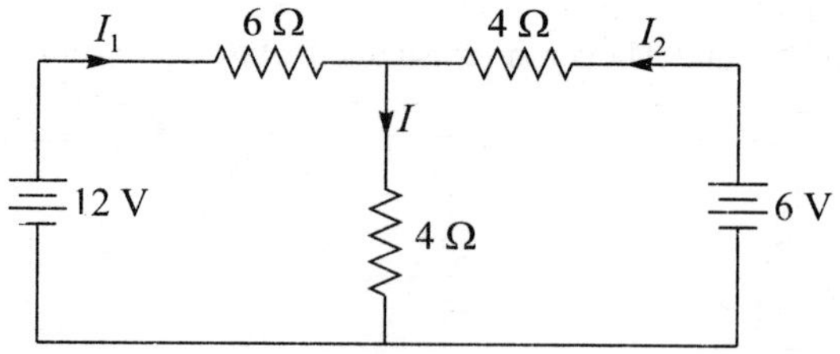

Fig. 3.61

13. Find the branch currents in the following network using superposition theorem.

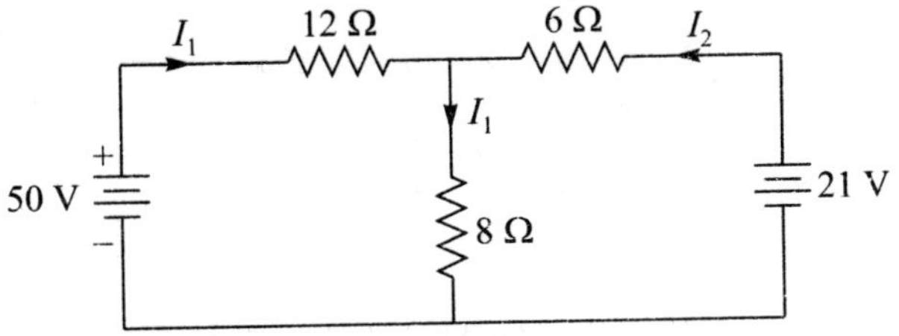

Fig. 3.62

QUESTIONS

A. Short Answer Questions:

1. State and explain Thevenin's Theorem.
2. Define the following terms:
   (*i*) Network (*ii*) Node (*iii*) Branch
3. Give the statements of Thevenin's theorem and Norton's theorem.
4. Give the procedure of superposition theorem to a simple network.
5. Distinguish between Linear and Non Linear networks.
6. State and prove reciprocity theorem.
7. Give the main point of difference between Thevenin's and Norton's Theorems.
8. Give the procedure to Thevenise a network.
9. Give the procedure to Nortonize a network.
10. State and prove Thevenin's Theorem.

B. Long Answer Questions:

1. State and prove superposition theorem and reciprocity theorem.
2. State and prove maximum power transfer theorem. Give one application of this theorem.
3. State and prove Millman Theorem.
4. State and explain Norton's network theorem. How is Norton's equivalent circuit related with the Thevenin's equivalent circuit.
5. State and prove Thevenin's and maximum power transfer theorem.
6. State and prove Thevenin's theorem. Give Millman's statement.
7. State and explain maximum power transfer theorem of A.C. sources.
8. State and prove Millman's reciprocity theorems.
9. State and prove maximum power transfer theorem and obtain true expression for the maximum power delivered to it.
10. State and prove superposition theorem. Explain the method to Thevenise a network.

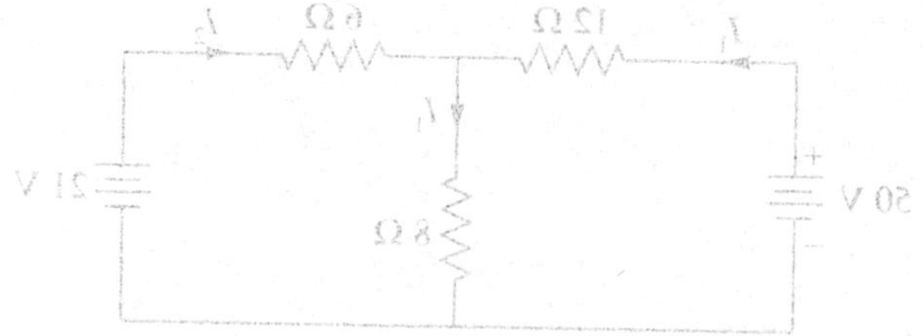

UNIT–II

CHAPTER 4

Cathode Ray Oscilloscope

4.1 INTRODUCTION

Cathode rays are streams of electrons observed in vacuum tubes, *i.e.* evacuated glass tubes that are equipped with at least two electrodes, a cathode (negative electrode) and an anode (positive electrode) in a configuration known as a diode.

When the cathode is heated, it emits radiation, which travels to the anode. If the inner glass walls behind the anode are coated with a phosphorescent material the incident electrons induce a glow. The presence of cathode rays was first postulated in early studies in vacuum tubes by placing metal shapes between the electrodes, thereby casting a shadow on the phosphorescent coating. This suggested that the cause of the light emission was due to rays emitted by the cathode and hitting the coating. They travel towards the anode in straight lines and continue past it for some distance. Because of the following properties they are used in a CRO.

Properties of Cathode Rays

Like a wave :

1. they travel in straight lines
2. Produce a shadow when obstructed by objects
3. could pass through thin metal foils without disturbing them

Like a particle :

Left the surface of the cathode at 90 degrees, not radiating like a wave

1. Deflected by magnetic fields
2. Could turn a wheel in the path of the ray
3. Travel far slower than light

4.2 THE IMPORTANT PARTS OF A CRO OR CRT

The cathode ray tube (CRT) is a vacuum tube containing an electron gun (a source of electrons) and a fluorescent screen, with internal or external means to accelerate and deflect the electron beam, used to form images in the form of light emitted from the fluorescent screen. The image may represent electrical waveforms (oscilloscope), pictures (television, computer monitor), radar targets and others.

The single electron beam can be processed in such a way as to display moving pictures in natural colors.

The CRT uses an evacuated glass envelope which is large, deep, heavy, and relatively fragile. Display technologies without these disadvantages, such as flat plasma screens, liquid crystal displays, have replaced CRTs in many applications and are becoming increasingly common as costs decline.

General description

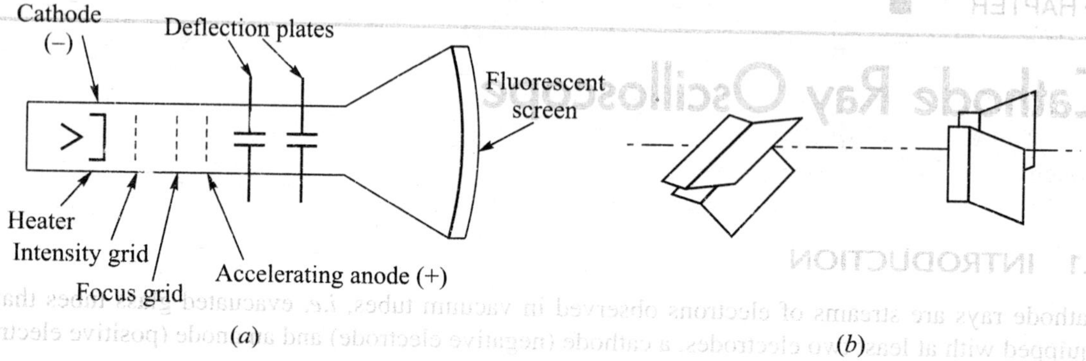

Fig. 4.1. Cathode-ray tube : (*a*) schematic, (*b*) detail of the deflection plates.

The earliest version of the CRT was invented by the German physicist Ferdinand Braun in 1897 and is also known as the 'Braun tube'. It was a cold-cathode diode, a modification of the Crookes tube with a phosphor-coated screen. The first version to use a hot cathode was developed by John B. Johnson (who gave his name to the term Johnson noise) and Harry Weiner Weinhart of Western Electric, and became a commercial product in 1922. The cathode rays are now known to be a beam of electrons emitted from a heated cathode inside a vacuum tube and accelerated by a potential difference between this cathode and an anode. The screen is covered with a phosphorescent coating (often transition metals or rare earth elements), which emits visible light when excited by high-energy electrons. The beam is deflected either by a magnetic or an electric field to move the bright dot to the required position on the screen.

In television sets and computer monitors the entire front area of the tube is scanned systematically in a fixed pattern called a raster. An image is produced by modulating the intensity of the electron beam with a received video signal (or another signal derived from it). In all CRT TV receivers except some very early models, the beam is deflected by magnetic deflection, a varying magnetic field generated by coils (the magnetic yoke), driven by electronic circuits, around the neck of the tube.

4.3 ELECTRON GUN

An electron gun is an electrical component that produces an electron beam that has a precise kinetic energy and is most often used in televisions and monitors which use cathode ray tube technology, as well as in other instruments, such as electron microscopes and particle accelerators. Electron guns may be classified in several ways: by the type of electric field generation (DC or RF), by emission mechanism (thermionic, photocathode, cold emission, plasma source), by focusing (pure electrostatic or with magnetic fields), or by the number of electrodes.

Most color cathode ray tubes - such as those used in color televisions - are made up of three electron guns, each one producing a different stream of electrons. Each stream travels through a shadow mask where the electrons will impinge upon either a red, green or blue phosphor to light up a color pixe on the screen. The resultant color that is seen by the viewer will be a combination of these three.

Fig. 4.2

The source of the electron beam is the electron gun, which produces a stream of electrons through thermionic emission, and focuses it into a thin beam. The gun is located in the narrow, cylindrical neck at the extreme rear of a CRT and has electrical connecting pins, usually arranged in a circular configuration, extending from its end. These pins provide external connections to the cathode, to various grid elements in the gun used to focus and modulate the beam, and, in electrostatic deflection CRTs, to the deflection plates. Since the CRT is a hot-cathode device, these pins also provide connections to one or more filament heaters within the electron guns. When a CRT is operating, the heaters can often be seen glowing orange through the glass walls of the CRT neck. The need for these heaters to 'warm up' causes a delay between the time that a CRT is first turned on, and the time that a display becomes visible. In older tubes, this could take fifteen seconds or more; modern CRT displays have fast-starting circuits which produce an image within about two seconds, using either briefly increased heater current or elevated cathode voltage. Once the CRT has warmed up, the heaters stay on continuously. The electrodes are often covered with a black layer, a patented process used by all major CRT manufacturers to improve electron density.

When electrons strike the poorly-conductive phosphor layer on the glass CRT, it becomes electrically charged, and tends to repel electrons, reducing brightness this effect is known as "sticking". To prevent this the interior side of the phosphor layer can be covered with a layer of aluminium connected to the conductive layer inside the tube, which disposes of this charge. It has the additional advantages of increasing brightness by reflecting towards the viewer light emitted towards the back of the tube, and protecting the phosphor from ion bombardment.

4.4 THE GLASS ENVELOPE

The outer glass allows the light generated by the phosphor out of the monitor, but it must block dangerous X-rays generated by high energy electrons impacting the inside of the CRT face. For this reason, the glass is leaded. Color tubes require significantly higher anode voltages than monochrome tubes (as high as 32,000 volts in large tubes), partly to compensate for the blockage of some electrons by the aperture mask or grille; the amount of X-rays produced increases with voltage. Because of leaded glass, other shielding, and protective circuits designed to prevent the anode voltage from rising too high in case of malfunction, the X-ray emission of modern CRTs is well within approved safety limits.

CRT displays accumulate a static electrical charge on the screen, unless preventive measures are taken. This charge does not pose a safety hazard, but can lead to significant degradation of image quality through attraction of dust particles to the surface of the screen. Unless the display is regularly cleaned with a dry cloth or special cleaning tissue (using ordinary household cleaners may damage anti-glare protective layer on the screen), after a few months the brightness and clarity of the image drops significantly.

The high voltage (EHT) used for accelerating the electrons is provided by a transformer. For CRTs used in televisions, this is usually a flyback transformer that steps up the line (horizontal) deflection supply to as much as 32,000 volts for a color tube, although monochrome tubes and speciality CRTs may operate at much lower voltages. The output of the transformer is rectified and the pulsating output voltage is smoothed by a capacitor formed by the tube itself (the accelerating anode being one plate, the glass being the dielectric, and the grounded (earthed) Aquadag coating on the outside of the tube being the other plate). Before all-glass tubes, the structure between the screen and the electron gun was made from a heavy metal cone which served as the accelerating anode. Smoothing of the EHT was then done with a high voltage capacitor, external to the tube itself. In the earliest televisions, before the invention of the flyback transformer design, a linear high-voltage supply was used; because these supplies were capable of delivering much more current at their high voltage than flyback high voltage systems - in the case of an accident they proved extremely dangerous. The flyback circuit design addressed this: in the case of a fault, the flyback systems deliver relatively little current, improving a person's chance of surviving a direct shock from the high voltage anode.

High Voltage

CRTs operate at very high voltages, which can persist long after the device containing the CRT has been switched off and/or unplugged. Residual charges of hundreds of volts can also remain in large capacitors in the power supply circuits of the device containing the CRT; these charges may persist. Modern circuits contain bleeder resistors, to ensure that the high-voltage supply is discharged to safe levels within a couple of minutes at most. These discharge devices can fail even on a modern unit and leave these high voltage charges present. The final anode connector on the bulb of the tube carries this high voltage.

4.5 ELECTROSTATIC DEFLECTION

Electrostatic deflection refers to a technique for modifying the path of a stream of charged particles by the use of an electric field applied transverse to the path of the particles.

The Lorentz force acts on any charged particle in an electromagnetic field. Electrostatic deflection uses a special, simplified case of this general effect by limiting the field to an electric field. An electric field applies a force on a particle that is proportional to the strength of the field and to the charge on the particle. The direction of the applied force is the same as the direction of the electric field. For electrostatic deflection, the applied electric field is arranged so that it lies in the plane perpendicular to the initial direction of the stream. The particles are accelerated by this force in proportion to the charge of the particles. The path the particles follow depends on their sideways acceleration and their velocity when they enter the deflecting field. Therefore it is important for good control of the direction that the particles in the stream have a uniform charge to mass ratio and that they move at a uniform speed. The technique is called electrostatic because the strength and direction of the applied field changes slowly relative to the time it takes for the particles to transit the field, and thus can be considered not to change (be static) for any particular particle.

The most common use for this technique is controlling the path of a stream of electrons in a vacuum. One application is in small cathode ray tubes for oscilloscopes. In these tubes the electric field is created by two sets of paired electrodes, mounted at right angles, that the electron stream flows between. This arrangement allows independent deflection of the beam in two dimensions (usually perceived as up/down (vertical) and right/left (horizontal)). The electrodes are commonly called deflection plates. Traditionally, the electrons pass through the vertical deflection plates first, yielding slightly higher sensitivity because of the longer travel time from the vertical deflection plates to the phosphor screen as compared to the horizontal deflection plates. In very high speed oscilloscopes, the deflection plates were often complex structures, combining a series of sub-plates with an electrical delay line. By matching the propagation speed of the electrical signal with the transit speed of the electrons, maximum bandwidth (frequency response) was achieved.

The technique works well whenever a sufficiently uniform stream can be created, as discussed above. Therefore, it has been used in controlling macroscopic particle streams in areas such as flow cytometry as well.

Electrostatic deflection is very useful for small deflection angles but is well known to be inferior to magnetic deflection for deflecting a charged particle beam into large angles - say over 10 degrees. The reason is that deflection aberrations become large as the deflection angle increases. This reduces the ability to finely focus the beam. Also in electrostatic deflection it has long been the practice to inject the beam midway between the charged deflection plates so as to avoid the fringe fields as much as possible. However it was found by computation methods that deflection aberrations would be significantly reduced if the beam were injected offset toward the attracting plate. That way the beam tends to follow equipotentials and the deflection force is normal to the beam direction. Thus offset, all the electrons in the beam are deflected into the same angle. There is an induced astigmatism that is correctable. This deflection idea has been tested and verified. Deflection angles of 50 degrees are reportedly possible without measurable deflection aberration. Optimal injection offset is approximately 1/3 of the plate gap toward the deflecting plate. The useful beam diameter is also approximately 1/3 of the gap.

4.6 FOCUSSING

The convergent beam from accelerating electrode has a tendency to spread because of the mutual repulsion between the electrons. Hence, some focussing device is required to bring the beam to a sharp focus at the screen. Two methods of focussing are (*i*) Electrostatic focussing; (*ii*) Magnetostatic focussing.

(i) Electrostatic Focussing

The final two electrodes of the electron gun are the first anode and the second or final anode. These provide an electron lens system for focussing the beam into a spot on the fluorescent screen.

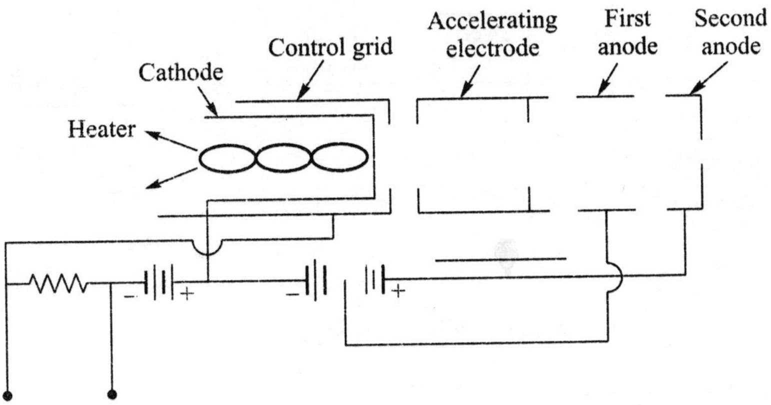

Fig. 4.3

Both these anodes are kept highly positive w.r.t. the cathode. The second or final anode is kept either at the same potential as accelerating electrode or higher, whereas the first anode potential is kept appreciably lower than the second anode potential. The final anode also has a central aperture so as to permit only a well defined beam to pass through it. The first anode generally does not have a transverse plate with an aperture. The final anode potential determines the velocity with which an electron in the beam travels to the screen on leaving the electron gun.

(ii) Magnetostatic Focussing

Magnetostatic focussing may be achieved either by permanent magnet or by passing current through a coil of wire, the axis of which coincides with the beam axis. The coil may extend the whole length of the electron beam or may be concentrated in a smaller region.

The principle of magnetic focussing is shown in figure below. The focussing coil surrounds the tube. The lines of magnetic field are uniformly distributed and are parallel to the direction of the electron motion.

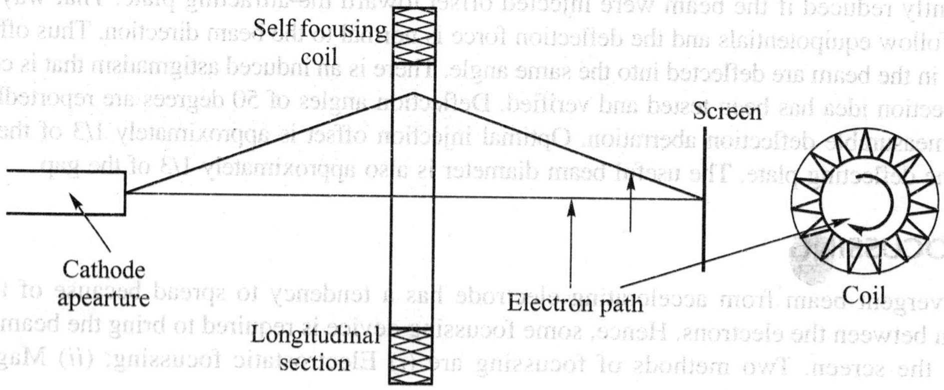

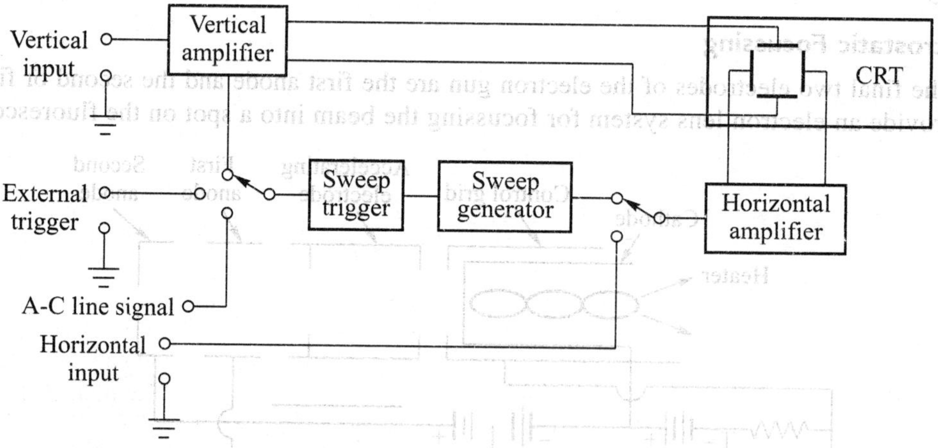

Fig. 4.4. Block diagram of a typical oscilloscope.

4.7 CATHODE-RAY OSCILLOSCOPE (CRO)

An oscilloscope (commonly abbreviated to scope or *O*-scope) is a type of electronic test equipment that allows signal voltages to be viewed, usually as a two-dimensional graph of one or more electrical potential differences (vertical axis) plotted as a function of time or of some other voltage (horizontal axis). The oscilloscope is one of the most versatile and widely-used electronic instruments.

Oscilloscopes are widely used when it is desired to observe the exact wave shape of an electrical signal. In addition to the amplitude of the signal, an oscilloscope can measure the frequency, show distortion, and show the relative timing of two related signals. Oscilloscopes are used in the sciences, medicine, engineering, telecommunications, and industry. General-purpose instruments are used for maintenance of electronic equipment and laboratory work. Special-purpose oscilloscopes may be used for such purposes as adjusting an automotive ignition system, or to display the waveform of the heartbeat.

Originally all oscilloscopes used cathode ray tubes as their display element, but modern digital oscilloscopes use high-speed analog-to-digital converters and computer-like display screens and processing of signals. Oscilloscope peripheral modules for general purpose laptop or desktop personal computers can turn them into useful and flexible test instruments.

The earliest the simplest type of oscilloscope consisted of a cathode ray tube, a vertical amplifier, a timebase, a horizontal amplifier and a power supply. These are now called 'analogue' scopes to distinguish them from the 'digital' scopes that become common in the 1990s and 2000s.

Before the introduction of the CRO in its current form, the cathode ray tube had already been in use as a measuring device. The cathode ray tube is an evacuated glass envelope, similar to that in a black-and-white television set, with its flat face covered in a phosphorescent material (the phosphor). The screen is typically less than 20 cm in diameter, much smaller than the one in a television set.

In the neck of the tube is an electron gun, which is a heated metal plate with a wire mesh (the grid) in front of it. A small grid potential is used to block electrons from being accelerated when the electron beam needs to be turned off, as during sweep retrace or when no trigger events occur. A potential difference of at least several hundred volts is applied to make the heated plate (the cathode) negatively charged relative to the deflection plates. For higher bandwidth oscilloscopes where the trace may move more rapidly across the phosphor target, a positive post-deflection acceleration voltage of over 10,000 volts is often used, increasing the energy (speed) of the electrons that strike the phosphor. The kinetic energy of the electrons is converted by the phosphor into visible light at the point of impact. When switched on, a CRT normally displays a single bright dot in the center of the screen, but the dot can be moved about electrostatically or magnetically. The CRT in an oscilloscope uses electrostatic deflection.

Between the electron gun and the screen are two opposed pairs of metal plates called the deflection plates. The vertical amplifier generates a potential difference across one pair of plates, giving rise to a vertical electric field through which the electron beam passes. When the plate potentials are the same, the beam is not deflected.

When the top plate is positive with respect to the bottom plate, the beam is deflected upwards; when the field is reversed, the beam is deflected downwards. The horizontal amplifier does a similar job with the other pair of deflection plates, causing the beam to move left or right. This deflection system is called electrostatic deflection, and is different from the electromagnetic deflection system used in television tubes. In comparison to magnetic deflection, electrostatic deflection can more readily follow random changes in potential, but is limited to small deflection angles.

The timebase is an electronic circuit that generates a ramp voltage. This is a voltage that changes continuously and linearly with time. When it reaches a predefined value the ramp is reset, with the voltage reestablishing its initial value. When a trigger event is recognized the reset is released, allowing the ramp to increase again. The timebase voltage usually drives the horizontal amplifier. Its effect is to sweep the electron beam at constant speed from left to right across the screen, then quickly return the beam to the left in time to begin the next sweep. The timebase can be adjusted to match the sweep time to the period of the signal.

Meanwhile, the vertical amplifier is driven by an external voltage (the vertical input) that is taken from the circuit or experiment that is being measured. The amplifier has a very high input impedance, typically one megohm, so that it draws only a tiny current from the signal source. The amplifier drives the vertical deflection plates with a voltage that is proportional to the vertical input. Because the electrons have already been accelerated by hundreds of volt, this amplifier also has to deliver almost hundred volts, and this with a very high bandwidth. The gain of the vertical amplifier can be adjusted to suit the amplitude of the input voltage. A positive input voltage bends the electron beam upwards, and a negative voltage bends it downwards, so that the vertical deflection of the dot shows the value of the input.

The response of this system is much faster than that of mechanical measuring devices such as the multimeter, where the inertia of the pointer slows down its response to the input.

When all these components work together, the result is a bright trace on the screen that represents a graph of voltage against time. Voltage is on the vertical axis, and time on the horizontal.

Observing high speed signals, especially non-repetitive signals, with a conventional CRO is difficult, due to non-stable or changing triggering threshold which makes it hard to "freeze" the waveform on the screen. This often requires the room to be darkened or a special viewing hood to be placed over the face of the display tube. To aid in viewing such signals, special oscilloscopes have borrowed from night vision technology, employing a microchannel plate in the tube face to amplify faint light signals.

Although a CRO allows one to view a signal, in its basic form it has no means of recording that signal on paper for the purpose of documentation. Therefore, special oscilloscope cameras were developed to photograph the screen directly. Early cameras used roll or plate film, while in the 1970s Polaroid instant cameras became popular.

The vertical amplifier and timebase controls are calibrated to show the vertical distance on the screen that corresponds to a given voltage difference, and the horizontal distance that corresponds to a given time interval.

The power supply is an important component of the scope. It provides low voltages to power the cathode heater in the tube, and the vertical and horizontal amplifiers. High voltages are needed to drive the electrostatic deflection plates. These voltages must be very stable. Any variations will cause errors in the position and brightness of the trace.

Later analogue oscilloscopes added digital processing to the standard design. The same basic architecture - cathode ray tube, vertical and horizontal amplifiers - was retained, but the electron beam was controlled by digital circuitry that could display graphics and text mixed with the analogue waveforms. The extra features that this system provides include:

(1) on-screen display of amplifier and timebase settings;

(2) voltage cursors-adjustable horizontal lines with voltage display;

(3) time cursors-adjustable vertical lines with time display;

(4) on-screen menus for trigger settings and other functions.

X-Y mode

Most modern oscilloscopes have several inputs for voltages, and thus can be used to plot one varying voltage versus another. This is especially useful for graphing I-V curves (current versus voltage characteristics) for components such as diodes, as well as Lissajous patterns. Lissajous figures are an example of how an oscilloscope can be used to track phase differences between multiple input signals. This is very frequently used in broadcast engineering to plot the left and right stereophonic channels, to ensure that the stereo generator is calibrated properly.

The cathode-ray oscilloscope (CRO) is a common laboratory instrument that provides accurate time and amplitude measurements of voltage signals over a wide range of frequencies. Its reliability, stability, and ease of operation make it suitable as a general purpose laboratory instrument. The heart of the CRO is a cathode-ray tube shown schematically in Fig. 4.1.

The cathode ray is a beam of electrons which are emitted by the heated cathode (negative electrode) and accelerated toward the fluorescent screen. The assembly of the cathode, intensity grid, focus grid, and accelerating anode (positive electrode) is called an electron gun. Its purpose is to generate the electron beam and control its intensity and focus. Between the electron gun and the fluorescent screen are two pair of metal plates - one oriented to provide horizontal deflection of the beam and another pair oriented to give vertical deflection to the beam. These plates are thus referred to as the horizontal and vertical deflection plates. The combination of these two deflections allows the beam to reach any portion of the fluorescent screen. Whenever the electron beam hits the screen, the phosphor is excited and light is emitted from that point. This conversion of electron energy into light allows us to write with points or lines of light on an otherwise darkened screen.

In the most common use of the oscilloscope the signal to be studied is first amplified and then applied to the vertical (deflection) plates to deflect the beam vertically and at the same time a voltage that increases linearly with time is applied to the horizontal (deflection) plates thus causing the beam to be deflected horizontally at a uniform (constant) rate. The signal applied to the vertical plates is thus displayed on the screen as a function of time. The horizontal axis serves as a uniform time scale.

The linear deflection or sweep of the beam horizontally is accomplished by use of a sweep generator that is incorporated in the oscilloscope circuitry. The voltage output of such a generator is that of a sawtooth wave as shown in Fig. 4.5. Application of one cycle of this voltage difference, which increases linearly with time, to the horizontal plates causes the beam to be deflected linearly with time across the tube face. When the voltage suddenly falls to zero, as at points (*a*) (*b*) (*c*), ... etc., the end of each sweep - the beam flies backs to its initial position. The horizontal deflection of the beam is repeated periodically, the frequency of this periodicity is adjustable by external controls.

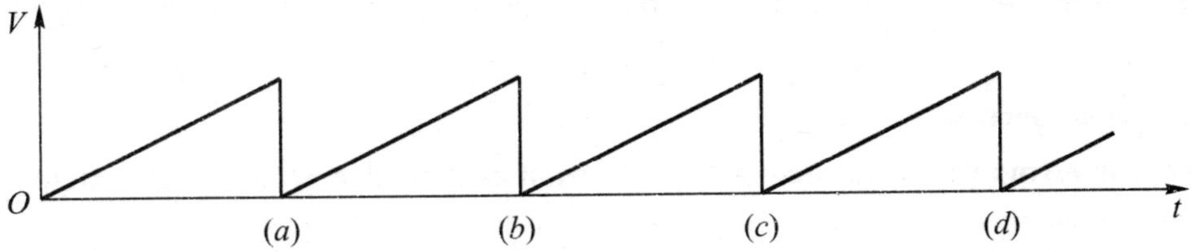

Fig. 4.5. Voltage difference V between horizontal plates as a function of time *t*.

To obtain steady traces on the tube face, an internal number of cycles of the unknown signal that is applied to the vertical plates must be associated with each cycle of the sweep generator. Thus, with such a matching of synchronization of the two deflections, the pattern on the tube face repeats itself and hence appears to remain stationary. The persistence of vision in the human eye and of the glow of the fluorescent screen aids in producing a stationary pattern. In addition, the electron beam is cut off (blanked) during flyback so that the retrace sweep is not observed.

4.8 CRO OPERATION

A simplified block diagram of a typical oscilloscope is shown in Fig. 4.4. In general, the instrument is operated in the following manner. The signal to be displayed is amplified by the vertical amplifier and applied to the vertical deflection plates of the CRT. A portion of the signal in the vertical amplifier is applied to the sweep trigger as a triggering signal. The sweep trigger then generates a pulse coincident with a selected point in the cycle of the triggering signal. This pulse turns on the sweep generator, initiating the sawtooth wave form. The sawtooth wave is amplified by the horizontal amplifier and applied to the horizontal deflection plates. Usually, additional provisions signal are made for applying an external triggering signal or utilizing the 60 Hz line for triggering. Also the sweep generator may be bypassed and an external signal applied directly to the horizontal amplifier.

4.9 CRO CONTROLS

The controls available on most oscilloscopes provide a wide range of operating conditions and thus make the instrument especially versatile. Since many of these controls are common to most oscilloscopes a brief description of them follows.

Cathode-raytube

Power and Scale Illumination: Turns instrument on and controls illumination of the graticule.

Focus: Focus the spot or trace on the screen.

Intensity: Regulates the brightness of the spot or trace.

Vertical amplifier section

Position: Controls vertical positioning of oscilloscope display.

Sensitivity: Selects the sensitivity of the vertical amplifier in calibrated steps.

Variable Sensitivity: Provides a continuous range of sensitivities between the calibrated steps. Normally the sensitivity is calibrated only when the variable knob is in the fully clockwise position.

AC-DC-GND: Selects desired coupling (ac or dc) for incoming signal applied to vertical amplifier, or grounds the amplifier input. Selecting dc couples the input directly to the amplifier; selecting ac sends the signal through a capacitor before going to the amplifier thus blocking any constant component.

Horizontal-sweep section

Sweep time/cm: Selects desired sweep rate from calibrated steps or admits external signal to horizontal amplifier.

Sweep time/cm Variable: Provides continuously variable sweep rates. Calibrated position is fully clockwise.

Position: Controls horizontal position of trace on screen.

Horizontal Variable: Controls the attenuation (reduction) of signal applied to horizontal amplifier through Ext. Horiz. connector.

Trigger

The trigger selects the timing of the beginning of the horizontal sweep.

Slope: Selects whether triggering occurs on an increasing (+) or decreasing (–) portion of trigger signal.

Coupling: Selects whether triggering occurs at a specific dc or ac level.

Source: Selects the source of the triggering signal.

INT - (internal) - from signal on vertical amplifier

EXT - (external) - from an external signal inserted at the **EXT. TRIG. INPUT.**

LINE - 60 cycle trigger

Level: Selects the voltage point on the triggering signal at which sweep is triggered. It also allows automatic (auto) triggering of allows sweep to run free (free run).

4.10 CONNECTIONS FOR THE OSCILLOSCOPE

Vertical Input: A pair of jacks for connecting the signal under study to the *Y* (or vertical) amplifier. The lower jack is grounded to the case.

Horizontal Input: A pair of jacks for connecting an external signal to the horizontal amplifier. The lower terminal is grounded to the case of the oscilloscope.

External Trigger Input: Input connector for external trigger signal.

Cal. Out: Provides amplitude calibrated square waves of 25 and 500 millivolts for use in calibrating the gain of the amplifiers.

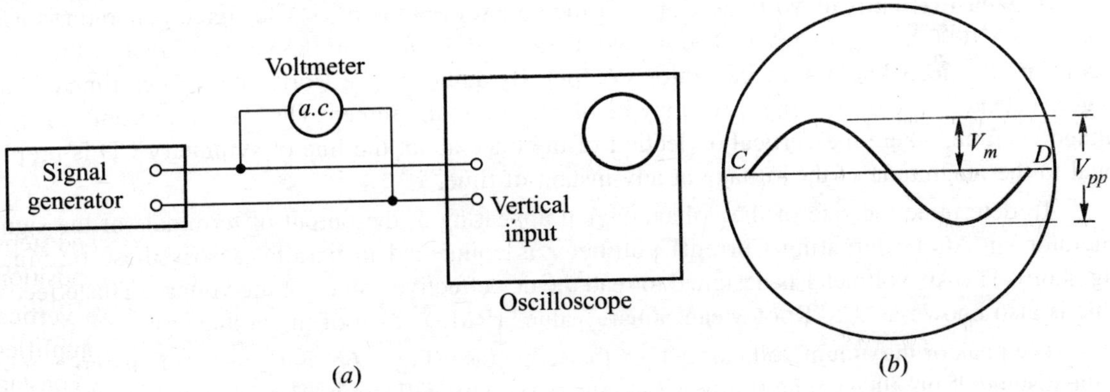

Fig. 4.6. (*a*) Circuit for procedure II. (*b*) Trace seen on scope.

Accuracy of the vertical deflection is +3%. Sensitivity is variable.

Horizontal sweep should be accurate to within 3%. Range of sweep is variable.

Operating Instructions: Before plugging the oscilloscope into a wall receptacle, set the controls as follows:

(*a*) Power switch at off
(*b*) Intensity fully counter clockwise
(*c*) Vertical centering in the center of range
(*d*) Horizontal centering in the center of range
(*e*) Vertical at 0.2
(*f*) Sweep times 1

Plug line cord into a standard ac wall recepticle. Turn power on. Do not advance the Intensity Control.

Allow the scope to warm up for approximately two minutes, then turn the Intensity Control until the beam is visible on the screen.

Warning: Never advance the Intensity Control so far that an excessively bright spot appears. Bright spots imply burning of the screen. A sharp focused spot of high intensity (great brightness) should never be allowed to remain fixed in one position on the screen for any length of time as damage to the screen may occur.

Adjust Horizontal and Vertical Centering Controls. Adjust the focus to give a sharp trace. Set trigger to internal, level to auto.

Procedure

I. Set the signal generator to a frequency of 1000 cycles per second. Connect the output from the generator to the vertical input of the oscilloscope. Establish a steady trace of this input signal on the scope. Adjust all of the scope and signal generator controls until you become familiar with the function of each. The purpose of such "playing" is to allow the student to become so familiar with the oscilloscope that it becomes an aid (tool) in making measurements in other experiments and not as a formidable obstacle. Note: If the vertical gain is set too low, it may not be possible to obtain a steady trace.

II. Measurements of Voltage: Consider the circuit in Fig. 4.6(*a*). The signal generator is used to produce a 1000 hertz sine wave. The AC voltmeter and the leads to the vertical input of the oscilloscope are connected across the generator's output. By adjusting the Horizontal Sweep time/cm and trigger, a steady trace of the sine wave may be displayed on the screen. The trace represents a plot of voltage vs. time, where the vertical deflection of the trace about the line of symmetry CD is proportional to the magnitude of the voltage at any instant of time.

To determine the size of the voltage signal appearing at the output of terminals of the signal generator, an AC (Alternating Current) voltmeter is connected in parallel across these terminals (Fig. 4.6*a*). The AC voltmeter is designed to read the dc "effective value" of the voltage. This effective value is also known as the "Root Mean Square value" (RMS) value of the voltage.

The peak or maximum voltage seen on the scope face (Fig. 4.6*b*) is V_m volts and is represented by the distance from the symmetry line *CD* to the maximum deflection. The relationship between the magnitude of the peak voltage displayed on the scope and the effective or RMS voltage (VRMS) read on the AC voltmeter is

$$V_{RMS} = 0.707\ V_m \text{ (for a sine or cosine wave).}$$

Thus

$$V_m = \frac{V_{RMS}}{0.707}$$

Agreement is expected between the voltage reading of the multimeter and that of the oscilloscope. For a symmetric wave (sine or cosine) the value of V_m may be taken as 1/2 the peak to peak signal V_{pp}.

The variable sensitivity control signal may be used to adjust the display to fill a convenient range of the scope face. In this position, the trace is no longer calibrated so that you can not just read the size of the signal by counting the number of divisions and multiplying by the scale factor. However, you can figure out what the new calibration is and use it as long as the variable control remains unchanged.

Caution: The mathematical prescription given for RMS signals is valid only for sinusoidal signals. The meter will not indicate the correct voltage when used to measure non-sinusoidal signals.

III. Frequency Measurements: When the horizontal sweep voltage is applied, voltage measurements can still be taken from the vertical deflection. Moreover, the signal is displayed as a function of time. If the time base (*i.e.* sweep) is calibrated, such measurements as pulse duration or signal period can be made. Frequencies can then be determined as reciprocal of the periods.

Set the oscillator to 1000 Hz. Display the signal on the CRO and measure the period of the oscillations. Use the horizontal distance between two points such as *C* to *D* in Fig. 4.6*b*.

Set the horizontal gain so that only one complete wave form is displayed.

Then reset the horizontal until 5 waves are seen. Keep the time base control in a calibrated position. Measure the distance (and hence time) for 5 complete cycles and calculate the frequency from this measurement. Compare your result with the value determined above.

Repeat your measurements for other frequencies of 150 Hz, 5 kHz, 50 kHz as set on the signal generator.

IV. Lissajous Figures: When sine-wave signals of different frequencies are input to the horizontal and vertical amplifiers a stationary pattern is formed on the CRT when the ratio of the two frequencies is an integral fraction such as 1/2, 2/3, 4/3, 1/5 etc. These stationary patterns are known as Lissajous figures and can be used for comparison measurement of frequencies.

Use two oscillators to generate some simple Lissajous figures like those shown in Fig. 4.7. You will find it difficult to maintain the Lissajous figures in a fixed configuration because the two oscillators are not phase and frequency locked. Their frequencies and phase drift slowly causing the two different signals to change slightly with respect to each other.

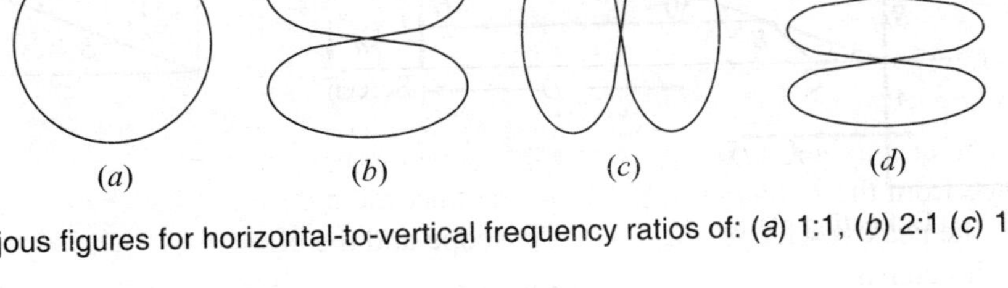

Fig. 4.7. Lissajous figures for horizontal-to-vertical frequency ratios of: (*a*) 1:1, (*b*) 2:1 (*c*) 1:2, and (*d*) 3:1

4.11 FLUORESCENT SCREEN

The Fluorescent materials are compound of light metals such as Zinc, Cadmium, Beryllium and Calcium, with non-metals such as Oxygen, Silicon and Sulphur. The CRT screen is coated with a Fluorescent material.

Approximately 15 different types of phosphorus have been standardised for various *C-R* tube applications, including television, radar and oscilloscope. The two phosphorus most commonly used are designated as P_4 and P_6 . P_4 phosphor is a general purpose material for black and white display, whereas P_6 type is a more expensive material specially suited to the reproduction of display in colour.

The important characteristics of phosphor are : (1) Colour, (2) After glow (or) decay characteristics, (3) Luminescence (or) Liminous efficiency.

Luminous efficiency is the ratio of the visible power in candles generated by the phosphor to the electrical power in watts delivered by the beam. It ranges normally from 1 to 5 candles per watt for tubes used in Television.

When a beam moves past a particular screen element the excitation is removed from that portion of the phosphor and the light begins to decay along a power curve (t^{-h}) (or) an exponential curve e^{-t} depending upon the nature of the phosphor material.

It is essential that this decay should substantially complete itself within a time comparable with the effective duration of persistence of vision.

Uses of Cathode Ray Oscilloscope

Cathode Ray Oscilloscope and C.R. tube are used for a variety of purposes.

(1) In television receiver, a CRT along with associated sweep circuits is used for the recreation of the images.

(2) CRT is used to give visual indication of the position of the target in RADAR.

(3) For measurement of voltage or current which may be direct or alternating.

(4) CRO's are used to study various types of waveforms.

(5) CRO's are used for measurement of frequency and phase angle.

4.12 DEFLECTION SYSTEM

There are two methods of deflecting the electron beam. (1) Electrostatic deflection system and (2) Magnetic deflection system.

(1) Electrostatic Deflection System - Deflection Sensitivity

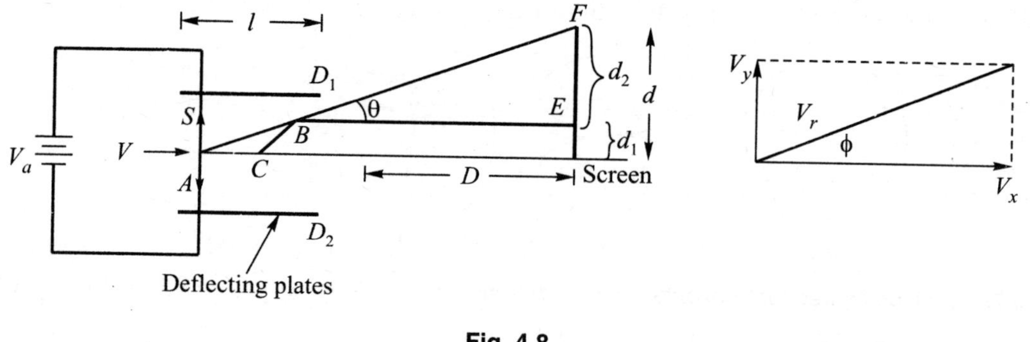

Fig. 4.8

Electrostatic deflection sensitivity of a pair of deflection plates of a C.R.O. is defined as the amount of deflection (in mm or inch) of electron spot produced when a voltage of one volt d.c. is applied between the corresponding deflection plates. The deflection sensitivity is expressed in terms of mm/volt (or) inch/volt. This deflection sensitivity may be different of *x-x* and *y-y* plates. The amount of deflection produced may be calculated knowing (1) The dimensions of the tube components and (2) Deflecting voltage.

The above figure shows the configuration of the electrostatic deflection system. Two plates of length '*l*', and spacing '*s*' are kept at a distance *D* from the screen. Let the voltage applied between the plates be 'V_d' volts and '*V*' be the velocity of an electron on entering the field of the deflection plates.

Then $$V = \sqrt{\left(\frac{2Q_e V_a}{m}\right)} = 5.9 \times 10^5 \sqrt{V_a} \text{ metres/Sec.} \quad ...(1)$$

Here V_a = The voltage of the final anode in volts

Q_e = charge of an electron in coulombs = – 1.6 × 10^{-19} coulombs

m = mass of an electron in kilograms = 9.1 × 10^{-31} kg

As the beam passes through the field of the deflection plates, electrons are attracted towards the positive plate by a force which is given by

$$\text{Force} = \text{mass} \times \text{acceleration} = Q_e E$$

where *E* = electric field

$$m\frac{d^2y}{dt^2} = -Q_e\frac{V_d}{S} \quad \text{where} \quad E = \frac{V_d}{S}$$

(∵ Potential between the plates D_1 and D_2 is V_d)

$$\frac{d^2y}{dt^2} = -\frac{Q_e V_d}{mS} \quad ...(2)$$

where *m* is mass of electron = 9.11 × 10^{-31} kg. and Q_e = – 1.6 × 10^{-19} coulomb. With the help of the equation (2) the velocity and displacement of the electron can be calculated provided the initial boundary conditions are known. It is assumed that the electron reaches at *A* whose coordinates are $x = 0$ and $y = 0$ at $t = 0$.

At the point *A*, $v_y = 0$ and $v_x = \sqrt{\left(\frac{2Q_e V_a}{m}\right)}$

Integrating equation (2), we get

$$\frac{dy}{dt} = v_y = \frac{-Q_e V_d}{m\,.\,S}\,t + C$$

where *C* is the constant of integration. When $t = 0$, $v_y = 0$, hence $C = 0$. Therefore

$$\therefore \quad \frac{dy}{dt} = \frac{-Q_e V_d}{m\,.\,S}\,.\,t \quad ...(3)$$

Integrating again equation (3), we have $y = \frac{-Q_e V_d}{2m\,.\,S}\,.\,t^2 + D$

But $D = 0$ as $y = 0$ when $t = 0$ so that $y = \frac{-Q_e V_d}{2m\,.\,S}\,.\,t^2$ ...(4)

Equation (4) gives the displacement of the electron along *Y* axis at any instant.

The distance travelled by the electron in time *t* along *X* axis is

$$x = \text{velocity} \times \text{time} = \left(\frac{-2Q_e V_a}{m}\right)^{1/2}.\,t \quad ...(5)$$

Eliminating t from equations (4) and (5), we have

$$y = \frac{-Q_e V_d}{2mS} \cdot \frac{x^2 \cdot m}{2(-Q_e) \cdot V_a} = \frac{V_d \cdot x^2}{4 \cdot S \cdot V_a} \qquad ...(6)$$

This is the equation of parabola and hence the electron moves in a parabolic path between the plates. After the point B, it follows a straight path and strikes the screen at a point F, the tangent FB makes an angle θ with the X-axis.

From figure, $\tan\theta = \dfrac{d}{(D + 1/2)}$ ...(7)

We know that $\tan\theta = \left.\dfrac{dy}{dx}\right|_{x=1}$

From equation (6) $\tan\theta = \left.\dfrac{dy}{dx}\right|_{x=1} = \left.\dfrac{V_d \cdot x}{2 \cdot S \cdot V_a}\right|_{x=1}$ or $\tan\theta = \dfrac{V_d \cdot 1}{2 \cdot S \cdot V_a}$ ...(8)

From equations (7) and (8), we have $\dfrac{d}{(D + 1/2)} = \dfrac{V_d \cdot 1}{2 \cdot S \cdot V_a}$

$\therefore$ Deflection sensitivity $= \dfrac{d}{V_d} = \dfrac{1}{2 \cdot S \cdot V_a}(D + 1/2)$ ...(9)

(2) Magnetic deflection sensitivity (or) Magnetic deflection system

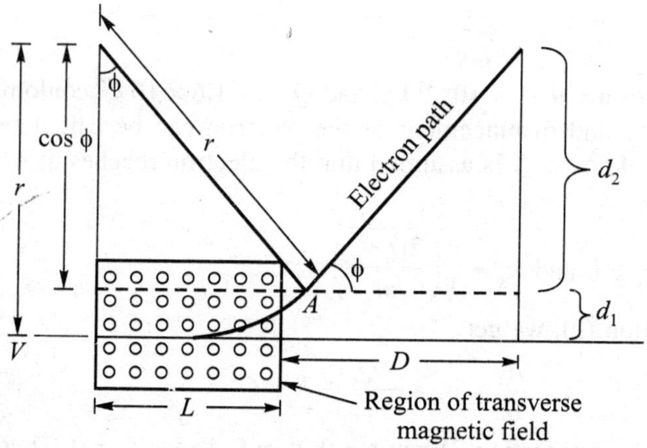

Fig. 4.9

In this case magnetic field is set up by passing current through a coil whose axis is normal to the direction of electron beam. The magnetic field, normal to the beam, exerts a force on electrons urging them to move in a direction at right angles to both the magnetic field and the direction of motion of electrons.

The magnetic field continues to deflect the electron beam at right angles to its line of motion. Hence the path of electron in the magnetic field is a circle. The radius of this circle depends on forward velocity 'V' of the electron, and on the strength of the magnetic flux 'B'.

The force exerted on the electron by the magnetic field is given by $F = q\ v\ b \sin\theta$. ...(1)

In the case under consideration, $\phi = 90°$. Equating the centrifugal force to the centripetal force.

$$B \cdot qv = \frac{mv^2}{r}$$

$$r = \frac{mv}{qB} \qquad \text{...(2)}$$

Also $$\tan\theta = \frac{d^3}{D}$$

$$d_2 = D\tan\theta = D\frac{\sin\theta}{\cos\theta} = \frac{D \cdot \frac{l}{r}}{\sqrt{(1-\sin^2\theta)}} = \frac{Dl}{r\sqrt{\left(1-\frac{l^2}{r^2}\right)}} = \frac{Dl}{\sqrt{(r^2-l^2)}}$$

Now $$d = r - r\sqrt{\left(1-\frac{l^2}{r^2}\right)} + \frac{D_1}{\sqrt{(r^2-l^2)}}$$

If $l << r$, then l and $\frac{l^2}{r^2}$ can be neglected. $\therefore \quad d = (Dl/r)$

But $$r = \frac{mv}{qB}$$

Hence, $$d = \frac{Dlq \cdot B}{m \cdot v} \qquad \text{...(4)}$$

Substituting the value of 'V' in terms of final anode voltage, we get

$$d = \frac{Dl \cdot B\sqrt{qlm}}{\sqrt{2va}} \qquad \text{...(5)}$$

$\therefore$ Hence Magnetic Deflection Sensitivity $\frac{d}{B} = \frac{Dl\sqrt{qlm}}{\sqrt{2Va}}$.

Alternatively magnetic deflection sensitivity may be defined as the amount of deflection (in mm or inches) caused when a current of one milliampere flows through the deflection coil. Magnetic deflection sensitivity is then expressed in mm/mA (or) inch/mA.

Defects or Deflection are (1) Defocusing, (2) Ion spot and Ion trap.

4.13 TIME BASE

When we apply an a.c. voltage to the vertical plates without any voltage across the horizontal plates, the spot moves up and down and we observe a vertical line on the screen. Thus we observe a vertical line even though we have applied an a.c. voltage. This is due to the fact that we have not provided a time axis to the voltage versus time graph.

To provide the time axis, we have to move the spot horizontally from one end of the screen to the other end at a rate proportional time. Immediately after it reaches the other end, the spot should return back to the starting point. Thus, we provide time base (or) time axis.

Time - base may be defined as a circuit for causing the deflection of the spot in the horizontal (or vertical or sometimes circular) direction to vary in a known manner, usually linearly with time. The name "time-base" is given, as the deflection is proportional to time and this deflection is along the x-axis or base line of the coordinate system. The wave form of this time Base voltage is of the shape of saw-tooth and hence this voltage is often called "Saw-tooth Voltage".

The common time-base circuit (or) voltage sweep generators are

(1) BJT Miller voltage sweep generator

(2) Bootstrap voltage sweep generator

(3) JFET miller voltage sweep generator

(4) Regenerative (or) Self gating miller sweep generator

(5) Double bootstrap JFET voltage sweep generator

SOLVED EXAMPLES

Example 1: *In a Cathode Ray tube, the length of deflecting plates in the direction of the beam is 4 cm, the spacing of the plates is 0.5 cm and the distance of the Fluorescent Screen from the centre of the plates is 18 cm. Calculate the deflection sensitivity in cm/volt, if the final anode voltage is (a) 200 Volts, (b) 400 Volts, (c) 600 Volts.*

Solution: Deflection sensitivity $= \dfrac{lD}{2SV}$

Where $l = 4$ cm, $D = 18$ cm, $S = 0.5$ cm

(*i*) For $V_a = 200$ volts

$$\text{Deflection sensitivity} = \frac{4 \times 8}{2 \times 0.5 \times 200} = \frac{72}{200} = 0.36 \text{ cm/volt}$$

(*ii*) For $V_a = 400$ volts

$$\text{Deflection sensitivity} = \frac{4 \times 8}{2 \times 0.5 \times 400} = \frac{72}{400} = 0.18 \text{ cm/volt}$$

(*iii*) For $V_a = 600$ volts

$$\text{Deflection sensitivity} = \frac{4 \times 8}{2 \times 0.5 \times 600} = \frac{72}{600} = 0.12 \text{ cm/volt}$$

Example 2: *In a CRT pair of deflecting plates are 2.0 cm long and are spaced 1 cm apart. The distance from the centre of the plates to the screen is 24 cm. The final anode voltage is 500 volts. Calculate (a) the displacement produced by deflecting voltage of 30 volts ; (b) the angle which the beam makes with the axis of the tube on emerging from the field and (c) Velocity of the beam on emerging from the field.*

Solution: (*a*) Deflection produced

$$d = \frac{V_d l D}{2SV_a} = \frac{130 \times 2 \times 24}{2 \times 1 \times 500} = \frac{6240}{1000} = 6.24 \text{ cm}$$

(*b*)
$$\tan \phi = \frac{d}{D} = \frac{V_d l}{2SV_a} = \frac{30 \times 2}{2 \times 1 \times 500} = \frac{60}{1000} = 0.06$$

$$\phi = \tan^{-1} = (0.06) = 3^0\, 26'$$

(*c*)
$$V = \sqrt{\frac{2q \cdot V_a}{m}} = 5.94 \times 10^5 \sqrt{V_a} \text{ metres/Sec}$$

$$= 5.94 \times 10^5 \sqrt{500} = 13.28 \times 10^6 \text{ metres/sec.}$$

Example 3: *A Cathode Ray Tube has a magnetic deflection field of flux density 10^{-6} webers/ sq. meter for a distance of 30 m along the tube axis. The distance from the end of the field to the screen is 20 cm. The final anode voltage is 800 volts. Calculate the deflection of electron spot in cm. Also find the displacement of a charged particle of charge twice that of an electron and mass 7344 times as large.*

Solution:
$$d = \frac{DlB\sqrt{q/m}}{\sqrt{2V}}$$

For an electron

$$d = \frac{20 \times 10^{-2} \times 3 \times 10^{-4} \times 10^{-2}\sqrt{1.76 \times 10^{11}}}{\sqrt{2 \times 800}} = 0.629 \text{ cm}$$

(*ii*) For the charged particle

$$d = \frac{20 \times 10^{-2} \times 3 \times 10^{-2} \times 10^{-4}\sqrt{1.76 \times 10^{11} \times 2/7344}}{\sqrt{2 \times 800}}$$

$$d = 0.0104 \text{ cm}$$

Example 4: *Two voltages of same frequency V_1 and V_2 are applied to the horizontal and vertical inputs of a C.R.O. The resulting trace is an ellipse as shown below.*

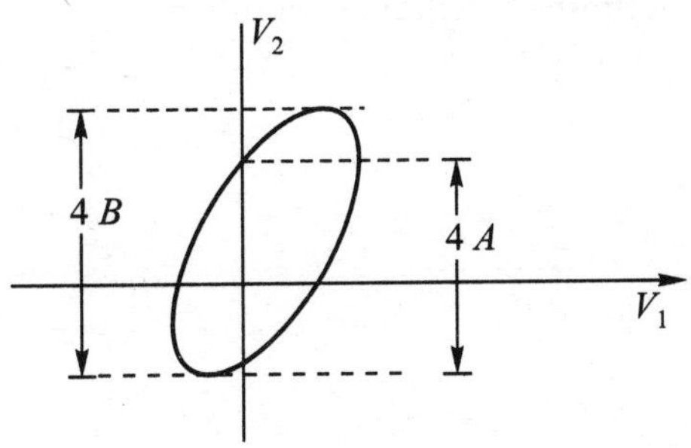

Fig. 4.10

A = 2.0 cm and B = 4.0 cm. Estimate phase difference between V_1 and V_2.

Solution:
$$\sin \phi = \frac{A}{B} = \frac{2}{4} = \frac{1}{2}$$
$$\phi = 30°$$

EXERCISE

1. In a C.R. Tube the final anode voltage is 1800 volts and the distance from the final anode to screen is 40 cm. It is kept with its axis vertical. Calculate the deflection of the spot due to earth's magnetic field. The earth's magnetic field is 0.2 AT/Cm and angle of dip is 55 degrees.
2. A cathode ray tube base *x*-plate deflection sensitivity of 0.6 mm/volt. Find the type of pattern obtained on the screen and its dimensions for the following types of applied voltage at the *x* plates (*i*) alternating voltage 50 sin 2 π 50 *t*; (*ii*) 80 sin 40 *t*.
3. In a cathode ray tube, the distance of the screen from the middle of the plates is 24 cm and the length of the magnetic field along the axis is 4 cm. If the final anode voltage is 400 V, calculate magnetic field intensity required to produce a deflection of 1.2 cm on the screen, $e/m = 5 \times 10^{-11}$ esu/gm.

4. A cathode ray tube has final anode voltage of 1250 volts and the distance from the final anode to screen is 24 cm. The earth's magnetic field is 0.24 AT/Cm and the angle of dip is 50 degrees. Find the deflection of the luminous spot on the screen from the axis of the tube, when this axis is horizontal and (*i*) in the plane of the magnetic maridian, (*b*) at the right angles to the maridian and (*c*) at the angle of 60° degrees with the magnetic maridian.

QUESTIONS

1. What are the principle uses of a C.R.O.?
2. Describe the working of C.R.T.
3. Describe the method of Electrostatic focussing of a C.R.O.
4. Give the purpose of attenuator in C.R.O.
5. Write a short note on "Electron Gun".
6. Draw the diagram of C.R.T. and explain the function of each part.
7. Give the block diagram of C.R.O. and explain.
8. Describe the time base in C.R.O.
9. Explain the action of fluorescent screen of a C.R.O.
10. Mention the front panel controls of C.R.O.
11. Compare the electrostatic and magnetic deflection systems.
12. Derive the expression for electrostatic deflection sensitivity.

CHAPTER 5

R.L. and R.C. Circuits

5.1 INTRODUCTION

Consider a circuit having a d.c. source and an inductance or a capacitance in series with a resistance. When the source is switched on, the process of energy storage starts and the circuit voltage and current build up gradually rather than instantly. When the capacitor or inductor has stored the energy to its full value, the circuit voltage and current do not change with time any more and the circuit is said to have acquired steady state. The current in the circuit or the voltages across the elements under steady state condition is known as steady state response. When the source is switched off, the energy stored in the inductor or capacitor releases gradually and it does not permit the circuit voltage and current to fall abruptly to zero, but takes a finite time. The entire process is shown in Fig. 5.1. During the process of storage or decay, of energy the circuit is said to be in a transient state. The current in the circuit or the voltage across the elements under transient state is known as transient response of the circuit. During transient state the variation of voltage and current depends on circuit parameters.

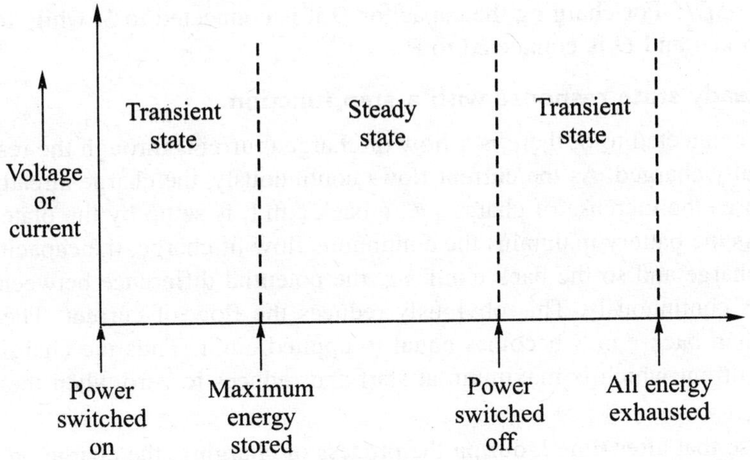

Fig. 5.1

Similarly if an a.c. source is applied to the electrical circuit, the response of the circuit consists of a steady-state response along with transient response. If $f(x)$ represents the total response of the circuit, the steady-state response $f_s(x)$ is defined as

$$f_s(x) = \underset{t \to \infty}{\text{Lt}} f(x)$$

and transient response will be the remaining part of total response other than the steady-state solutions.

Definition of Step Forcing Function

When the energy source which is suddenly applied to a circuit, is a d.c. source, then we say it is a step forcing function. The operation of a switch in series with a battery is equivalent to a step forcing function which is zero upto the instant the switch if off and is equal to the battery voltage after the switch is closed. A unit step forcing function is a function of time which has zero values for all values of its argument less than zero and which is unity for all positive values of its argument (Fig. 5.2). Mathematically we can write the unit step forcing function as

$$f(t) = 1 \text{ for } t \geq 0$$
$$f(t) = 0 \text{ for } t < 0$$

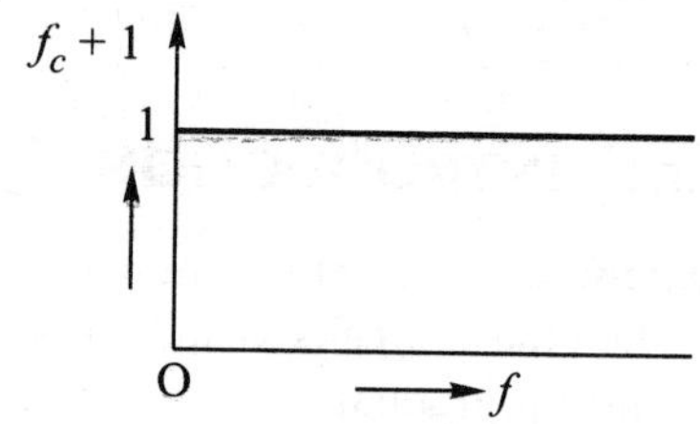

Fig. 5.2

When the battery of e.m.f. E is connected in the circuit, then we can write the step forcing function as

$$v(t) = E\, f(t)$$
$$v(t) = E \text{ for } t \geq 0$$
$$= 0 \text{ for } t < 0$$

5.2 TRANSIENT RESPONSE OF AN R-C CIRCUIT WHEN D.C. SOURCE IS CONNECTED

Consider a circuit consists of a resistance R, a capacitor of capacity C alongwith a battery of e.m.f. E and a two way key *SOP*. For charging the capacitor O it is connected to S, while for discharging the capacitor OS is broken and O is connected to P.

Transient and steady state response with a step function

When O is connected to S, there is a flow of charge (current) through the resistance R and the capacitor is gradually charged. As the current flows continuously, the charge already accumulated by the capacitor opposes the increase of charge *i.e.*, a back e.m.f. is setup by the plates of the capacitor during charging. As the battery maintains the continuous flow of charge, the capacitor gradually goes on accumulating charge and so the back e.m.f. *i.e.* the potential difference between the plates of the capacitor increases continuously. This obviously reduces the flow of current. The charging current reduces to zero when back e.m.f. becomes equal to applied e.m.f. Thus the charging of a capacitor involves varying current which is maximum at start and reduces to zero when the capacitor is fully charged.

Now suppose that after time t, during the process of charging, the charge on the capacitor is q and the current through resistance R is i. Then

The potential difference across resistance $R = R_i$

The potential difference across condenser $= q/C$

Applying Kirchoff's second law to the closed circuit, the equation of e.m.f.s is

$$Ri + \frac{q}{C} = E \qquad ...(1)$$

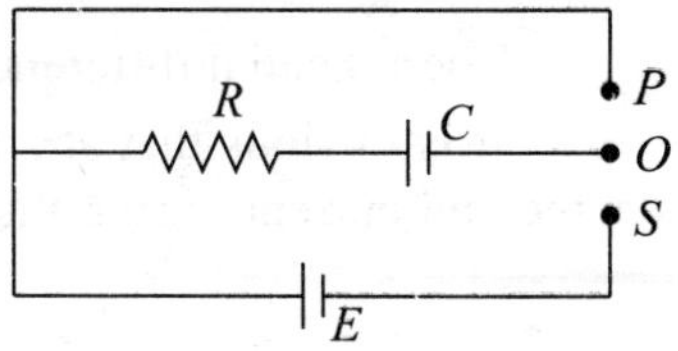

Fig. 5.3

But $\qquad i = \frac{dq}{dt}$ = rate of flow of charge

Therefore, $\qquad R = \frac{dq}{dt} + \frac{q}{C} = E$

or $\qquad R\frac{dq}{dt} = E - \frac{q}{C}$

$$dt = \frac{Rdq}{E - \frac{q}{C}} = -RC\frac{-{}^{1}/_{C}\,dq}{E - {}^{q}/_{C}}$$

on integrating, we get

$$t = -RC \log_e\left(E - \frac{q}{C}\right) + A \qquad ...(2)$$

where A is constant of integration.

From initial conditions, we know that at $t = 0$, $q = 0$

$\therefore \qquad O = -RC \log_e E + A$

$$A = RC \log_e E$$

substituting value of A in eqn. (2) we get

$$t = -RC \log_e\left(E - \frac{q}{C}\right) + RC \log_e E$$

$$= -RC \log_e E - \frac{q/C}{E} \qquad ...(3)$$

$$\log_e \frac{E - q/C}{E} = -\frac{t}{RC}$$

$$= \frac{E - q/C}{E} = e^{-t/RC}$$

which yields, $\qquad q = CE(1 - e^{-t/RC}) \qquad ...(4)$

But $\qquad CE$ = maximum value of charge = q_o (say)

$\therefore \qquad q = q_o(1 - e^{-t/RC}) \qquad ...(5)$

The factor RC is called the time constant τ of the circuit. The second part $(-q_o e^{-t/RC})$ represents the transient response, whereas the first part $(-q_0)$ represents the steady state response. Putting $RC = \tau = t$ in eqn. (5) we get

$$q = q_0(1 - e^{-1}) = q_0\left(1 - \frac{1}{2.72}\right) = 0.638q_o$$

Thus the time constant (RC) of R-C circuit is defined as the time in which the charge of condenser reaches to 0.638 times its final value. The rate of charging or current after time t will be

$$i = \frac{dq}{dt} = \frac{d}{dt}\,[q_o(1 - e^{-t/RC})]$$

$$= \frac{q_o}{dt}.e^{-t/RC} = \frac{E}{R}\,e^{-t/RC} = i_o e^{-t/RC} \qquad ...(6)$$

where $i_o = \dfrac{E}{R}$ = maximum or initial value of current.

The potential difference across R is given by $v_R = Ri = Ee^{-t/RC}$.

It is obvious that at $t = 0$, $i = i_o$ and at $t = \infty$, $i = 0$. The graphs representing the variation of charge and current during the process of charge are shown.

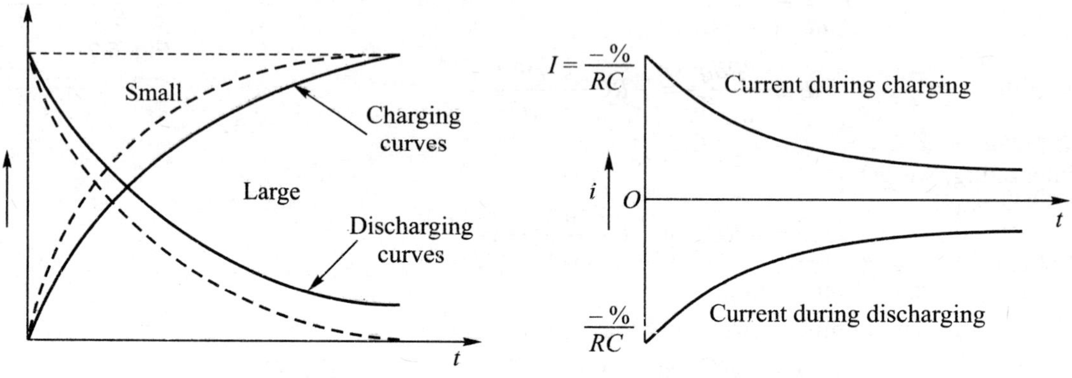

(a) (b)

Fig. 5.4

5.3 TRANSIENT RESPONSE OF R-C CIRCUIT WHEN NO SOURCE IS CONNECTED

Capacitor be charged to a value q_0 by connecting OS. Then connection OS is broken and O is connected to P. Now the capacitor of initial charge q_o is being discharged through the resistance R.

Suppose after time t during the process of discharging, the charge on the capacitor is q and the current through the resistance R is i.

Now since the battery is not in the circuit, application of Kirchoff's second law yields,

$$Ri + \frac{q}{C} = 0$$

$$R\frac{dq}{dt} + \frac{q}{c} = 0$$

Since $$i = \frac{dq}{dt}$$

$$\frac{dq}{dt} = \frac{-dt}{RC}$$

integrating, we get $$\log_e q = \frac{-t}{RC} + \text{B} \qquad ...(8)$$

where B is constant of integration.

From initial conditions we have at $t = 0$, $q = q_{oi}$ therefore equation (8) gives $B = \log_e q_o$ Substituting value of B in (8), we get

$$\log_e q = \frac{-t}{RC} + \log_e q_o \quad .$$

$$\log_e \frac{q}{q_0} = \frac{-t}{RC}$$

$$\therefore \quad q = q_o\, e^{-t/RC} \quad ...(9a)$$

If we write $q = cv$ and $q_o = CE,$ where v is the potential difference across C at time t and E is the initial maximum potential difference across C, then,

$$v = Ee^{t-/RC} \quad ...(9b)$$

If $t = \tau = RC$ = time constant, then

$$q = q_o\, e^{-1} = \frac{q_o}{E} = 0.362 q_0 \quad ...(10)$$

i.e., time constant of *R-C* circuit may be defined as the time in which the charge of the capacitor (or the potential difference across the capacitor) falls to $\frac{1}{e}$ (= 0.362) times its initial value during the process of discharging.

The charge decreases exponentially during discharging process and its value depends upon the value of time constant RC *e.g.* (9) represents the transient response, the steady state component being zero.

The current at any instant t is given by

$$i = \frac{dq}{dt} = \frac{d}{dt}\,[q_o\, e^{-t/RC}]$$

$$= \frac{q_o}{RC}\, e^{-t/RC} = \frac{-E}{R}\, e^{-t/RC} = -\, i_o e^{-t/RC} \quad ...(11)$$

where $i_o = E/R$ = maximum or initial value of current. It is obvious that at $t = 0$, $i = -\, i_0$ and $t = \infty$, $i = 0$.

The graphs representing the variation of charge and current during the process of discharging are shown in figure 5.4 (*a*), (*b*).

5.4 TRANSIENT RESPONE OF AN R-L CIRCUIT WHEN D.C. SOURCE IS CONNECTED

Figure 5.5 represents a circuit containing a coil of self-inductance L, a non-inductive resistance R and a battery of constant e.m.f. E. The battery E may be connected in or removed from the circuit by the help of switch S.

Transient and steady state response with a step function

The battery E is included in the circuit by throwing the switch P to O. When the circuit is open there is no flux linked with the coil, hence as soon as the battery is included in the circuit the current begins to flow and a magnetic flux is linked with the coil. As a result, a self-induced e.m.f. sets up in the coil which opposes the rise of current in the circuit. Hence the current does not attain its final steady value i_O instantaneously; but grows at a rate depending upon the inductance and resistance of circuit.

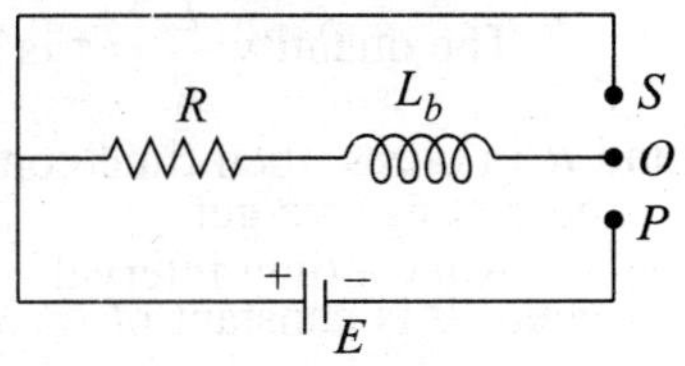

Fig. 5.5

During the variable state of growth of current if i is the instantaneous current in the circuit, the e.m.f. induced in the circuit is $-L\frac{di}{dt}$. Hence the effective e.m.f. driving the current in the circuit is $E - L\frac{di}{dt}$. Therefore from Ohm's law, this must be equal to Ri. Thus

$$E - L\frac{di}{dt} = Ri \qquad \text{...(1)}$$

$$L\frac{di}{dt} = E - Ri \quad i.e. \quad \frac{di}{E - Ri} = \frac{dt}{L}$$

Integrating, we get $\int \frac{di}{E - Ri} = \int \frac{dt}{L}$

i.e., $$-\frac{1}{R}\log_e(E - Ri) = \frac{t}{L} + A_1 \qquad \text{...(2)}$$

where A_1 is constant of integration.

Since at $t = 0$, $i = 0$ therefore

$$A_1 = -\frac{1}{R}\log_e E$$

Substituting the value of A_1 in equation (2) we get

$$-\frac{1}{R}\log_e(E - Ri) = \frac{t}{L} - \frac{1}{R}\log_e E$$

i.e. $$\log_e \frac{E - Ri}{E} = \frac{-R}{L}t$$

$$1 - \frac{Ri}{E} = e^{-(R/tL)t}$$

$$i = \frac{E}{R}(i - e^{(R/L)t}) \qquad \text{...(3)}$$

The maximum steady value of current i is obtained when $t = \infty$, from equation (3) this value is

$$i_0 = \frac{E}{R} \qquad \text{...(4)}$$

Hence equation (3) may be written as

$$i = i_0(1 - e^{-(R/L)t}) \qquad \text{...(5)}$$

From this equation it is clear that the current in the circuit rises according to an exponential law, figure 5.6 represents the growth (or rise) of current i with time t according to equation (5).

The quantity $\frac{L}{R} = \tau$ is known as the time constant of L-R circuit, when L is measured in henry and R in ohms, then time constant will be in second.

After a time interval $\frac{L}{R}$ second, the current in the circuit is given by

$$i = i_o\,(1 - e^{-(R/L.L/R)})$$

$$= i_o(1 - e^{-1}) = 0.63i_o$$

Therefore time constant of L-R circuit may be defined as the time in which the current rises to nearly two-third of its maximum steady value.

In equation (5) the component $i_o e^{-(R/L)t}$ represents the transient current, while i_o is the steady state current. The transient and steady state currents are flowing in opposite directions.

From equation (5), the rate of growth of current in the circuit is given by

$$\frac{di}{dt} = i_o \frac{R}{L} e^{-(R/L)t}$$

$$= i_o \frac{R}{L}\left(1 - \frac{i}{i_o}\right) \text{ using equation (5)}$$

$$= \frac{R}{L}(i_o - i) \qquad ...(6)$$

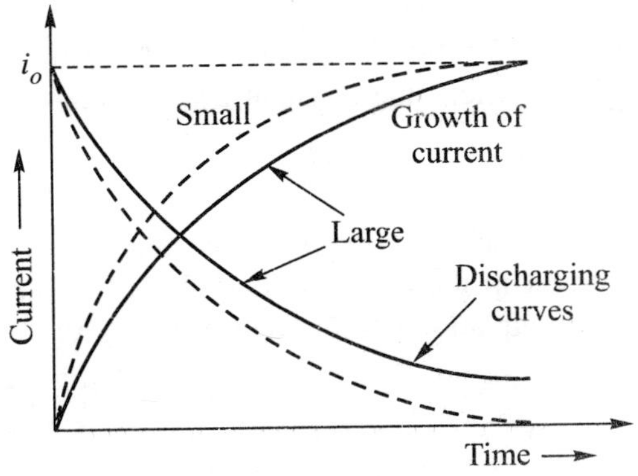

Fig. 5.6

From this equation it is clear that for larger $\frac{R}{L}$ or smaller $\frac{L}{R}$, the rate of growth of current in the circuit is greater. Hence for smaller time constant $\frac{L}{R}$ of a circuit, the current in the circuit attains its maximum steady value more rapidly.

5.5 TRANSIENT RESPONSE OF R-L CIRCUIT WHEN NO SOURCE IS CONNECTED

Now in figure 5.5 the battery is removed from the circuit by throwing switch O from position (S) to position (P). In this the e.m.f. applied in the circuit is zero, the resistance of the circuit remains unchanged and the current in the circuit decays from maximum value $i_o\left(=\frac{E}{R}\right)$ to zero value. During the decay of current, the flux linked with the coil changes. Hence an induced e.m.f. is developed in the coil which opposes the decay of current in the circuit. Thus the current takes some time to attain its value from maximum value i_o.

During the variable state of decay of current if i is the instantaneous current in the circuit, the e.m.f. induced in the circuit is $-L\frac{di}{dt}$. Hence from ohm's law, we have

$$-L\frac{di}{dt} = Ri$$

$$\frac{di}{i} = -\frac{R}{L}dt \qquad ...(7)$$

integrating, we get

$$\log_e i = \frac{-R}{L}t + A_2 \qquad ...(8)$$

where A_2 is a constant of integration

Since at $t = 0$, $i = i_0$ therefore

$$A_2 = \log_e i_0$$

Substituting this value of A_2 in equation (8), we get

$$\log_e \frac{i}{i_0} = \frac{-R}{L} t$$

$$\therefore \qquad i = i_0 e - (R/L)t \qquad \text{...(9)}$$

From this equation it is clear that the current in the circuit decays exponentially. Figure 5.6 represents the decay of current i with time t according to equation (9). During decaying process of current flowing is the transient one, the steady component being zero.

After time constant $L/R = \tau$ the current in the circuit is given by

$$i = i_0 e^{-(R/L)(L/R)} = i_0 e^{-1}$$

$$= 1/e.i_0 \qquad \text{...(10)}$$

Therefore the time constant L/R of an R-L circuit may also be defined as the time in which the current in the circuit falls to $1/e$ of its maximum value when external source of e.m.f. is removed. From equation (9), the rate of decay of current in the circuit is given by

$$\frac{di}{dt} = -i_0 . \frac{R}{L} e^{-(R/L)t}$$

$$= -i_0 . \frac{R}{L} . \frac{i}{i_0} \quad \text{(using equation 9)}$$

$$= \frac{-R}{L} i \qquad \text{...(11)}$$

Thus it is clear that the rate of decay of current in the circuit is greater for greater R/L or smaller L/R (time constant). Hence for smaller time constant L/R of a circuit, the current in the circuit decays more rapidly.

5.6 FREQUENCY RESPONSE OF RC CIRCUITS

Passive Filters

An electrical filter is a circuit that can be designed to modify, reshape or reject all unwanted frequencies of an electrical signal and accept or pass only those signals wanted by the circuits designer. In low frequency applications (up to 100 kHz), passive filters are usually made from simple RC (Resistor-Capacitor) networks while higher frequency filters (above 100 kHz) are usually made from RLC (Resistor-Inductor-Capacitor) components.

Simple first order filters can be made by connecting together a single resistor and a single capacitor in series across an input signal (V_{in}), with the output signal (V_{out}) taken from the junction of these two components and depending on which way around the resistor and capacitor are connected determines the type of filter, *i.e.* either Low Pass or High Pass. The function of any filter is to allow signals of a given band of frequencies to pass unaltered while attenuating the others. As there are two passive components within this type of filter design the output signal has a smaller amplitude than its corresponding input signal, therefore it has a gain of less than one (unity).

The Low Pass RC Filter

A simple passive RC Low Pass Filter or LPF, can be made by connecting together in series a single Resistor and a single Capacitor as shown below. In this type of filter arrangement the input signal (V_{in}) is applied to the series combination *i.e.* both the Resistor and Capacitor but the output

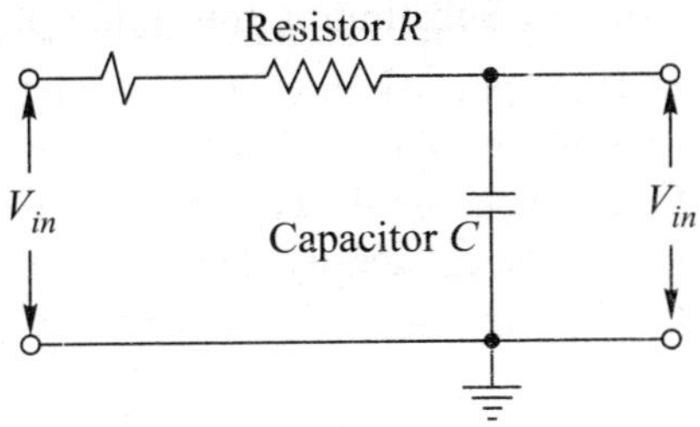

Fig. 5.7

signal V_{out} is taken across the capacitor only. This type of filter is known generally as a 1st order Filter, because it has only "one" reactive component in the circuit, the capacitor.

The reactance of the capacitor varies inversely with frequency, while the resistor remains constant as the frequency changes. At low frequencies the capacitive reactance (X_c) of the capacitor will be very large compared to the resistance value of the resistor, R and as a result the voltage across the capacitor V_c will also be large while the voltage drop across the resistor V_r will be much lower. At high frequencies the reverse is true with V_c being small and V_r being large. So, by using the potential divider equation of two resistances in series, we can calculate the output voltage of the filter for any given frequency.

For example

Given that $R = 47\ \text{k}\Omega$, $C = 47\text{nf}$ and $V_{in} = 10$ V, find the output voltage (V_{out}) at a frequency of 100 Hz and again at frequency of 10,000 Hz or 10 kHz.

At a frequency of 100 Hz.

$$X_C = \frac{1}{2\pi fC} = \frac{1}{2\pi \times 100 \times 47 \times 10^{9}} = 33{,}863\ \Omega$$

$$V_{OUT} = V_{IN} \times \frac{X_C}{\sqrt{R^2 + X_C^2}} 10 \times \frac{33863}{\sqrt{4700^2 + 33863^2}} = 9.9\ \text{V}$$

At a frequency of 10 kHz.

$$X_C = \frac{1}{2\pi fC} = \frac{1}{2\pi \times 10{,}000 \times 47 \times 10^{-5}} = 338.6\ \Omega$$

$$V_{OUT} = V_{IN} \times \frac{X}{\sqrt{R^2 + X_C^2}} = 10 \times \frac{338.6}{\sqrt{4700^2 + 338.6^2}} = 0.718\ \text{V}$$

It can be seen that as the frequency increases from 100 Hz to 10 kHz, the output voltage V_{out} decreases from 8.78 V to 0.672 V. By plotting the output voltage against the input frequency, the Frequency Response Curve or Bode Plot function of the low pass filter can be found, as shown below.

Frequency Response of a 1st Order Low Pass Filter

The Bode Plot shows the Frequency Response of the filter to be nearly flat for low frequencies and all of the input signal is passed directly to the output, resulting in a gain of nearly 1 until it reaches the cut-off frequency point (f_c). This is because the reactance of the capacitor is high at low frequencies and blocks any current flow through the capacitor. After this point the response of the circuit decreases giving a slope of – 20dB/Decade or (– 6dB/Octave) "roll-off" as signals above this frequency become greatly attenuated, until at very high frequencies the reactance of the capacitor becomes so low that it gives the effect of a short circuit condition on the output terminals resulting in zero output.

For this type of low pass filter circuit, all the frequencies below this cut-off f_C point that are unaltered with little or no attenuation are said to be in the filters passband zone, while frequencies above this point that are unwanted or attenuated are generally said to be in the filters Stopband zone.

This "Cut-off", "Corner" or "Breakpoint" frequency is defined as being the frequency point where the capacitive reactance and resistance are equal, $R = X_c = 47\ \text{k}\Omega$. When this occurs the output signal is attenuated to 70.7% of the input signal value of the input. Although $R = X_c$, the output is not half of the input signal. This is because it is equal to the vector sum of the two and is therefore 0.707 of the input. As the filter contains a capacitor, the Phase Angle (ϕ) of the output signal lags behind that of the input and at the – 3dB cut-off frequency (f_C) is 45° out of phase. This is due to the time taken to charge the capacitor as the input voltage changes, resulting in the output voltage (the voltage across the capacitor) "lagging" behind that of the input signal. The higher the input frequency applied to the filter the more the capacitor lags and the circuit becomes more and more "out of phase".

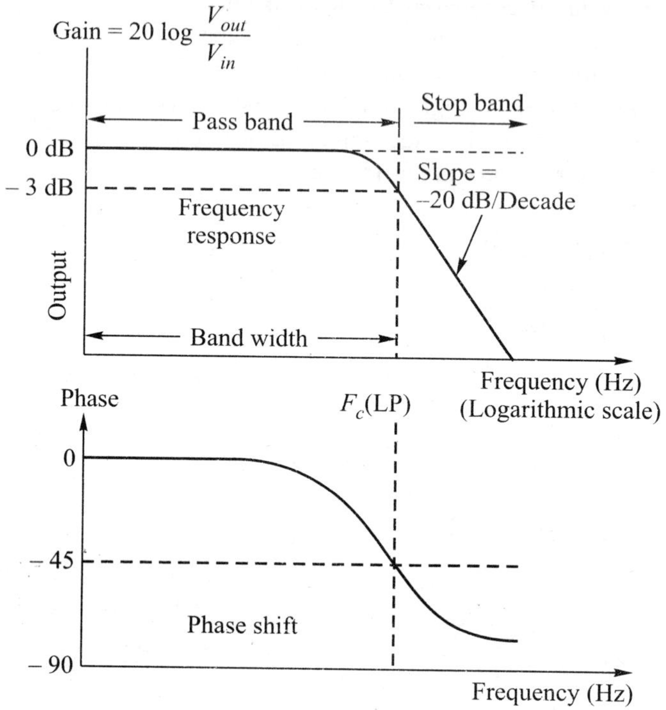

Fig. 5.8

This cut-off frequency and phase shift can be found by using the equation:

$$f_C = \frac{1}{2\pi RC} = \frac{1}{2\pi \times 4700 \times 47 \times 10^{-2}} = 720 \text{ Hz}$$

Phase Shift ϕ = – arctan ($2\pi f\ RC$)

Then for our simple example of a "Passive Low Pass RC Filter" above, the cut-off frequency (f_C) is given as 720 Hz with an output voltage of 70.7% of the input voltage value and a phase shift angle of 45°.

Time Constant

We know from above, that the filters cut-off frequency (f_C) is the product of the resistance (R) and the capacitance (C) in the circuit with respect to some specified frequency point and that by altering any one of the two components alters this cut-off frequency point by either increasing it or decreasing it. We also know that the phase shift of the circuit lags behind that of the input signal due to the time required to charge and then discharge the capacitor as the sinewave changes. This combination of R and C produces a charging and discharging effect on the capacitor known as its Time Constant (τ) of the circuit as seen in the RC Circuits section.

This time constant, tau (τ), is related to the cut-off frequency f_C as

$$\tau = RC = \frac{1}{2\pi f_C}$$

or expressed in terms of the cut-off frequency, f_C as

$$f_C = \frac{1}{2\pi RC} \text{ or } \frac{1}{2\pi\tau}.$$

The output voltage V_{out} depends upon the time constant and the frequency of the input signal. With an a.c. sinusoidal signal the circuit behaves as a simple 1st order low pass filter. But what if we were to change the input signal to that of a "square wave" shaped signal that has an almost vertical step input, the response of the circuit changes dramatically and produces a circuit known commonly as an Integrator.

So to summarize, the Low Pass Filter has a constant output voltage from D.C. (0Hz), up to a specified cut-off frequency (f_C). This cut-off frequency point is 0.707 or 3-dB (dB = – 20 log V_{out}/V_{in}) of the voltage gain allowed to pass. The frequency range "below" this cut-off point f_C is generally known as the Pass Band and the frequency range "above" this cut-off point is generally known as the Stop Band. A simple 1st order low pass filter can be made using a single resistor in series with a single non-polarised capacitor (or any single reactive component) across an input signal V_{in}, whilst the output signal V_{out} is taken from across the capacitor. The cut-off frequency of – 3dB point, can be found using the formula, $f_C = 1/(2\pi RC)$. The phase angle of the output signal at f_C *is* – 45°.

The gain of the filter or any filter for that matter, is generally expressed in Decibels and is a function of the output value divided by its corresponding input value and is given as:

$$\text{Gain in dB} = 20 \log \frac{V_{out}}{V_{in}}$$

Applications of Low Pass RC Filters are in Audio Amplifiers and Speaker systems to direct the lower frequency pass signals to the larger pass speakers or to reduce any high frequency noise or "hiss" type distortion. When used like this in audio applications the low pass filter is sometimes called a "high-cut", or "treble cut" filter.

If we were to reverse the positions of the resistor and capacitor in the circuit so that the output voltage is now taken from across the resistor, we would have a circuit that produces an output frequency response curve similar to that of a Passive High Pass Filter, and this is discussed in the next section.

The RC Integrator

The integrator is a low pass filter type circuit that converts a square wave step response input into a triangular waveform output as the capacitor charges and discharges. A Triangular waveform consists of alternate but equal positive and negative ramps. As seen below, if the RC time constant is long compared to the time period of the input waveform the resultant output waveform will be triangular in shape and the higher the input frequency the lower will be the output amplitude compared to that of the input.

Consider the output across the capacitor at high frequency *i.e.*

$$\omega >> \frac{1}{RC}$$

This means that the capacitor has insufficient time to charge up and so its voltage is very small. Thus the input voltage approximately equals the voltage across the resistor. To see this, consider the expression for I given above:

$$I = \frac{V_{in}}{R + 1/j\omega C}$$

but note that the frequency condition described means that

$$\omega C >> \frac{1}{R}$$

so $$I \approx \frac{V_{in}}{R} \text{ which is just Ohm's Law.}$$

Now, $$V_C = \frac{1}{C}\int_0^t I dt$$

so $$V_C \approx \frac{1}{RC}\int_0^t V_{in} dt$$

which is an integrator across the capacitor.

This then makes this type of circuit ideal for converting one type of electronic signal to another for use in wave-generating or wave-shaping circuits.

High Pass RC Filter

A High Pass Filter or HPF, is the exact opposite to that of the previously seen Low Pass RC Filter Circuit as now the output signal V_{out} is now taken from across the resistor as shown below. The high pass filter circuit as its name implies, attenuates or rejects all low frequency signals and passes only high frequency signals above the selected cut-off point, f_c.

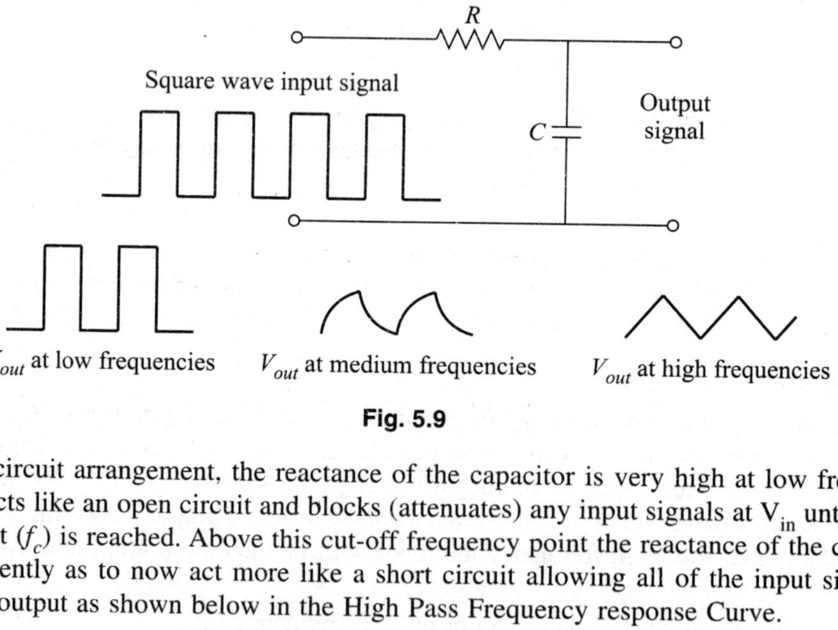

Fig. 5.9

In this circuit arrangement, the reactance of the capacitor is very high at low frequencies so the capacitor acts like an open circuit and blocks (attenuates) any input signals at V_{in} until the cut-off frequency point (f_c) is reached. Above this cut-off frequency point the reactance of the capacitor has reduced sufficiently as to now act more like a short circuit allowing all of the input signal to pass directly to the output as shown below in the High Pass Frequency response Curve.

Frequency Response of a 1st Order High Pass Filter

The Bode Plot or frequency response curve shown is for a High Pass filter which as mentioned previously, is the exact opposite to that of a low pass filter. Here the signal is attenuated at low frequencies and the output increases at 20dB/Decade (6dB/Octave) until the frequency reaches the

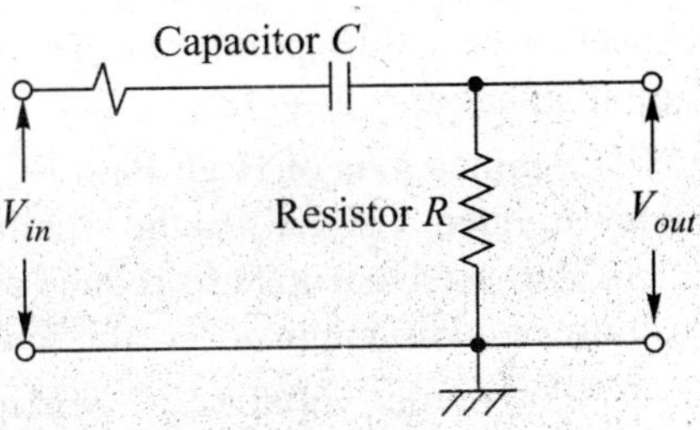

Fig. 5.10

cut-off point (f_c) where again $R = X_c$. At this frequency the output voltage is 70.7% of the input signal value or – 3dB (20 log (V_{out}/V_{in})) of the input. The phase angle of the output signal leads that of the input and is equal to + 45o at frequency f_c.

The cut-off frequency point for a 1st order high pass filter can be found using the same equation as that of the low pass filter, but the equation for the phase shift is modified slightly to account for the positive phase angle as shown below.

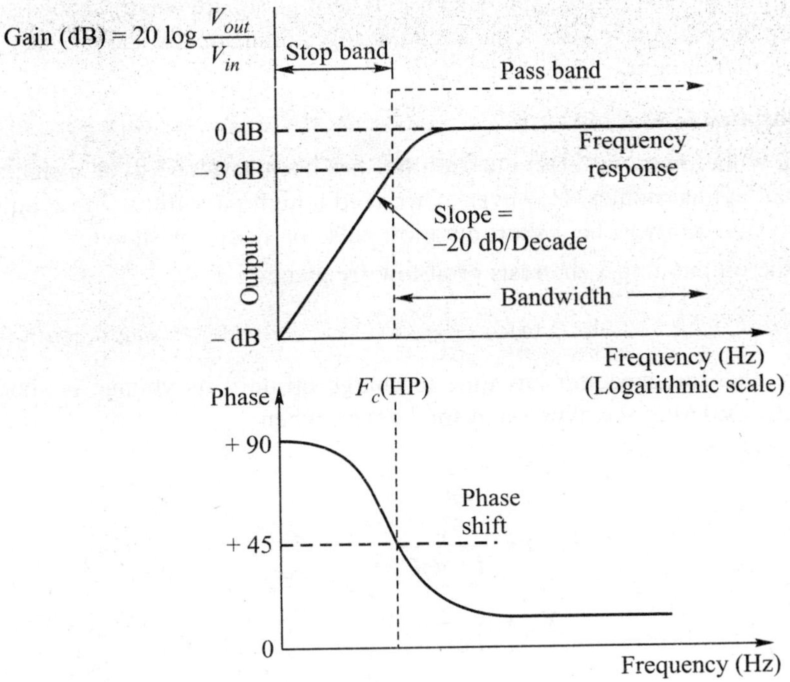

Fig. 5.11

$$f_c = \frac{1}{2\pi RC}$$

Phase Shift $$\phi = \arctan \frac{1}{2\pi fRC}$$

Example:

Calculate the cut-off or "breakpoint" frequency (f_c) for a simple High Pass RC Filter that has an 82 pF capacitor connected in series with a 240 kΩ resistor.

$$f_c = \frac{1}{2\pi RC} = \frac{1}{2\pi \times 240.000 \times 82 \times 10^{a}} = 8,087 \text{ He or BVRz}$$

The High Pass Filter has no output voltage from D.C. (0Hz), up to a specified cut-off frequency f_c. This cut-off frequency point is 70.7% or – 3 dB (dB = – 20 log V_{out}/V_{in}) of the voltage gain allowed to pass. The frequency range "below" this cut-off point fc is generally known as the Stop Band and the frequency range "above" this cut-off point is generally known as the Pass Band. The cut-off frequency or – 3dB point, can be found using the formula, $f_c = 1/(2\pi RC)$. The phase angle of the output signal at f_c is + 45°.

Applications of High Pass RC filters are in audio amplifiers and speaker systems to direct the higher frequency signals to the smaller "tweeter" type speakers while blocking the lower pass signals or are also used as filters to reduce any low frequency noise or "rumble" type distortion. When used like this in audio applications the low pass filter is sometimes called a "low-cut", or "bass cut" filter.

The output voltage V_{out} depends upon the time constant and the frequency of the input signal as seen previously. With an a.c. sinusoidal signal applied to the circuit it behaves as a simple 1st order high pass filter. But if we change the input signal to that of a "square wave" shaped signal that has an almost vertical step input, the response of the circuit changes dramatically and produces a circuit known commonly as an Differentiator.

The RC Differentiator

Up until now the input waveform to the filter has been assumed to be a sine wave consisting of the fundamental and harmonics. However, if we feed a high pass filter with a Square Wave input signal the output waveform will be a short duration pulse or spikes as shown.

Consider the output across the resistor at low frequency *i.e.*,

$$\omega << \frac{1}{RC}$$

This means that the capacitor has time to charge up until its voltage is almost equal to the source's voltage. Considering the expression for I again, when

$$R << \frac{1}{\omega C}$$

so

$$I \approx \frac{V_{in}}{I / j\omega C}$$

$$V_{in} \approx \frac{1}{j\omega C} \approx V_C$$

Now,

$$V_R = IR = C\frac{dV_C}{dt}R$$

$$V_R \approx RC\frac{dV_{in}}{dt}$$

which is a differentiator across the resistor.

Each cycle of the square wave input waveform produces two spikes at the output, one positive and another negative and whose amplitude is equal to that of the input. The rate of decay of the spikes depends upon the time constant of the *RC* components, ($t = C \times R$) and the input frequency. The output pulses resemble more and more the shape of the input signal as the frequency increases.

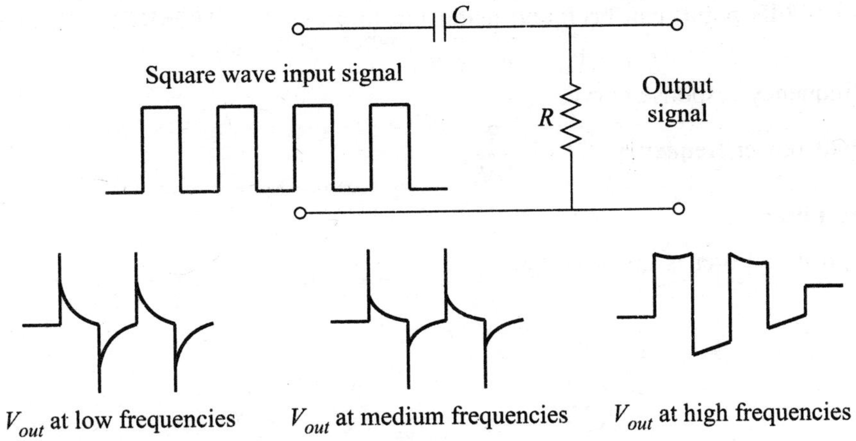

Fig. 5.12

5.7 FREQUENCY RESPONSE OF RL CIRCUITS

(*i*) Low Pass Filter

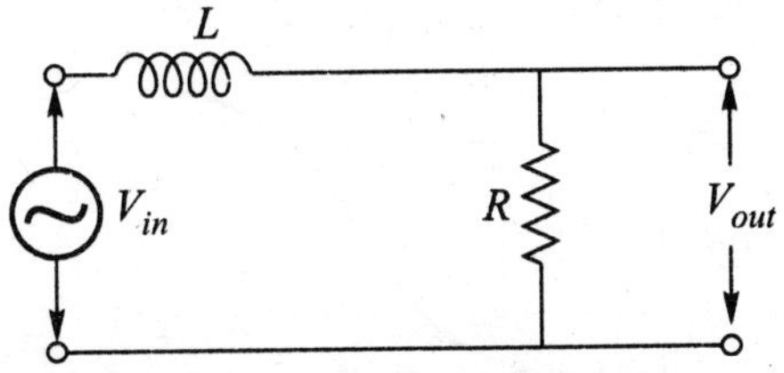

Fig. 5.13

The above figure shows the Low Pass Filter Circuit with R-L combination. Its frequency response curve will be as following. It is excited by constant amplitude a.c. voltage whose frequency can be varied.

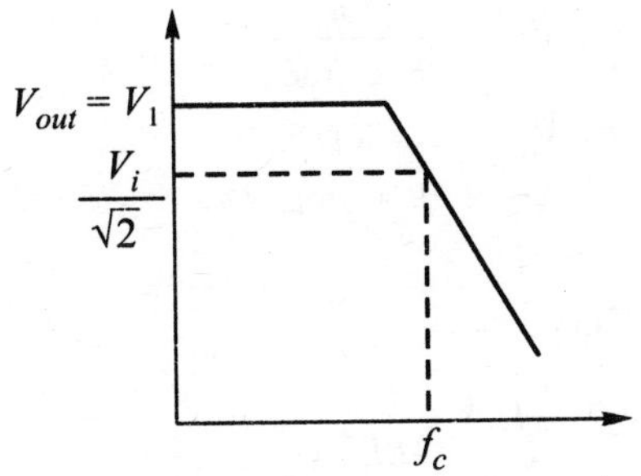

Fig. 5.14

Frequency Response

The output voltage is given by $V_{out} = \dfrac{V_{in} R}{R + j\omega L}$

$$\frac{V_{out}}{V_{in}} = \frac{R}{R + j\omega L}, \text{ where } \omega = 2\pi f$$

From frequency response curve,

upper half-power frequency $f_c = \dfrac{R}{2\pi L}$

(*ii*) High-Pass Filter

R-L High Pass Filter is given below

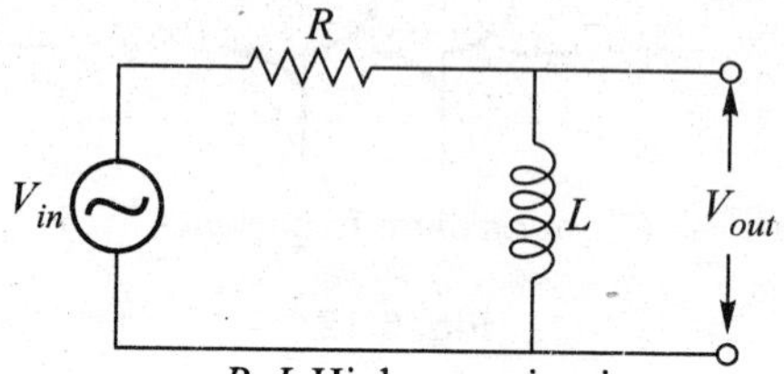

R–L High pass circuit

Fig. 5.15

This filter's frequency response curve will be as following

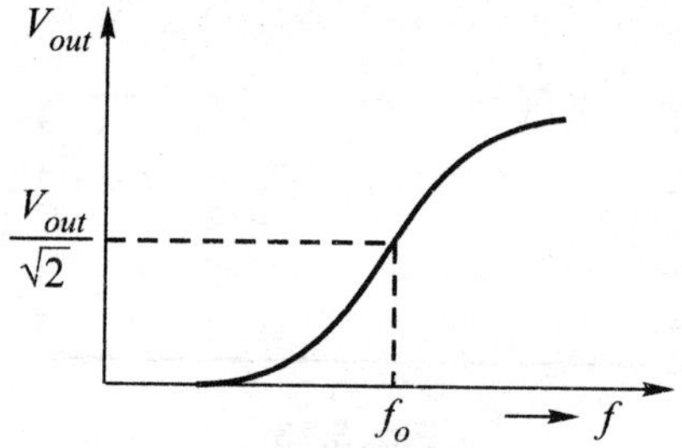

Frequency response

Fig. 5.16

The output voltage is given by

$$V_{out} = \frac{j\omega L}{R + j\omega L} V_{in}$$

$$\frac{V_{out}}{V_{in}} = \frac{j\omega L}{R + j\omega L}$$

where $2\pi f = \omega$

From Frequency Response Curve,

Lower cut off frequency $f_c = \dfrac{R}{2\pi L}$

(*iii*) High-Pass R-L Circuit as Differentiator

High-Pass R-L circuit is given below

Now, let us apply KVL to this R-L circuit.

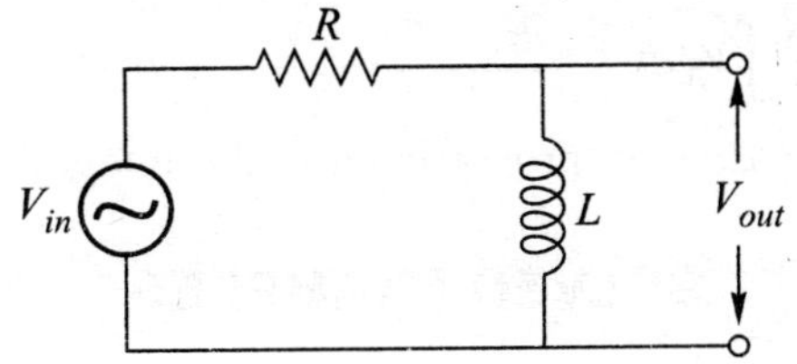

Fig. 5.17

$$V_{in} = iR + V_{out}, \text{ where } V_{out} = L\frac{di}{dt} \qquad ...(2)$$

On integrating equation (2), we get

$$\int V_{out}\, dt = Li$$

$$i = \int \frac{V_{out}}{L} dt \qquad ...(3)$$

Substituting equation (3) in (1), we get

$$V_{in} = R\int \frac{V_{out}}{L}\, dt + V_{out}$$

If $\frac{L}{R} << T$, V_o can be neglected, then

$$V_{in} = \frac{R}{L} \int V_{out}\, dt$$

on differentiating both sides,

$$V_{out} = \frac{L}{R}\frac{dV_i}{dt} \qquad ...(4)$$

Thus, the o/p voltage is directly proportional to the derivative of the input.

(*iv*) Low Pass R-L Circuit as Integrator

Form Low Pass R-L Circuit

Now, on applying KVL to this, we get

$$V_{in} = L\frac{di}{dt} + V_{out} \qquad ...(1)$$

and

$$V_{out} = iR \qquad ...(2)$$

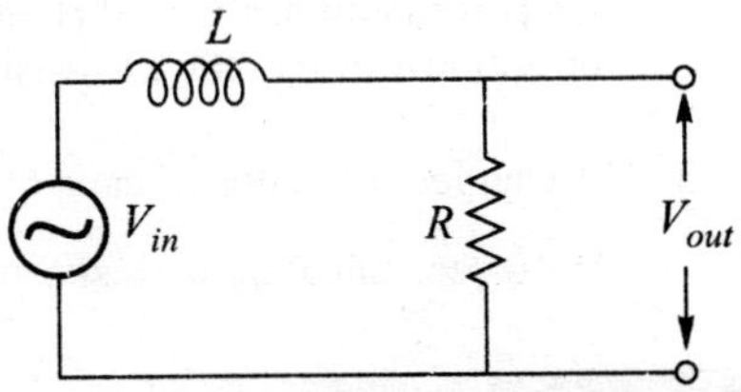

Fig. 5.18

On integrating equation (1), we get

$$\int V_{in} dt = Li + \int V_{out} dt$$

$$\int V_{in} dt = \frac{L}{R} V_{out} + \int V_{out} dt \qquad \left[\text{From equation (2)} \quad \frac{V_{out}}{R} = i\right]$$

If $\frac{L}{R} >> T$, $\int V_{out}$ can be neglected in comparison with $\frac{L}{R}$ V_O. Therefore

$$\int V_{in} dt = \frac{L}{R} V_{out}$$

$$\int V_{out} = \frac{R}{L} \int V_{in} dt$$

Thus the output voltage is directly proportional to the integral of the input.

SOLVED EXAMPLES

Example 1: *A coil of self inductance of 100 Henry and resistance 100 Ω are joined to a source of 4 volts in series. Calculate the time constant and maximum current.*

Solution: Time constant $= \frac{L}{R} = \frac{100}{100} = 1$ sec

Maximum current $= \frac{E}{R} = \frac{4}{100} = 0.04$ amp.

Example 2: *An inductor of self inductance 5H and resistance 5Ω are connected in series to a source. Calculate the time in which the current will attain half its steady value.*

Solution:

$$i = i_0[1 - e^{Rt/L}]$$

$$L = 5H,\ R = 5\Omega,\ i = i_0/2$$

$$\frac{i_o}{2} = i_0[1 - e^{-5t/5}]$$

$$\frac{1}{2} = 1 - e^{-t}$$

$$e^{-t} = 1 - \frac{1}{2} = \frac{1}{2}$$

$$e^{t} = 2$$

$$t = 2.3\ \Omega\ \log_{10}{}^{2} = 0.693 \text{ sec}$$

EXERCISE

1. A 25 mF capacitor charged to 100V is discharged through a 0.1 MΩ resistance. Find the time taken by the current to decrease to 50% of its initial value.
2. An L-R circuit has $L = 50$ H and $R = 50$ Ω. An applied d.c. voltage of 10 V will produce a steady current of 200 mA in the 200 Ω resistor. How much is the current after 0.1S?
3. A charged capacitor of capacity 10 mF is associated by a short parallel resistance. If $\frac{1}{4}$ of charge leaks in 10 sec, calculate its resistance?

QUESTIONS

1. Describe the step response of R-L circuit.
2. Distinguish between high pass and low pass filters.
3. Discuss the growth of charge in C-R circuit.
4. Discuss the differentiation of signal by a filter.
5. Describe the working of low pass R-L circuit as integrator.
6. Exhibit the high pass R-L circuit as differentiator of signal.
7. Distinguish between transient response and frequency response of circuit.
8. Explain RC high pass filter.
9. Describe RC low pass filter.

CHAPTER 6

Resonance

6.1 INTRODUCTION

Resonance in AC circuits implies a special frequency determined by the values of the resistance, capacitance, and inductance. For series resonance the condition of resonance is straightforward and it is characterized by minimum impedance and zero phase. Parallel resonance, which is more common in electronic practice, requires a more careful definition.

6.2 RLC CIRCUIT

An RLC circuit (also known as a resonant circuit, tuned circuit, or LCR circuit) is an electrical circuit consisting of a resistor (R), an inductor (L), and a capacitor (C), connected in series or in parallel. This configuration forms a harmonic oscillator.

Tuned circuits have many applications particularly for oscillating circuits and in radio and communication engineering. They can be used to select a certain narrow range of frequencies from the total spectrum of ambient radio waves. For example, AM/FM radios with analog tuners typically use an RLC circuit to tune a radio frequency. Most commonly a variable capacitor is attached to the tuning knob, which allows you to change the value of C in the circuit and tune to stations on different frequencies.

Series RLC

In this circuit, the three components are all in series with the voltage source.

Series RLC Circuit notations:

v = the voltage of the power source (measured in volts V)

i = the current in the circuit (measured in amperes A)

R = the resistance of the resistor (measured in ohms)

L = the inductance of the inductor (measured in henrys)

C = the capacitance of the capacitor (measured in farads)

q = the charge across the capacitor (measured in coulombs C)

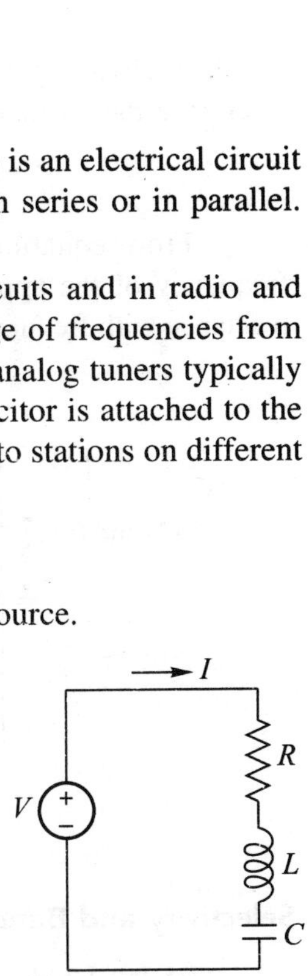

Fig. 6.1

6.3 SERIES RESONANCE DERIVATION

The impedance of the circuit is given by

$$z = R + j\omega l - \frac{j}{\omega c}$$

$$= R + j\left(\omega L - \frac{1}{\omega c}\right) \quad \text{...(1)}$$

At resonance the reactive component in the Impedance is zero *i.e.* $\omega L - \dfrac{1}{\omega c} = 0$

$$\omega L = \frac{1}{\omega c}$$

$$\omega = \omega_0 = \frac{1}{\sqrt{LC}}$$

Hence Resonant frequency $f_0 = \dfrac{\omega_0}{2\pi} = \dfrac{1}{2\pi\sqrt{LC}}$ ...(2)

At this frequency, the reactive term disappears and $|Z| = R$. Thus at resonance, the Impedance is equal to R while at other frequencies, the magnitude of 'Z' is

$$|Z| = \sqrt{R^2 + \left(\omega L - \frac{1}{\omega C}\right)^2}$$

which is always greater than R. Hence in series circuit, the impedance is minimum at resonance. Therefore the current will be maximum at resonance and its value is

$$I_o = \frac{E}{R} \quad \text{...(3)}$$

From equation (2) it is obvious that the resonance may be obtained by either varying the frequency of the applied voltage or by varying either L or C. The variation of Impedance (Z) and reactance with frequency and the variation of current with frequency in a series circuit are shown below.

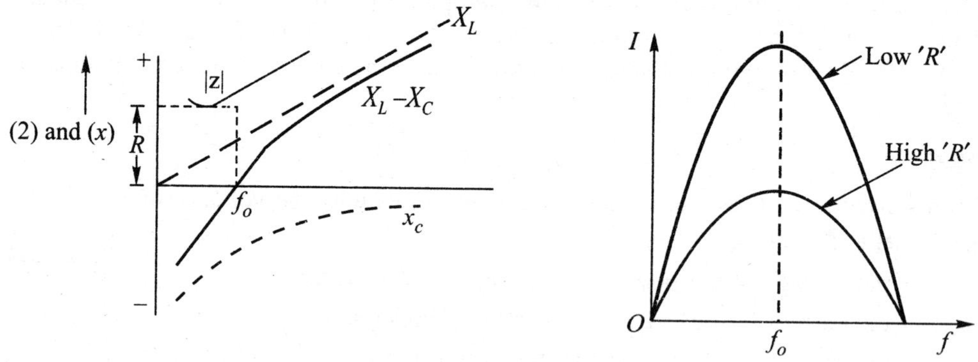

Fig. 6.2

Selectivity and Bandwidth

As we know that in a series LCR circuit the Impedance is minimum at resonant frequency. Therefore it passes frequencies near the resonant frequency more readly than any other frequencies. Thus at resonance frequency it is called as acceptor circuit and has selective properties. If the resistance of the circuit is low, then the curve of current against frequent has a sharp peak and the circuit is said to be sharply resonant or highly selective. If the resistance is high the peak is broadened and the selectivity is poor.

To compare the selectivity or sharpness of different circuits, a band of frequencies is chosen at which current falls to $\frac{1}{\sqrt{2}}$ times of its maximum value of E/R as shown in the figure.

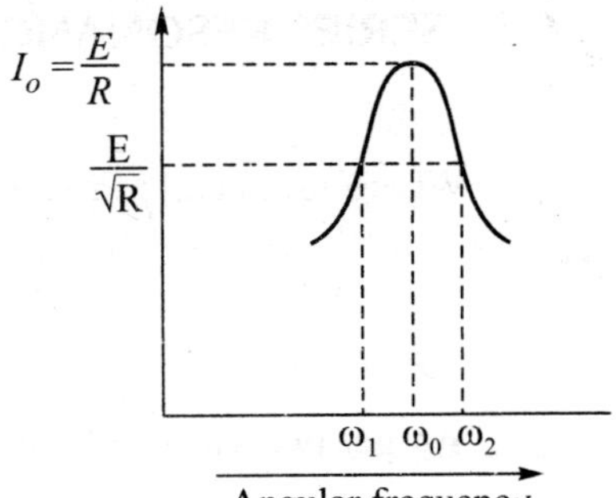

Fig. 6.3

The degree of selectivity of sharpness of resonance is expressed in term of 'Q' known as Q–factor or quality factor. Mathematically defined as

$$Q = \frac{\omega_0}{\omega_2 - \omega_1} = \frac{\omega_0}{\Delta\omega} = \frac{f_0}{\Delta f} \qquad ...(1)$$

Consider W_a be the angular frequency corresponding to the points where $I = E/\sqrt{2}R$, and these points are called as half power points.

In a LCR circuit the current at $f_a = \frac{\omega_a}{2\pi}$ is

$$\frac{E}{\sqrt{2}R} = I = \frac{E}{\sqrt{R^2 + \left(\omega_a L - \frac{1}{\omega_a C}\right)^2}}$$

$$\sqrt{2}R = \sqrt{R^2 + \left(\omega_a L - \frac{1}{\omega_a C}\right)^2}$$

$$\therefore \qquad \left(\omega_a L - \frac{1}{\omega_a C}\right)^2 = R^2$$

$$\omega_a L - \frac{1}{\omega_a C} = \pm R$$

$$\therefore \qquad \omega_a^2 \pm \frac{R}{L}\omega_a - \frac{1}{LC} = 0$$

It is a quadratic equation in W_a, therefore the roots are

$$\omega_a = \frac{\pm \frac{R}{L} \pm \sqrt{\frac{R^2}{L^2} + \frac{4}{LC}}}{2}$$

$$= \pm \frac{R}{2L} \pm \sqrt{\frac{R^2}{4L^2} + \frac{1}{LC}}$$

As 'R' is minimum when the circuit is most selective, neglecting $\left(\frac{R}{2L}\right)^2$ as compared with $\frac{1}{LC}$ we get

$$\omega_a = \pm\frac{R}{2L} + \sqrt{\frac{1}{LC}} = \pm\frac{R}{2L} \pm \omega_0$$

Where ω_0 is called resonant frequency which is equal to $\frac{1}{\sqrt{LC}}$. Taking only positive value of ω_a *i.e.*

$$\omega_a = \omega_0 \pm \frac{R}{2L}$$

We get two values of 'ω' as

$$\omega_1 = \omega_0 - \frac{R}{2L} \text{ and } \omega_2 = \omega_0 + \frac{R}{2L}$$

$\therefore$ The allowed values of frequencies can be obtained by taking $\omega_2 - \omega_1 = \Delta\omega$ which is called as Bandwidth

$$\therefore \qquad \Delta\omega = \omega_2 - \omega_1 = \frac{R}{L} \text{ Radiation/Sec.}$$

$$\therefore \qquad \Delta f = f_2 - f_1 = \frac{\Delta\omega}{2\pi} = \frac{R}{2\pi L} \text{ C/Sec.}$$

Here f_1 and f_2 are called half power frequencies at which the current will be 70.7% of its maximum resonance. The frequency f_1 is called as lower cut-off and f_2 is called as upper cut-off frequency.

The quality factor of the circuit is

$$Q = \frac{\omega_0}{\Delta\omega} = \frac{\omega_0}{R/L} = \frac{\omega_0 L}{R}$$

At resonance we have $w_0 L = \frac{1}{\omega C}$

$$\therefore \qquad Q = \frac{1}{\omega_0 CR} = \frac{1}{\frac{1}{\sqrt{LC}} \cdot CR} = \frac{1}{R}\sqrt{\frac{L}{C}}$$

Sharpness of Resonance Vs Bandwidth

A series resonant circuit looks like a resistance at the resonant frequency. (Figure 6.4) Since the definition of resonance is $X_L = X_C$, the reactive components cancel, leaving only the resistance to contribute to the impedance. The impedance is also at a minimum at resonance. (Figure 6.5) Below the resonant frequency, the series resonant circuit looks capacitive since the impedance of the capacitor increases to a value greater than the decreasing inductive reactance, leaving a net capacitive value. Above resonance, the inductive reactance increases, capacitive reactance decreases, leaving a net inductive component.

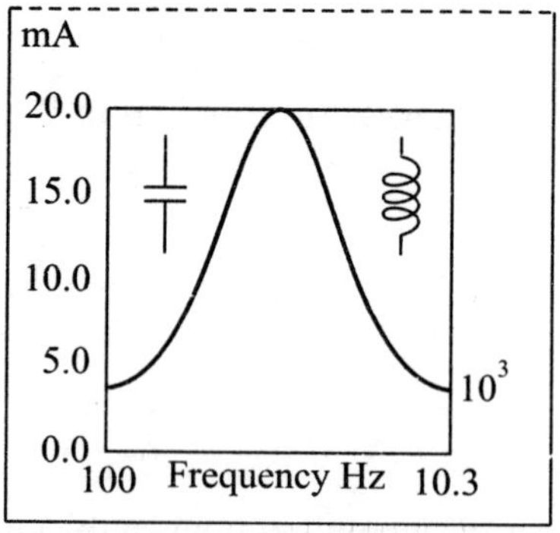

Fig. 6.4. At resonance the series resonant circuit appears purely resistive. Below resonance it looks capacitive. Above resonance it appears inductive.

Current is maximum at resonance, impedance at a minimum. Current is set by the value of the resistance. Above or below resonance, impedance increases.

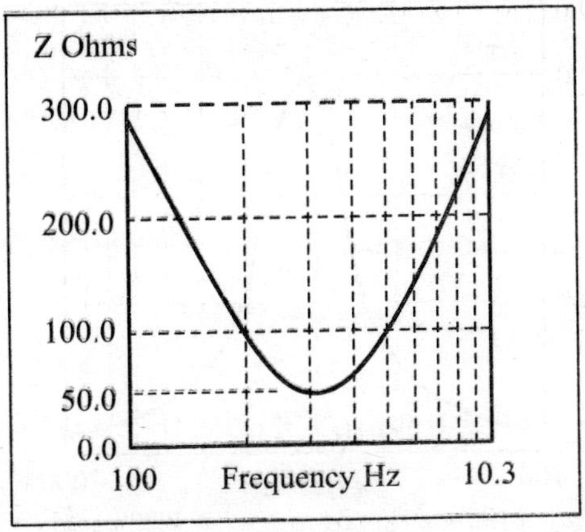

Fig. 6.5. Impedance is at a minimum at resonance in a series resonant circuit.

The resonant current peak may be changed by varying the series resistor, which changes the Q. (Figure 6.6) This also affects the broadness of the curve. A low resistance, high Q circuit has a narrow bandwidth, as compared to a high resistance, low Q circuit. Bandwidth in terms of Q and resonant frequency:

$$BW = f_c/Q$$

where f_c = resonant frequency

Q = quality factor

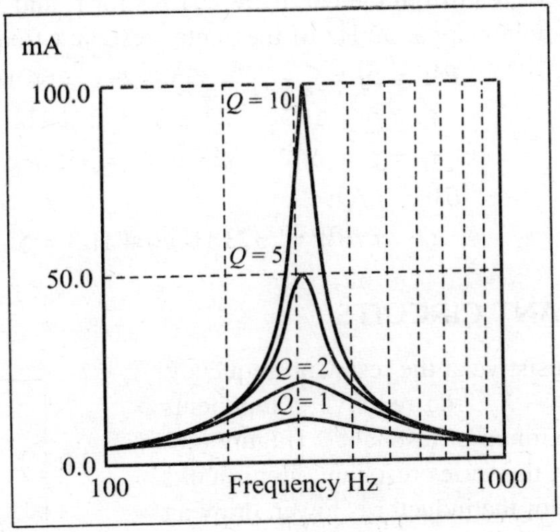

Fig. 6.6. A high Q resonant circuit has a narrow bandwidth as compared to a low Q.

Bandwidth is measured between the 0.707 current amplitude points. The 0.707 current points correspond to the half power points since $P = I^2R$, $(0.707)^2 = (0.5)$. (Figure 6.7)

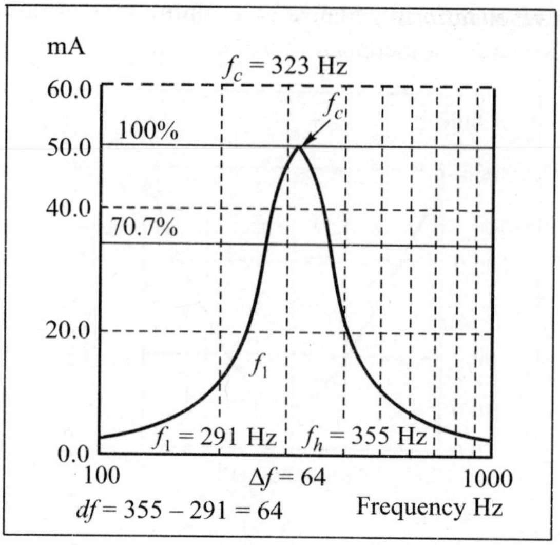

Fig. 6.7. *Bandwidth, Δf is measured between the 70.7% amplitude points of series resonant circuit.*

$$BW = \Delta f = f_h - f_1 = f_c/Q$$

Where f_h = high band edge, f_1 = low band edge

$$f_1 = f_c - \Delta f/2$$
$$f_h = f_c + \Delta f/2$$

Where f_c = center frequency (resonant frequency)

In Figure 6.7, the 100% current point is 50 mA. The 70.7% level is 0707 (50mA) = 35.4 mA. The upper and lower band edges read from the curve are 291 Hz for f_1 and 355 for f_h. The bandwidth is 64 Hz, and the half power points are ± 32 Hz of the center resonant frequency:

$$BW = \Delta f = f_h - f_1 = 355 - 291 = 64 \text{ Hz}$$
$$f_1 = f_c - \Delta f/2 = 323 - 32 = 291 \text{ Hz}$$
$$f_h = f_c + \Delta f/2 = 323 + 32 = 355 \text{ Hz}$$

Since $BW = f_c/Q$:

$$Q = f_c/BW = (323 \text{ Hz})/(64 \text{ Hz}) = 5$$

6.4 PARALLEL RESONANT CIRCUITS

A parallel resonant circuit is resistive at the resonant frequency. (Figure 6.8) at resonance $X_L = X_C$, the reactive components cancel. The impedance is maximum at resonance. (Figure 6.8) Below the resonant frequency, the series resonant circuit looks inductive since the impedance of the inductor is lower, drawing the larger proportion of current. Above resonance, the capacitive reactance decreases, drawing the larger current, thus, taking on a capacitive characteristic.

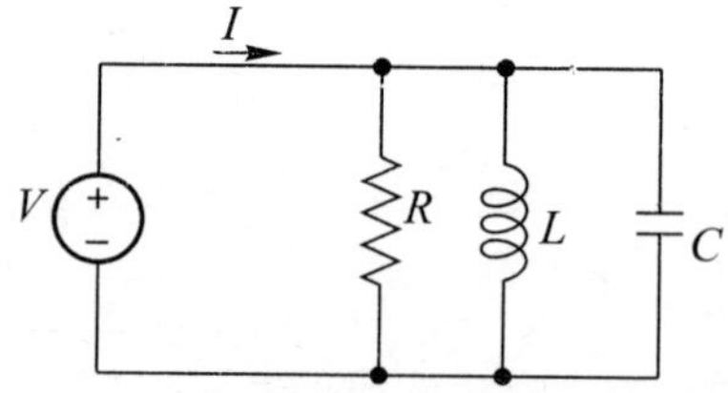

Parallel RLC circuit

Fig. 6.8

Parallel RLC Circuit notations

V = the voltage of the power source (measured in volts V)

I = the current in the circuit (measured in amperes A)

R = the resistance of the resistor (measured in ohms)

L = the inductance of the inductor (measured in henrys)

C = the capacitance of the capacitor (measured in farads)

The complex impedance of this circuit is given by adding up the impedances in parallel.

Parallel Resonance in RLC Circuit

Figure (6.9) depicts a parallel resonant circuit in which one branch consists of an inductance L with associated resistance R and the other a capacitor C. The resistance associated with capacitance is assumed to be negligible. An A.C. source of e.m.f. $E = E_0 \sin \omega t$ is connected across this parallel arrangement.

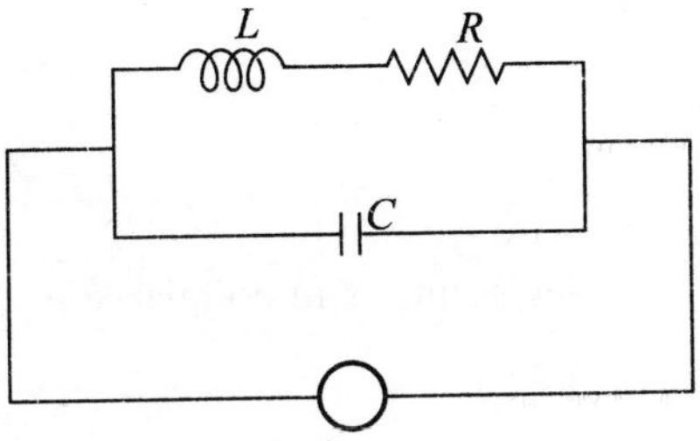

Fig. 6.9. $E = E_0 \sin \omega t$

The combined impedance of z of the two parallel branches is given by

$$\frac{1}{Z} = \frac{1}{R + j\omega L} + \frac{1}{1/j\omega C} = \frac{1}{R + j\omega L} + j\omega C$$

$$= \frac{(R - j\omega L)}{(R + j\omega L)(R - j\omega L)} + j\omega C$$

$$= \frac{R}{R^2 + \omega^2 L^2} + j\omega\left[C - \frac{L}{R^2 + \omega^2 L^2}\right] \quad \text{...(1)}$$

At resonance, the reactive component of impedance is zero. Hence resonance occurs when

$$j\omega\left(C - \frac{L}{R^2 + \omega^2 L^2}\right) = 0$$

$$\Rightarrow \quad C = \frac{L}{R^2 + \omega^2 L^2}$$

$$\Rightarrow \quad R^2 + \omega^2 L^2 = \frac{L}{C}$$

$$\Rightarrow \quad \omega^2 = \frac{1}{LC} - \frac{R^2}{L^2} \quad \text{...(2)}$$

$$\Rightarrow \quad \omega = \omega_0 = \sqrt{\left(\frac{1}{LC} - \frac{R^2}{L^2}\right)}$$

$\therefore$ Resonant frequency
$$f_0 = \frac{1}{2\pi}\sqrt{\left(\frac{1}{LC} - \frac{R^2}{L^2}\right)} \quad \text{...(3)}$$

If R is very small, $\frac{R^2}{L^2}$ is small as compared to $\frac{1}{LC}$ and the resonant frequency becomes

$$f_0 = \frac{1}{2\pi\sqrt{LC}} \quad \text{...(4)}$$

The impedance at resonance is then given by eqn. (1)

$$\frac{1}{Z} = \frac{R}{R^2 + \omega^2 + L^2} = \frac{R}{L/C} \qquad \left[\because\ R^2 + \omega^2 L^2 = \frac{L}{C}\right]$$

or

$$Z = \frac{L}{CR} \qquad ...(5)$$

This impedance at resonance is called the dynamic resistance. It is obvious that smaller the resistance R, larger is the impedance. If R is negligible, the impedance is infinite at resonance.

If R is small, a useful approximation for the value of Z may be made from

$$\frac{1}{Z} = \frac{R}{R^2 + \omega_o^{\ 2}\ L^2}$$

$$Z = R + \frac{\omega_o^{\ 2}\ L^2}{R}$$

Neglecting R in comparison to $\frac{\omega_o^{\ 2}\ L^2}{R}$

$$Z = \frac{\omega_o^{\ 2}\ L^2}{R} = \frac{\omega_0^{\ 2}}{R}\ .\ \omega_o L = Q.\omega_o L$$

where $\frac{\omega_o L}{R}$ = Q as for series resonance. Thus at resonance, the impedance of the circuit becomes ϑ times the impedance of the branch line.

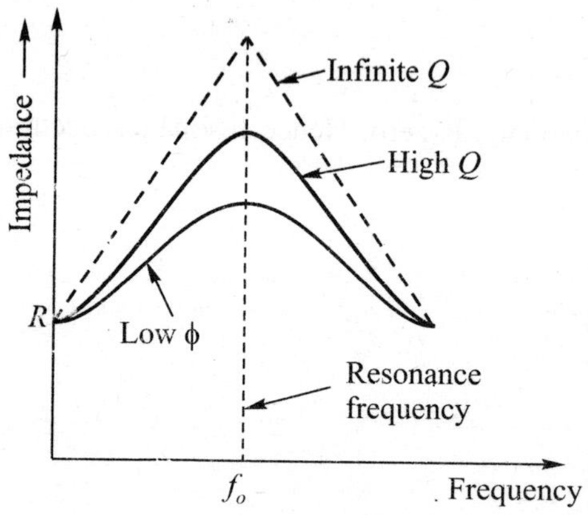

Fig. 6.10

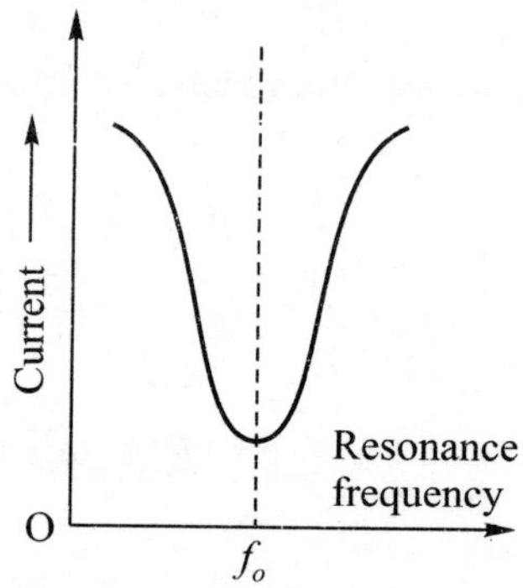

Fig. 6.11

If ϑ is large, impedance is maximum at resonance. Figure (6.10) depicts the variation of parallel circuit impedance as a function of frequency for different ϑ values. The current will be minimum at resonance which has been shown diagrammatically in figure (6.11). A parallel resonant circuit can, therefore, be used to suppress the current of a particular frequency out of current of many other frequencies. Hence the circuit is known as a rejector circuit.

Reactance Sketch: The variation of susceptance B (Inverse of reactance X) in a parallel circuit with frequency is shown in figure (6.12). It may be seen that the nature of variation of suspectance in a parallel circuit is similar to that of reactance in a series circuit. The reactance curve can be obtained as the reciprocal of the suspectance curve and has been depicted in figure (6.12).

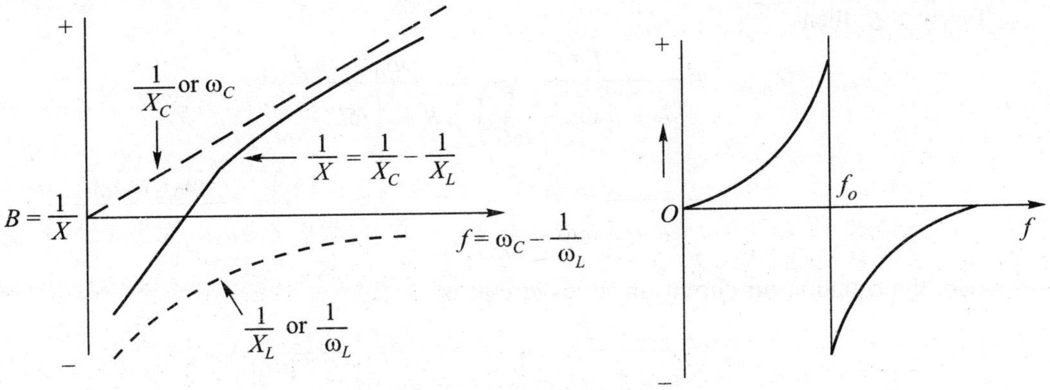

Fig. 6.12

Rejector Circuit: A parallel resonant circuit has maximum impedance at resonance and does not allow a frequency equal to the natural frequency of the circuit. Thus it suppresses (or rejects) a particular frequency from several other frequencies. Hence it is called a rejector or "Filter Circuit".

6.5 SELECTIVITY OF PARALLEL RESONANT CIRCUIT

Figure (6.13) shows a parallel resonant circuit connected to a generator of internal impedance Rg. Let E' be the voltage across the parallel circuit. If E is kept constant and frequency varied, then E' will be approximately equal to E when Rg is small at all frequencies compared with the impedance of parallel circuit. Then the circuit will be selective. If, however Rg is large compared with the impedance of parallel circuit, the current flowing through the circuit will approximately be equal to $\frac{E}{Eg}$ and E' will be proportional to the impedance of the parallel circuit.

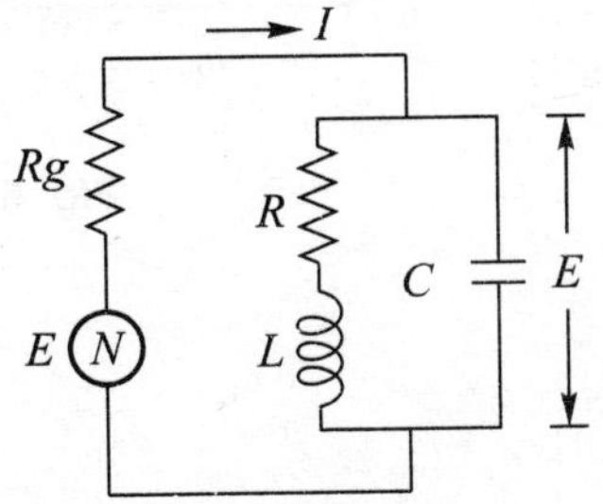

Fig. 6.13

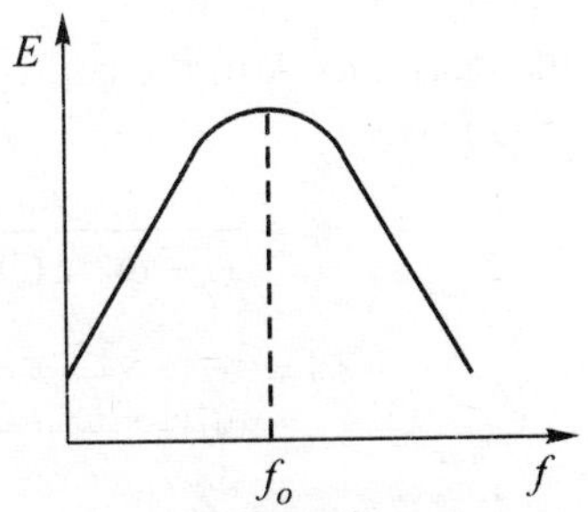

Fig. 6.14

Since impedance of parallel circuit is maximum at resonance, E' will also be a maximum at resonance as shown in figure 6.14. Thus parallel circuits are highly selective when energised from a high impedance generator.

Now, the expression for impedance Z of parallel resonant circuit is

$$\frac{1}{Z} = \frac{1}{R + j\omega L} + \frac{1}{1/j\omega C} = \frac{\frac{1}{1j\omega C} + R + j\omega L}{(R + j\omega L)\frac{1}{j\omega L}} \quad \text{or} \quad Z = \frac{(R + j\omega L)\frac{1}{j\omega C}}{R + j\left(\omega L - \frac{1}{\omega C}\right)}$$

If $\omega L >> R$, then

$$Z_{\text{parallel}} = \frac{L/C}{R + j\left(\omega L - \frac{1}{\omega C}\right)} = \frac{\omega_0 L/\omega_0 C}{R + j\left(\omega L - \frac{1}{\omega C}\right)}$$

$$= \frac{(\omega_0 L)^2}{R + j\left(\omega L - \frac{1}{\omega C}\right)} \left(\therefore\ \omega_0 L = \frac{1}{\omega_0 C}\right) = \frac{\omega_0 L/\omega_0 C}{R + j\left(\omega L - \frac{1}{\omega C}\right)}$$

Also, the equation for current in series circuit is

$$I_{\text{Series}} = \frac{E}{R + j\left(\omega L - \frac{1}{\omega C}\right)} = \frac{\omega_0 L/\omega_0 C}{R + j\left(\omega L - \frac{1}{\omega C}\right)}$$

A comparison of equations (2), (3) shows that both the parallel impedance and the series current are equal to a constant divided by at some impedance. The resonance curve of the parallel impedance can, therefore, be considered to have the same shape as the resonance curve of the series current and the same working rules can be applied for estimating the sharpness of parallel resonance which were applied to the series resonance.

$$(Z_o = L/CR)$$

For high Q values, Bw corresponding to impedance $\frac{Z_0}{\sqrt{2}}$ is called the pass band width. The degree of selectivity Q of the parallel resonant circuit is defined in a similar manner as for series resonant circuit *i.e.*, $Q = \frac{\omega_0}{\Delta\omega}$. The sharpness of the curve can be increased by increasing the circuit Q.

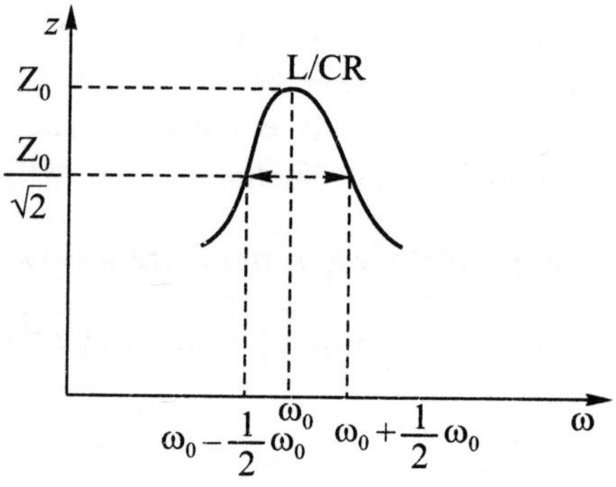

Fig. 6.15

Since the resistance R of the coil is negligibly small, then equation (1) gives

$$Z = \frac{j\omega L}{(1 - \omega^2 LC) + j\omega CR}$$

or

$$|Z| = \frac{\omega L}{\sqrt{[(1 - \omega^2 LC)^2 + (\omega^2 C^2 R^2)]}}$$

For $R \to 0$ and $|Z| = \frac{|Z_0|}{\sqrt{2}}$; $\omega = \omega_0 \pm \frac{1}{2}\Delta\omega$, the terms ωL and $\omega^2 C^2 R^2$ of equation (4) will approximately be equal to $\omega_0 L$ and $\omega_0^2 C^2 R^2$ but the terms $1 - \omega^2 LC$ will no longer be zero (Since when $R \to 0$, $\omega_0 = \frac{1}{\sqrt{[LC]}}$ and hence for this term we put $\omega = \omega_0 + \frac{1}{2}\Delta\omega$. Thus

$$\frac{|Z_0|}{\sqrt{2}} = \frac{\omega_0 L}{\sqrt{\left\{1 - \left[\omega_0 + \frac{1}{2}\Delta\omega\right]^2 LC\right\}^2 + \omega_0^2 C^2 R^2}}$$

Dividing numerator and denominator by $\omega_0\, CR$.

$$\frac{|Z_0|}{\sqrt{2}} = \frac{L/CR}{\dfrac{1}{\omega_0\, CR}\sqrt{\left[\left\{1-\left(\omega_0+\dfrac{1}{2}\Delta\omega\right)^2 LC\right\}^2 + \omega_0^{\,2}C^2R^2\right]}}$$

But $L/CR = Z_0$, hence

$$\frac{|Z_0|}{\sqrt{2}} = \frac{Z_0}{\dfrac{1}{\omega_0\, CR}\sqrt{\left[\left\{1-\left(\omega_0+\dfrac{1}{2}\Delta\omega\right)^2 LC\right\}^2 + \omega_0^{\,2}C^2R^2\right]}}$$

Therefore

$$\sqrt{2}\,\omega_0 CR = \sqrt{\left[\left\{1-\left(\omega_0+\frac{1}{2}\Delta\omega\right)^2 LC\right\}^2 + \omega_0^{\,2}C^2R^2\right]}$$

or

$$2\omega_0^{\,2}C^2R^2 = \omega_0^{\,2}C^2R^2 = \left[1-\left(\omega_0+\frac{1}{2}\Delta\omega\right)^2 . LC\right]^2$$

$$\omega_0^{\,2}C^2R^2 = \left[1-\left(\omega_0^{\,2}+\omega_0 . \Delta\omega+\frac{1}{4}\Delta\omega^2\right) . \frac{1}{\omega_0^{\,2}}\right]^2 = \left(\frac{D\omega}{\omega_0}\right)^2$$

$$\therefore \qquad \frac{\Delta\omega}{\omega_0} = \pm\, \omega_0 CR$$

Therefore, the selectivity of the parallel resonant circuit

$$Q = \frac{\omega_0}{\Delta\omega} = \frac{1}{\omega_0\, CR} = \frac{\omega_0 L}{R}$$

$$= \frac{1}{R}\frac{1}{\sqrt{(LC)}} = \frac{1}{R}\sqrt{\left(\frac{L}{C}\right)} \qquad \left(\therefore\ \omega_0 L = \frac{1}{\omega_0 C}\right)$$

Thus the selectivity of the circuit may be increased by increasing the value of (L/C) on decreasing the value of R.

6.6 COMPARISON BETWEEN SERIES AND PARALLEL RESONANT CIRCUITS

Series Resonant Circuit	Parallel Resonant Circuit
1. Resonant frequency f_o is independent of resistance R, $\left(f_o = \dfrac{1}{2\pi\sqrt{LC}}\right)$	1. Resonant frequency depends on resistance $f_0 = \dfrac{1}{2\pi}\sqrt{\left(\dfrac{1}{LC}-\dfrac{R^2}{L^2}\right)}$ but for small resistance in the circuit, the divergence from the series resonant value $\dfrac{1}{2\pi\sqrt{(LC)}}$ is not very much.

2. Impedance Z at resonance is minimum and purely resistive.
3. Current at resonance is inversely proportional to resistance of the circuit ($I = E/R$) and is independent of capacity C and inductance L.
4. The series resonant circuit, called the acceptor circuit, offers an easy path for currents at resonant frequency (current maximum at resonance) and a more difficult path for another frequencies.
5. For high selectivity, the circuit should be energised by a generator of low internal impedance.
6. At resonance, exhibits a voltage.

2. Impedance at resonance is maximum *i.e.*, admittance ($1/Z$) at minimum value.
3. Current at resonance is directly proportional to resistance $\left(I = \frac{ECR}{L}\right)$ and an increase in the ratio of (C/L) increases the supply current.
4. The parallel resonant circuit, called the rejector circuit, offers a difficult path for current at resonant frequency (current minimum at a resonance) and an easier path for all other frequencies.
5. For high selectivity, the circuit should be energised by a generator having high internal impedance.
6. At resonance, exhibits a current.

Q and Bandwidth of a Resonant Circuit

The Q, quality factor, of a resonant circuit is a measure of the "goodness" or quality of a resonant circuit. A higher value for this figure of merit corresponds to a more narrow bandwidth, which is desirable in many applications. More formally, Q is the ratio of power stored to power dissipated in the circuit reactance and resistance, respectively:

$$Q = P_{stored}/P_{dissipated} = I^2X/I^2R$$

$$Q = X/R$$

where X = Capacitive or Inductive reactance at resonance

R = Series resistance.

This formula is applicable to series resonant circuits, and also parallel resonant circuits if the resistance is in series with the inductor. This is the case in practical applications, as we are mostly concerned with the resistance of the inductor limiting the Q. Note: Some text may show X and R interchanged in the "Q" formula for a parallel resonant circuit. This is correct for a large value of R in parallel with C and L. Our formula is correct for a small R in series with L.

A practical application of "Q" is that voltage across L or C in a series resonant circuit is Q times total applied voltage. In a parallel resonant circuit, current through L or C is Q times the total applied current.

Problem 1: *In a series RLC circuit, R = 100 Ω, L = 0.5H and C = 40 µf. Calculate the resonant, lower and upper half frequencies.*

Solution: Resonant frequency

$$f_o = \frac{1}{2\pi\sqrt{LC}} = \frac{1}{2 \times 3.14\sqrt{0.5 \times 40 \times 10^{-6}}}$$

$$= \frac{1}{6.28\sqrt{20 \times 10^{-6}}} = \frac{1}{6.25 \times 10^{-3} \times 4.97}$$

$$= 35.6 \text{ Hz}$$

Lower half power frequency

$$f_1 = \frac{\omega_1}{2\pi} = \frac{1}{2\pi}\left(\omega_o - \frac{R}{2L}\right)$$

$$= f_o - \frac{R}{4\pi L} = 35.6 - \frac{100}{4 \times 3.14 \times 0.5}$$

$$= 35.6 - 15.9 = 19.7 \text{ Hz}$$

Upper half power frequency

$$f_2 = f_o + \frac{R}{4\pi L} = 35.6 + 15.9 = 51.5 \text{ Hz.}$$

Problem 2: *Find the quality factor for an LCR series circuit with L = 0.25 mH, C = 25 μf and R = 10 Ω.*

Solution:

$$Q = \frac{1}{R}\sqrt{\frac{L}{C}} = \frac{1}{10}\sqrt{\frac{0.25 \times 10^{-3}}{25 \times 10^{-6}}}$$

$$= \frac{1}{10}\sqrt{10} = \frac{1}{\sqrt{10}} = 0.316.$$

Problem 3: *A series RLC circuit has R = 5 Ω, L = 40 mH and C = 1 μf. Calculate resonant frequency, Q of the circuit and bandwidth.*

Solution: Resonant frequency

$$f_o = \frac{1}{2\pi\sqrt{LC}}$$

$$= \frac{1}{2 \times 3.14\sqrt{40 \times 10^{-3} \times 1 \times 10^{-6}}} = 796.2 \text{ Hz}$$

Q of the circuit

$$Q = \frac{1}{R}\sqrt{\frac{L}{C}} = \frac{1}{5}\sqrt{\frac{40 \times 10^{-3}}{1 \times 10^{-6}}} = 40$$

Bandwidth

$$\Delta f = \frac{f_o}{Q} = \frac{796.2}{40} = 19.9 \text{ Hz}$$

Problem 4: *A series LCR circuit has Q = 120 at resonance, a capacitance 200 pf connected in series with an inductance of 150 μH. Calculate its bandwidth.*

Solution: Resonant frequency

$$f_o = \frac{1}{2\pi\sqrt{LC}}$$

$$= \frac{1}{6.28\sqrt{150 \times 10^{-6} \times 200 \times 10^{-12}}}$$

$$= 9.19 \times 10^5 \text{ Hz}$$

Now

$$Q = \frac{f_o}{f_2 - f_1} = \frac{f_o}{\Delta f}$$

Bandwidth

$$\Delta f = \frac{f_o}{Q} = \frac{9.19 \times 10^5}{120} = 7.66 \text{ Hz.}$$

REVIEW QUESTIONS

1. What is a series resonant circuit ? Derive the expression for the resonant frequency of the circuit ?
2. What is quality factor of a circuit ? Explain LCR parallel Resonance circuit with necessary theory.
3. Write a short note on Q-factor.
4. Define reactance and impedance. Give the theory of LCR series resonant circuit.
5. Explain Q-factor and power factor.
6. Give the detailed theory of L-C-R series circuit carrying A.C. and explain resonance condition.
7. Deduce the condition when the charge is oscillatory, if the circuit contains the combination L, C, R.
8. Give the necessary theory of L-C-R series circuit and exhibit it as a series resonant circuit. What is series resonance ?
9. What is parallel resonance ? Give the detailed theory of L-C-R parallel circuit and exhibit it as a parallel resonant circuit.

EXERCISE

1. In an oscillatory circuit $L = 0.5$ Henry, $C = 2$ MF, what is the maximum value of resistance so that the circuit may just oscillate ? **[Ans.** $1 \times 10^3\ \Omega$]
2. A parallel resonant circuit is formed with a capacitor of capacity 50 MF, a coil of inductance 0.8 mH and a resistance of 10 Ω. Calculate the frequency of resonance and impedance of resonance.
3. Calculate the resonant frequency of an LCR parallel resonant circuit with 100 m, $C = 10$ MF and $R = 1$ KΩ. **[Ans.** 159.2 Hz]
4. Given that L = 1 Henry, $C = 1$M Farad, $R = 10\ \Omega$ of an LCR series resonant circuit, find its resonant frequency, impedance, admittance and quality factor.
5. Find the value of an inductance which should be connected in series with a capacitor of 5MF, resistance of 100 Ω and A.C. source of 50 Hz so that the power factor is unity.
6. A capacitor of 0.1 MF capacity is discharged through an inductance of 10 mH having a resistance of 200 Ω. If the circuit is oscillatory, calculate H's frequency.

SOLVED EXAMPLES

Example 1: *A 1 MF capacitor is discharged through a 1 Henry inductor. Calculate the frequency of discharge.*

Solution: The frequency of discharge

$$f = \frac{1}{2\pi}\sqrt{\frac{1}{LC}}$$

Given that

$$C = 1 \text{ MF} = 1 \times 10^{-6} \text{ F}$$

$$L = 1 \text{ H}$$

$$f = \frac{1}{2\pi}\sqrt{\frac{1}{1\times1\times10^{-6}}}$$

$$f = \frac{1}{2\pi} \times 10^3 = 159.2 \text{ Hertz.}$$

Example 2: *In a parallel LCR circuit, R = 100 Ω, L = 1H, C = 100 MF. Calculate resonant frequency and Q-factor.*

Solution:

$$f = \frac{1}{2\pi}\sqrt{\frac{1}{LC} - \frac{R^2}{L^2}} \cong \frac{1}{2\pi}\sqrt{\frac{1}{LC}}$$

$$f = \frac{1}{2 \times 3.14}\sqrt{\frac{1}{1 \times 100 \times 10^{-6}}}$$

$$f = \frac{100}{6.28} = 15.92 \text{ Hertz}$$

$$Q = \frac{\omega L}{R} = \frac{2\pi f L}{R} = \frac{2 \times 3.14 \times 15.92 \times 1}{100} = 0.99.$$

Example 3: *In a series LCR circuit, R = 500 Ω, L = 2H, C = 5 MF. Calculate resonant frequency and Q-factor.*

Solution:

$$f = \frac{1}{2\pi\sqrt{LC}} = \frac{1}{2 \times 3.14\sqrt{2 \times 5 \times 10^{-6}}}$$

$$f = 50.35 \text{ Hertz}$$

$$\text{Quality factor} = Q = \frac{\omega L}{R} = \frac{2\pi f L}{R}$$

$$Q = \frac{2 \times 3.14 \times 50.35 \times 2}{500} = 1.264.$$

Example 4: *In an oscillatory circuit, L = 0.2 H, C = 0.00128 MF, what is the maximum value of resistance so that the circuit may just oscillate ?*

Solution:

$$R = 2\sqrt{\frac{L}{C}} = 2\sqrt{\frac{0.2}{0.00128 \times 10^{-6}}} = 25 \times 10^{-3}\ \Omega$$

Example 5: *Calculate f_o ?*

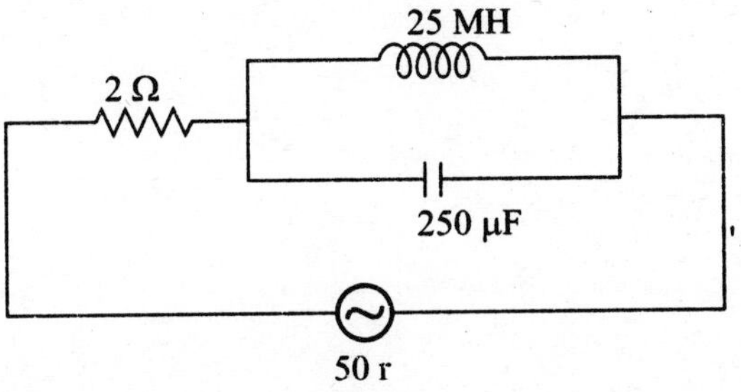

Fig. 6.16

Solution:

$$f_o = \frac{1}{2\pi\sqrt{LC}}$$

$$f_o = \frac{1}{2 \times 3.14\sqrt{25 \times 10^{-3} \times 250 \times 10^{-6}}}$$

$$f_o = \frac{31622.7766}{6.28 \times 79.05} = \frac{31622.7766}{496.47}$$

$$f_o = 63.69 \text{ Hz.}$$

UNIT–III

CHAPTER 7

PN Junction

7.1 CLASSIFICATION OF SOLIDS BASED ON ELECTRICAL PROPERTIES OF CONDUCTOR

In science and engineering, a conductor is a material which contains movable electric charges. In metallic conductors, such as Copper or aluminum, the movable charged particles are electrons. Positive charges may also be mobile in the form of atoms in a lattice missing electrons called "holes" or ions, such as in the electrolyte of a battery.

All conductors contain electric charges which will move when an electric potential difference is applied across separate points on the material. This flow of charge is what is meant by electric current. In most materials, the rate of current is proportional to the voltage provided the temperature remains constant and the material remains in the same shape and state. The ratio between the voltage and the current is called the resistance of the object between the points where the voltage was applied. The resistance across a standard mass and shape of a material at a given temperature is called the resistivity of the material. The inverse of resistance and resistivity is conductance and conductivity. Some good examples of conductors are metals.

Most familiar conductors are metallic. Copper is the most common material for electrical wiring, silver is the best but expensive, and gold for high-quality surface-to-surface contacts. However, there are also many non-metallic conductors, including graphite, solutions of salts, and all plasmas.

Non-conducting materials lack mobile charges, and so resist the flow of electric current, generating heat. In fact, all materials offer some resistance and warm up when a current flows. Thus, proper design of an electrical conductor takes into account the temperature that the conductor needs to be able to endure without damage, as well as the quantity of electrical current. The motion of charges also creates an electromagnetic field around the conductor that exerts a mechanical radial squeezing force on the conductor. A conductor of a given material and volume has no real limit to the current it can carry without being destroyed as long as the heat generated by the resistive loss is removed and the conductor can withstand the radial forces.

Since all conductors have some resistance; and all insulators will carry some current, there is no theoretical dividing line between conductors and insulators. However, there is a large gap between the conductance of materials that will carry a useful current at working voltages and those that will carry a negligible current for the purpose in hand, so the categories of insulator and conductor do have practical utility.

Insulator

An Insulator is a material that resists the flow of electric current. It is an object intended to support or separate electrical conductors without passing current through itself. An insulation material has atoms with tightly bonded valence electrons. The term electrical insulation has the same meaning as the term dielectric.

Some materials such as silicon dioxide or Teflon are very good electrical insulators. A much larger class of materials, for example rubber-like polymers and most plastics are still "good enough" to insulate electrical wiring and cables even though they may have lower bulk resistivity. These materials can serve as practical and safe insulators for low to moderate voltages (hundreds, or even thousands, of volts).

Semiconductor

A semiconductor is a solid material that has electrical conductivity in between a conductor and an insulator; it can vary over that wide range either permanently or dynamically.

Semiconductors are important in electronic technology. Semiconductor devices, electronic components made of semiconductor materials, are essential in modern consumer electronics, including computers, mobile phones, and digital audio players. Silicon is used to create most semiconductors commercially, but dozens of other materials are used as well.

Classification of Solids based on Energy Band Theory

The electrons of a single free-standing atom occupy atomic orbital and form a discrete set of energy levels. If several atoms are brought together into a molecule, their atomic orbitals split, as in a coupled oscillation. This produces a number of molecular orbital proportional to the number of atoms. When a large number of atoms (of order 1020 or more) are brought together to form a solid, the number of orbital becomes exceedingly large, and the difference in energy between them becomes very small, so the levels may be considered to form continuous bands of energy rather than the discrete energy levels of the atoms in isolation. However, some intervals of energy contain no orbital, no matter how many atoms are aggregated, forming band gaps. Within an energy band, energy levels are so numerous as to be a near continuum. First, the separation between energy levels in a solid is comparable with the energy that electrons constantly exchange with photons (atomic vibrations). Second, it is comparable with the energy uncertainty due to the Heisenberg uncertainty principle, for reasonably long intervals of time. As a result, the separation between energy levels is of no consequence.

Conduction Band and Valence Band

In solids, the valence band is the highest range of electron energies where electrons are normally present at absolute zero. In semiconductors and insulators, there is a band gap above the valence band, followed by a conduction band above that. In metals, the conduction band has no energy gap separating it from the valence band (basically, this is correct only for semimetals. All solids have forbidden energy levels between the energy bands). The rest of this discussion refers to the valence band in semiconductors and insulators.

Semiconductor Band Structure

Semiconductors and insulators owe their low conductivity to the properties of the valence band in those materials. It just so happens that the number of electrons is precisely equal to the number of states available up to the top of the valence band. There are no available states in the band gap. This means that when an electric field is applied, the electrons cannot increase their energy (*i.e.*, accelerate) because there are no states available to the electrons where they would be moving faster than they are already going.

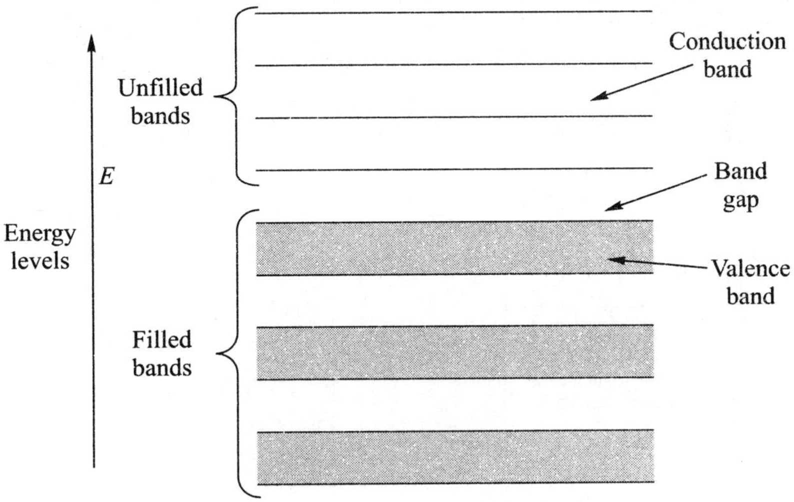

Fig. 7.1. Semiconductor band structure.

There is some conductivity in insulators, however. This is due to thermal excitation-some of the electrons get enough energy to jump the band gap in one go. Once they are in the conduction band, they can conduct electricity, as can the hole they left behind in the valence band. The hole is an empty state that allows electrons in the valence band some degree of freedom.

In the physics field of semiconductors and insulators, the conduction band is the range of electron energy, higher than that of the valence band, sufficient to make the electrons free to accelerate under the influence of an applied electric field and thus constitute an electric current. Semiconductors may cross this conduction band when they are excited.

Electronic Band Structure

In solid-state physics, the electronic band structure of a solid describes ranges of energy that an electron is "forbidden" or "allowed" to have. The band structure of a material determines several characteristics, in particular its electronic and optical properties.

Any solid has a large number of bands. In theory, it can be said to have infinitely many bands just as an atom has infinitely many energy levels. However, all but a few lie at energies so high that any electron that reaches those energies escapes from the solid. These bands are usually disregarded.

Bands have different widths, based upon the properties of the atomic orbitals from which they arise. Also, allowed bands may overlap, producing a single large band. Figure shows a simplified picture of the bands in a solid that allows the three major types of materials to be identified: metals, semiconductors and insulators.

Metals contain a band that is partly empty and partly filled regardless of temperature. Therefore they have very high conductivity.

The lowermost, almost fully occupied band in an insulator or semiconductor is called the valence band by analogy with the valence electrons of individual atoms. The uppermost, almost unoccupied band is called the conduction band because only when electrons are excited to the conduction band can current flow in these materials. The difference between insulators and semiconductors is only that the forbidden band gap between the valence band and conduction band is larger in an insulator, so that fewer electrons are found there and the electrical conductivity is lower. Because one of the main mechanisms for electrons to be excited to the conduction band is due to thermal energy, the conductivity of semiconductors is strongly dependent on the temperature of the material.

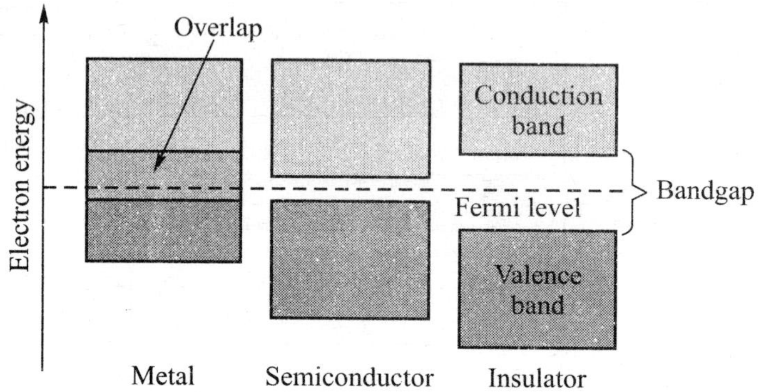

Fig. 7.2. Simplified diagram of the electronic band structure of metals, semiconductors, and insulator.

This band gap is one of the most useful aspects of the band structure, as it strongly influences the electrical and optical properties of the material. Electrons can transfer from one band to the other by means of carrier generation and recombination processes.

7.2 SEMI-CONDUCTORS

We have two types of semiconductors, intrinsic and extrinsic, an intrinsic semiconductor, also called an undoped semiconductor or *I*-type semiconductor, is a pure semiconductor without any significant dopant species present. The number of charge carriers is therefore determined by the properties of the material itself instead of the amount of impurities. In intrinsic semiconductors the number of excited electrons and the number of holes are equal.

The conductivity of intrinsic semiconductors can be due to crystal defects or thermal excitation. In an intrinsic semiconductor the number of electrons in the conduction band is equal to the number of holes in the valence band.

An extrinsic semiconductor is a semiconductor that has been doped, that is, into which a doping agent has been introduced, giving it different electrical properties than the intrinsic (pure) semiconductor. Doping involves adding dopant atoms to an intrinsic semiconductor, which changes the electron and hole carrier concentrations of the semiconductor at thermal equilibrium. Dominant carrier concentrations in an extrinsic semiconductor classify it as either an *n*-type or *p*-type semiconductor. The electrical properties of extrinsic semiconductors make them essential components of many electronic devices.

Semi-conductors materials such as Silicon and Germanium have electrical properties somewhere in the middle, between those of a "Conductor" and an "Insulator". They are not good conductors or good insulators (hence their name semi-conductors). They have very few "free electrons" because their atoms are closely grouped together in a crystalline pattern called a "Crystal Lattice". However, their ability to conduct electricity can be greatly improved by adding certain "Impurities" to this crystalline structure, thereby producing more free electrons than holes or vice versa. By controlling the amount of impurities added to the semiconductor material it is possible to control its conductivity. This process of adding impurity atoms to semiconductor atoms (the order of 1 impurity atom per 10 million (or more) atoms of the semiconductor) is called Doping.

Now we are going to talk about doping, in the process of doping we add a tiny amount of atoms from another material to the pure semiconductor. By doing so, you can drastically increase its ability to conduct a current. There are two forms of doping, *p* and *n*. *p* stands for positive and *n* for negative. Finally, two words that are good to know: a pure non-doped semiconductor is called intrinsic, while a doped semiconductor material is called extrinsic.

So, Semiconductors' intrinsic electrical properties are often permanently modified by introducing impurities by a process known as doping. Usually, it is sufficient to approximate that each impurity atom adds one electron or one "hole" that may flow freely. Upon the addition of a sufficiently large proportion of impurity dopants, semiconductors will conduct electricity nearly as well as metals. Depending on the kind of impurity, a doped region of semiconductor can have more electrons or holes, and is named *N*-type or *P*-type semiconductor material, respectively. Junctions between regions of *N*- and *P*-type semiconductors create electric fields, which cause electrons and holes to be available to move away from them, and this effect is critical to semiconductor device operation. Also, a density difference in the amount of impurities produces a small electric field in the region which is used to accelerate non-equilibrium electrons or holes.

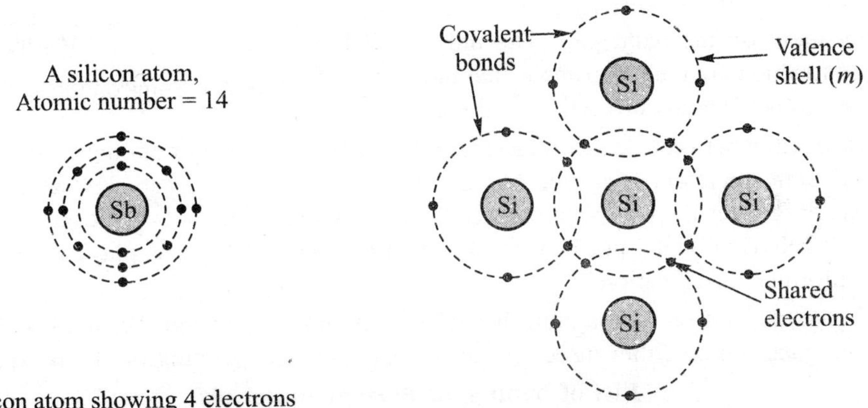

Silicon atom showing 4 electrons in its outer valence shell (*m*)

Silicon crystal lattice

Fig. 7.3. The diagram above shows the structure and lattice of a 'normal' pure crystal of Silicon.

In addition to permanent modification through doping, the resistance of semiconductors is normally modified dynamically by applying electric fields. The ability to control resistance/ conductivity in regions of semiconductor material dynamically through the application of electric fields is the feature that makes semiconductors useful. It has led to the development of a broad range of semiconductor devices, like transistors and diodes. Semiconductor devices that have dynamically controllable conductivity, such as transistors, are the building blocks of integrated circuits devices like the microprocessor. These "active" semiconductor devices (transistors) are combined with passive components implemented from semiconductor material such as capacitors and resistors, to produce complete electronic circuits.

The most commonly used semiconductor material is Silicon. It has four valence electrons in its outer most shell which it shares with its adjacent atoms in forming covalent bonds. The structure of the bond between two silicon atoms is such that each atom shares one electron with its neighbour making the bond very stable. As there are very few free electrons available to move from place to place producing an electrical current, crystals of pure silicon (or germanium) are therefore good insulators, or at the very least very high value resistors. Silicon atoms are arranged in a definite symmetrical pattern making them a crystalline solid structure. A crystal of pure silicon (silicon dioxide or glass) is generally said to be an intrinsic crystal.

Band Gap

In solid state physics and related applied fields, the band gap, also called an energy gap or stop band, is a region where a particle or Quasiparticle is forbidden from propagating. For insulators and semiconductors, the band gap generally refers to the energy difference between the top of the valence band and the bottom of the conduction band.

In semiconductors and insulators, electrons are confined to a number of bands of energy, and forbidden from other regions. The term "band gap" refers to the energy difference between the top of the valence band and the bottom of the conduction band; electrons are able to jump from one band to another.

The conductivity of intrinsic semiconductors is strongly dependent on the band gap. The only available carriers for conduction are the electrons which have enough thermal energy to be excited across the band gap.

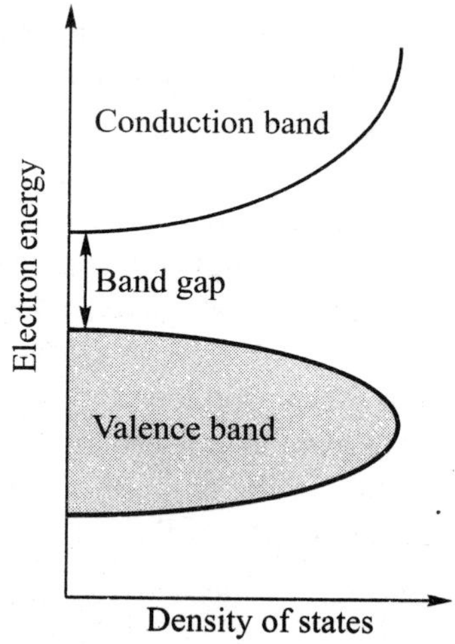

Fig. 7.4. Semiconductor band structure.

The distinction between semiconductors and insulators is a matter of convention. One approach is to think of semiconductors as a type of insulator with a low band gap. Insulators with a higher band gap, usually greater than 3 eV, are not considered semiconductors and generally do not exhibit semiconductive behaviour under practical conditions. Electron mobility also plays a role in determining a material's informal classification.

Band gaps depend on temperature because of thermal expansion. Band gaps also depend on pressure. Band gaps can be either direct or indirect bandgaps, depending on the band structure.

List of band gaps of some materials

Material	Symbol	Band gap (eV) at 300K
Silicon	Si	1.11 [1]
Germanium	Ge	0.67 [1]
Silicon carbide	SiC	2.86 [1]
Aluminum phosphide	AlP	2.45 [1]
Aluminium arsenide	AlAs	2.16 [1]
Aluminium antimonide	AlSb	1.6 [1]
Aluminium nitride	AlN	6.3
Diamond	C	5.5

Gallium(III) phosphide	GaP	2.26 [1]
Gallium(III) arsenide	GaAs	1.43 [1]
Gallium(III) nitride	GaN	3.4 [1]
Gallium(II) sulfide	GaS	2.5 (at 295 K)
Gallium antimonide	GaSb	0.7 [1]
Indium(III) phosphide	InP	1.35 [1]
Indium(III) arsenide	InAs	0.36 [1]
Zinc sulfide	ZnS	3.6 [1]
Zinc selenide	ZnSe	2.7 [1]
Zinc telluride	ZnTe	2.25 [1]
Cadmium sulfide	CdS	2.42 [1]
Cadmium selenide	CdSe	1.73 [1]
Cadmium telluride	CdTe	1.49 [2]
Lead(II) sulfide	PbS	0.37 [1]
Lead(II) selenide	PbSe	0.27 [1]
Lead(II) telluride	PbTe	0.29

N-type Semiconductor

In order for our silicon crystal to conduct electricity, we need to introduce an impurity atom such as Arsenic, Antimony or Phosphorus (V Group element) into the crystalline structure. These atoms have five outer electrons in their outermost covalent bond to share with other atoms and are commonly called "Pentavalent" impurities. This allows four of the five electrons to bond with its neighbouring silicon atoms leaving one "free electron" to move about when an electrical voltage is applied (electron flow). As each impurity atom "donates" one electron, pentavalent atoms are generally known as "Donors". Antimony (symbol Sb) is frequently used as a pentavalent additive as it has 51 electrons arranged in 5 shells around the nucleus. The resulting semiconductor material has an excess of current-carrying electrons, each with a negative charge, and is therefore referred to as "N-type" material with the electrons called "Majority Carriers" and the resultant holes "Minority Carriers".

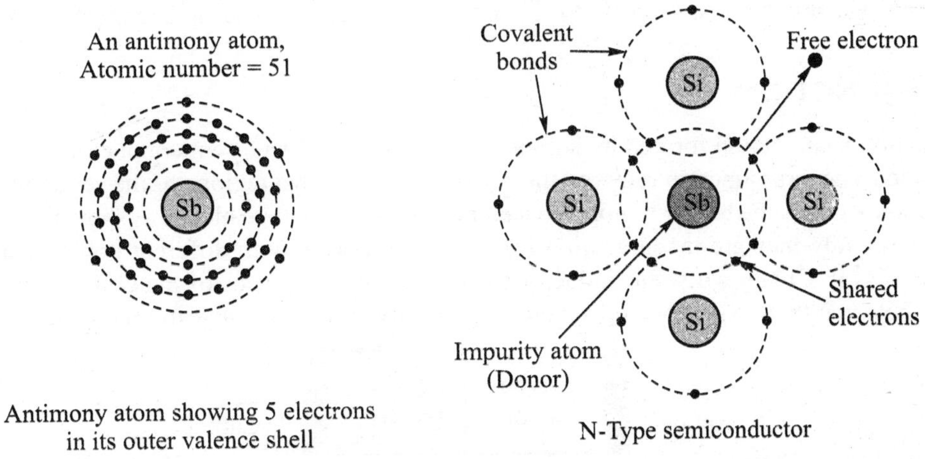

Fig. 7.5. The diagram above shows the structure and lattice of the donor impurity atom Antimony.

P-Type Semiconductor

If we go the other way, and introduce a "Trivalent" (3-electron) generally a III group element impurity into the crystal structure, such as Aluminum, Boron or Indium, only three valence electrons are available in the outermost covalent bond meaning that the fourth bond cannot be formed. Therefore, a complete connection is not possible, giving the semiconductor material an abundance of positively charged carriers known as "holes" in the structure of the crystal. As there is a hole an adjoining free electron is attracted to it and will try to move into the hole to fill it. However, the electron filling the hole leaves another hole behind it as it moves. This in turn attracts another electron which in turn creates another hole behind, and so forth giving the appearance that the holes are moving as a positive charge through the crystal structure (conventional current flow). As each impurity atom generates a hole, trivalent impurities are generally known as "Acceptors" as they are continually "accepting" extra electrons. Boron (symbol B) is frequently used as a trivalent additive as it has only 5 electrons arranged in 3 shells around the nucleus. Addition of Boron causes conduction to consist mainly of positive charge carriers that results in a "*P*-type" material and the positive holes are called "Majority Carriers" while the free electrons are called "Minority Carriers".

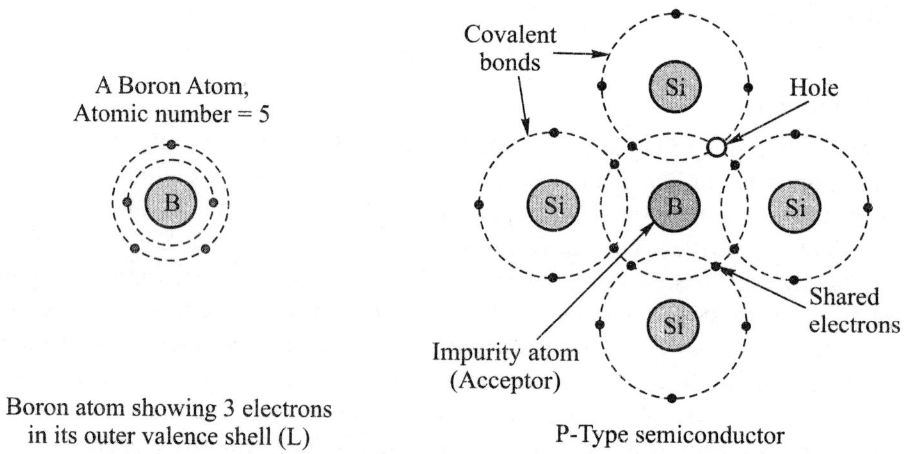

Fig. 7.6. The diagram above shows the structure and lattice of the acceptor impurity atom Boron.

7.3 P-N JUNCTION

A *p-n* junction is a junction formed by combining *P*-type and *N*-type semiconductors together in very close contact. The term junction refers to the region where the two regions of the semiconductor meet. It can be thought of as the border region between the *p*-type and *n*-type blocks as shown in the following diagram. Normally they are manufactured from a single crystal with different dopant concentrations diffused across it. Creating a semiconductor from two separate pieces of material introduces a grain boundary between them which would severely inhibit its utility by scattering the electrons and holes.

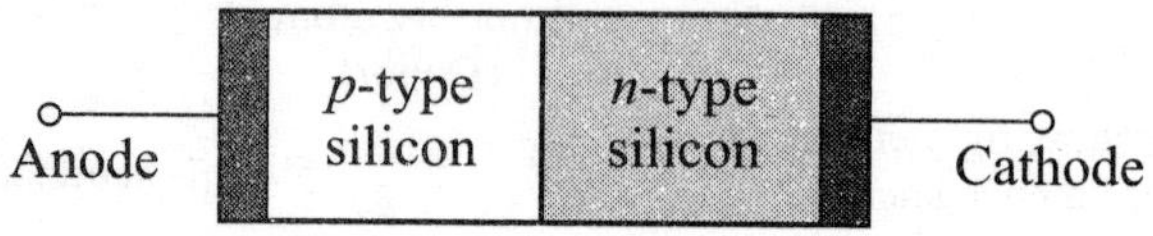

Fig. 7.7. A silicon p-n junction with no applied voltage.

Properties of a p-n Junction

The *p-n* junction possesses some interesting properties which have useful applications in modern electronics. A *p*-doped semiconductor is relatively conductive. The same is true of an *n*-doped semiconductor, but the junction between them is a nonconductor. This non conducting layer, called the depletion zone, occurs because the electrical charge carriers in doped *n*-type and *p*-type silicon (electrons and holes, respectively) attract and eliminate each other in a process called recombination. By manipulating this nonconductive layer, *p-n* junctions are commonly used as diodes: circuit elements that allow a flow of electricity in one direction but not in the other (opposite) direction. This property is explained in terms of the forward-bias and reverse-bias effects, where the term bias refers to an application of electric voltage to the *p-n* junction.

Equilibrium (zero bias)

In a *p-n* junction, without an external applied voltage, an equilibrium condition is reached in which a potential difference is formed across the junction. This potential difference is called built-in potential V_{bi}.

In an equilibrium *p-n* junction, electrons near the *p-n* interface tend to diffuse into the *p* region. As electrons diffuse, they leave positively charged ions (donors) on the *n* region. Similarly holes near the *p-n* interface begin to diffuse in the *n*-type region leaving fixed ions (acceptors) with negative charge. The regions nearby the *p-n* interfaces lose their neutrality and become charged, forming the space charge region or depletion layer see figure 7.8.

The electric field created by the space charge region opposes the diffusion process for both electrons and holes. There are two concurrent phenomena: the diffusion process that tends to generate more space charge, and the electric field generated by the space charge that tends to counteract the diffusion. The carrier concentration profile at equilibrium is shown in figure 7.9 with blue and red lines. Also shown are the two counterbalancing phenomena that establish equilibrium.

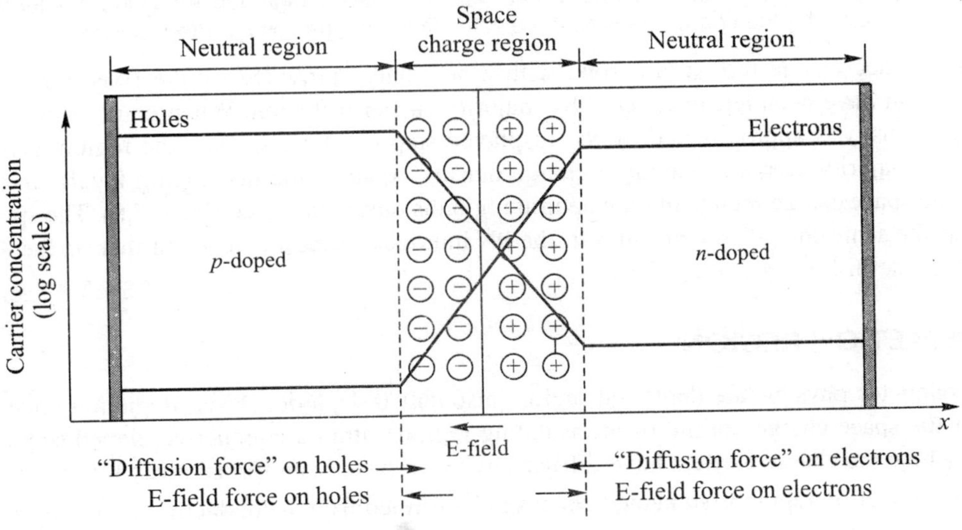

Fig. 7.8. A *p-n* junction in thermal equilibrium with zero bias voltage applied.

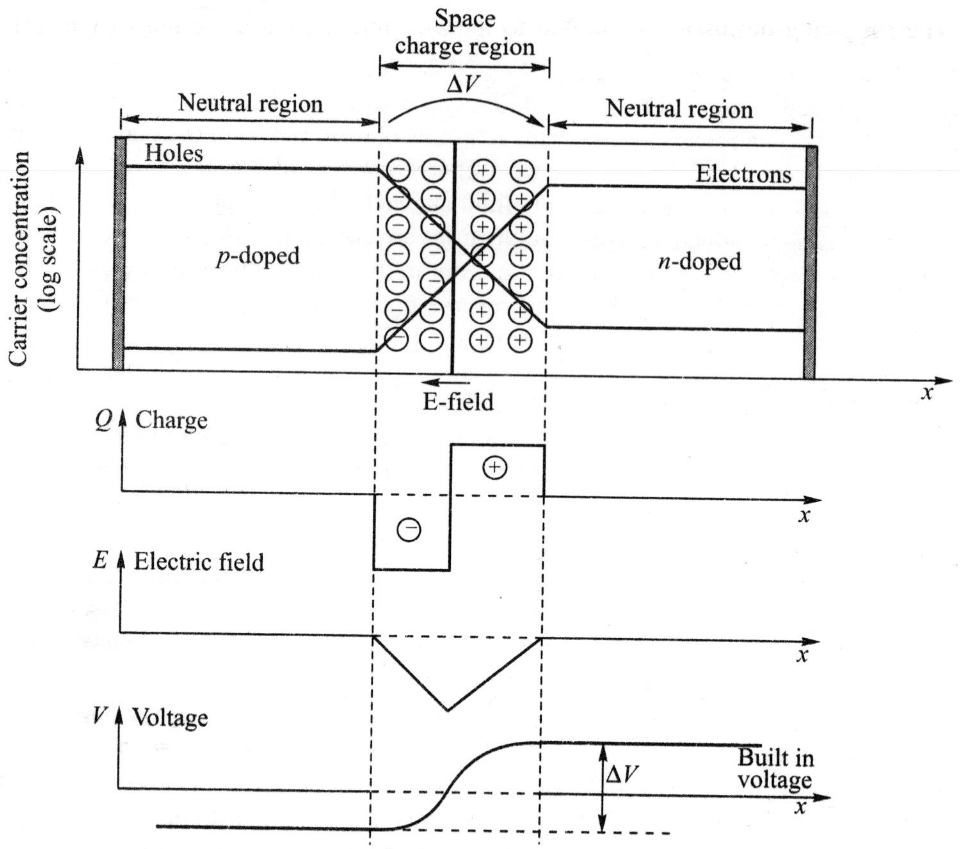

Fig. 7.9. A *p-n* junction in thermal equilibrium with zero bias voltage applied. Under the junction, plots for the charge density, the electric field and the voltage are reported.

The space charge region is a zone with a net charge provided by the fixed ions (donors or acceptors) that have been left uncovered by majority carrier diffusion. When equilibrium is reached, the charge density is approximated by the displayed step function. In fact, the region is completely depleted of majority carriers (leaving a charge density equal to the net doping level), and the edge between the space charge region and the neutral region is quite sharp (see figure 7.8). The space charge region has the same charge on both sides of the PN interfaces, thus it extends farther on the less doped side (the *n* side in figures 7.8 and 7.9).

7.4 DEPLETION REGION

In semiconductor physics, the depletion region, also called depletion layer, depletion zone, junction region or the space charge region, is an insulating region within a conductive, doped semiconductor material where the charge carriers have diffused away, or have been forced away by an electric field.

The depletion region is so named because it is formed from a conducting region by removal of all free charge carriers, leaving none to carry a current. Understanding the depletion region is key to explaining modern semiconductor electronics: diodes, bipolar junction transistors, field-effect transistors, and variable capacitance diodes all rely on depletion region phenomena.

The following discussion is limited to the *p-n* junction but depletion regions arise in all the devices mentioned above.

Formation of Depletion Region in a p-n Junction

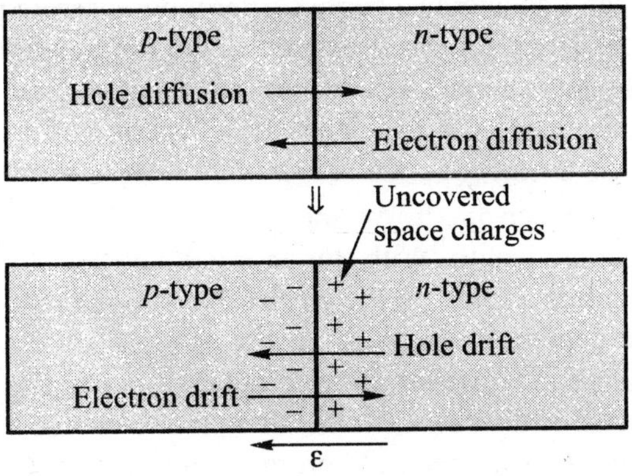

Fig. 7.10. Top: *p-n* junction before diffusion; Bottom: After equilibrium is reached.

A depletion region forms spontaneously across a *p-n* junction. It is most easily described when the junction is in thermal equilibrium or in a steady state: in both of these cases the properties of the system do not vary in time, they have been called dynamic equilibrium. Electrons and holes diffuse into regions with lower concentrations of electrons and holes, such as ink diffuses into water until it is uniformly distributed. By definition, *N*-type semiconductor has an excess of free electrons compared to the *P*-type region, and *P*-type has an excess of holes compared to the *N*-type region. Therefore when *N*-doped and *P*-doped pieces of semiconductor are placed together to form a junction, electrons diffuse into the *P*-side and holes diffuse into the *N*-side. Departure of an electron on the *N*-side for the *P*-side leaves a positive donor ion behind on the *N*-side, and likewise the hole leaves a negative acceptor ion on the *P*-side. Following transfer, the injected electrons come into contact with holes on the *P*-side and are eliminated by recombination. Likewise for the injected holes on the *N*-side. The net result is the injected electrons and holes are gone, leaving behind the charged ions adjacent to the interface in a region with no mobile carriers (called the depletion region). The uncompensated ions are positive on the *N* side and negative on the *P* side. This creates an electric field that provides a force opposing the continued diffusion of charge carriers. When the electric field is sufficient to arrest further transfer of holes and electrons, the depletion region has reached its equilibrium dimensions. Integrating the electric field across the depletion region determines what is called the built-in voltage also called the junction voltage or barrier voltage or contact potential.

Forward-bias

Forward-bias occurs when the *P*-type semiconductor material is connected to the positive terminal of a battery and the *N*-type semiconductor material is connected to the negative terminal, as shown below.

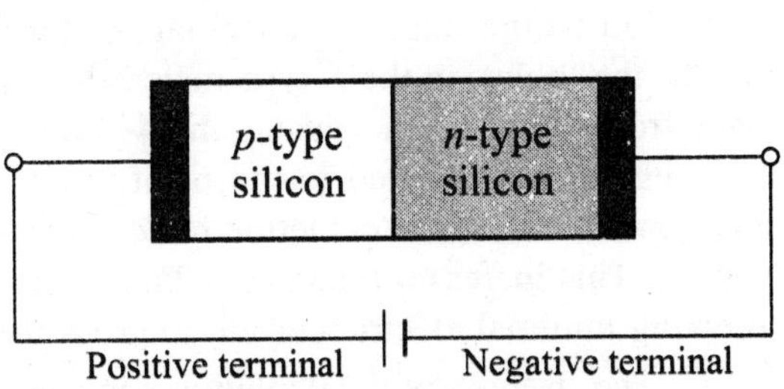

Fig. 7.11. A silicon p-n junction in Forward-bias.

With a battery connected this way, the holes in the *P*-type region and the electrons in the *N*-type region are pushed towards the junction. This reduces the width of the depletion zone. The positive charge applied to the *P*-type material repels the holes, while the negative charge applied to the *N*-type material repels the electrons. As electrons and holes are pushed towards the junction, the distance between them decreases. This lowers the barrier in potential. With increasing forward-bias voltage, the depletion zone eventually becomes thin enough that the zone's electric field can't counteract charge carrier motion across the *p-n* junction, consequently reducing electrical resistance. The electrons which cross the *p-n* junction into the *P*-type material (or holes which cross into the *N*-type material) will diffuse in the near-neutral region. Therefore, the amount of minority diffusion in the near-neutral zones determines the amount of current that may flow through the diode.

Only majority carriers (electrons in *N*-type material or holes in *P*-type) can flow through a semiconductor for a macroscopic length. With this in mind, consider the flow of electrons across the junction. The forward bias causes a force on the electrons pushing them from the *N* side toward the *P* side. With forward bias, the depletion region is narrow enough that electrons can cross the junction and inject into the *P*-type material. However, they do not continue to flow through the *P*-type material indefinitely, because it is energetically favorable for them to recombine with holes. The average length an electron travels through the *P*-type material before recombining is called the diffusion length, and it is typically on the order of microns.

Although the electrons penetrate only a short distance into the *P*-type material, the electric current continues uninterrupted, because holes (the majority carriers) begin to flow in the opposite direction. The total current (the sum of the electron and hole currents) is constant in space, because any variation would cause charge buildup over time. The flow of holes from the *P*-type region into the *N*-type region is exactly analogous to the flow of electrons from *N* to *P* (electrons and holes swap roles and the signs of all currents and voltages are reversed).

Therefore, the macroscopic picture of the current flow through the diode involves electrons flowing through the *N*-type region toward the junction, holes flowing through the *P*-type region in the opposite direction toward the junction, and the two species of carriers constantly recombining in the vicinity of the junction. The electrons and holes travel in opposite directions, but they also have opposite charges, so the overall current is in the same direction on both sides of the diode, as required.

Reverse-bias

Connecting the *P*-type region to the negative terminal of the battery and the *N*-type region to the positive terminal produces the reverse-bias effect. The connections are illustrated in the following diagram:

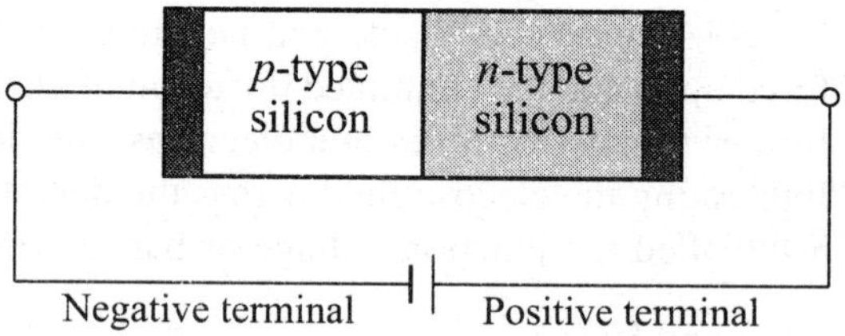

Fig. 7.12. A silicon p-n junction in Reverse-bias

Because the *P*-type material is now connected to the negative terminal of the power supply, the 'holes' in the *P*-type material are pulled away from the junction, causing the width of the depletion zone to increase. Similarly, because the *N*-type region is connected to the positive terminal, the electrons will also be pulled away from the junction. Therefore the depletion region widens, and does so increasingly with increasing reverse-bias voltage. This increases the voltage barrier causing a high resistance to the flow of charge carriers thus allowing minimal electric current to cross the *p-n* junction.

The strength of the depletion zone electric field increases as the reverse-bias voltage increases. Once the electric field intensity increases beyond a critical level, the *p-n* junction depletion zone breaks-down and current begins to flow, usually by either the Zener or avalanche breakdown processes. Both of these breakdown processes are non-destructive and are reversible, so long as the amount of current flowing does not reach levels that cause the semiconductor material to overheat and cause thermal damage.

The forward-bias and the reverse-bias properties of the *p-n* junction imply that it can be used as a diode. A *p-n* junction diode allows electric charges to flow in one direction, but not in the opposite direction; negative charges (electrons) can easily flow through the junction from *n* to *p* but not from *p* to *n* and the reverse is true for holes. When the *p-n* junction is forward-biased, electric charge flows freely due to reduced resistance of the *p-n* junction. When the *p-n* junction is reverse-biased, however, the junction barrier (and therefore resistance) becomes greater and charge flow is minimal.

7.5 DIODE

A diode offers a low resistance to electric current in one direction and a high resistance to it in the reverse direction. This property makes a diode useful as a rectifier, which can convert alternating current (AC) into direct current (DC). The term diode and rectifier will be used interchangeably; however, the term diode usually implies a small signal device with current typically in the milliamp range; and a rectifier, a power device, conducting from 1 to 1000 amps or even higher.

When the voltage applied in the reverse direction exceeds a certain value, a semiconductor diode "breaks down and conducts heavily in the direction of normally high resistance. When the reverse voltage at which breakdown occurs remains nearly constant for a wide range of currents, the phenomenon is called avalanching. A diode using this property is called a Zener diode. It can be used to regulate the voltage in a circuit.

Semiconductor diodes can be designed to have a variety of characteristics. One such diode, called a varactor, exhibits a capacitance that is dependent upon the voltage across it. In another kind, the tunnel diode, the current through the device decreases as the voltage is increased within a certain range; this property, known as negative resistance, makes it useful as an amplifier. Gunn diodes are negative-resistance diodes that are the basis of some microwave oscillators. Light-sensitive or photosensitive diodes can be used to measure illumination; the voltage drop across them depends on the amount of light that strikes them. A light-emitting diode (LED) produces light as current passes through it; some LED's can act as the light source of lasers.

However, some other forms of diode are created by depositing one material onto another e.g. Schottky diodes are made by placing some metal in contact with a semiconductor. In general, whenever we join two different, very pure, materials we're likely to make some sort of diode.

Diodes are referred to as non-linear circuit elements because of the characteristic curve. For most applications the non-linear region can be avoided and the device can be modeled by piece-wise linear circuit elements. Qualitatively we can just think of an ideal diode having two regions: a conduction region of zero resistance and an infinite resistance non-conduction region. Figure 7.13 shows a schematic symbol for a diode and the current-voltage curve for an ideal diode.

Diode is unidirectional, *i.e.* current flows in only one direction (anode to cathode internally). When a forward voltage is applied, the diode conducts; and when a reverse voltage is applied, there is no conduction.

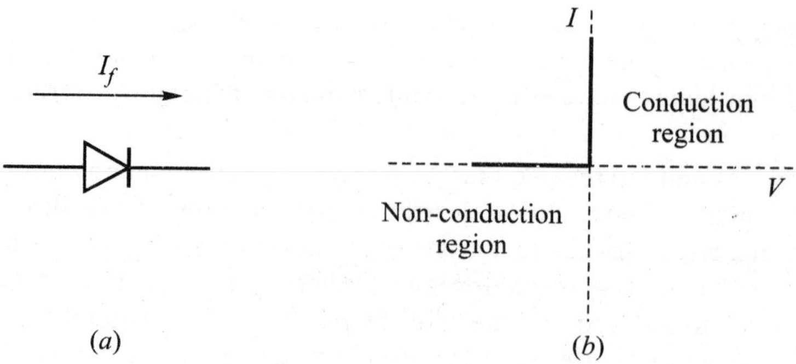

Fig. 7.13. (*a*) Schematic symbol for a diode and (*b*) current versus voltage for an ideal diode.

An ideal diode characteristic would be 7.14 (*a*): However, a typical diode characteristic is more like the following 7.14 (*b*):

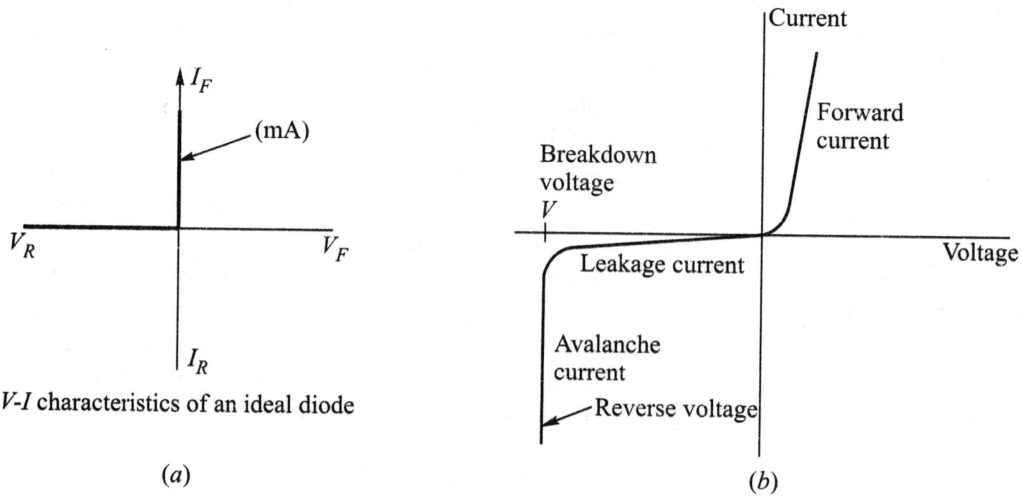

Fig. 7.14

Forward Voltage Drop, V_f

Notice that the diode conducts a small current in the forward direction up to a threshold voltage, 0.3 for germanium and 0.7 for silicon; after that it conducts as we might expect. The forward voltage drop, V_f, is specified at a forward current, I_f.

Leakage Current

In the reverse direction there is a small leakage current up until the reverse breakdown voltage is reached. This leakage is undesirable, obviously the lower the better; diodes are intended to operate below their breakdown voltage.

Switching

The switching speed of a diode depends upon its construction and fabrication. In general the smaller the chip the faster it switches, other things being equal. The chip geometry, doping levels, and the temperature at nativity determine switching speeds. The reverse recovery time, trr, is the time it takes a diode to switch from on to off.

Avalanche Breakdown

Avalanche breakdown is a phenomenon that can occur in both insulating and semiconducting materials. It is a form of electric current multiplication that can allow very large currents to flow within materials which are otherwise good insulators.

Explanation

Avalanche breakdown can occur within insulating or semiconducting solids, liquids, or gases when the electric field in the material is great enough to accelerate free electrons to the point that, when they strike atoms in the material, they can knock other electrons free: the number of free electrons is thus increased rapidly as newly generated particles become part of the process. This phenomenon is usefully employed in special purpose semiconductor devices such as the avalanche diode, the avalanche photodiode and the avalanche transistor, as well as in some gas filled tubes. In general purpose semiconductor devices such as common diodes, MOSFETs, transistors, it poses an upper limit on the operating voltages since the associated electric fields can start the process and cause excessive (if not unlimited) current flow and destruction of the device. When avalanche breakdown occurs within a solid insulating material it is almost always destructive. When an avalanche-like effect occurs without connecting two electrodes, it is referred to as an electron avalanche. Although there are some superficial similarities to Zener breakdown, the physical origins of the two phenomena are very different.

The Avalanche Process

Avalanche breakdown is a current multiplication process that occurs only in strong electric fields, which can be caused either by the presence of very high voltages, such as in electrical transmission systems, or by more moderate voltages which occur over very short distances, such as within semiconductor devices. The electric field strength necessary to achieve avalanche breakdown varies greatly between different materials: in air, 3 MV/m is typical, while in a good insulator such as some ceramics, fields in excess of 40 MV/m are required. Field strengths used in semiconductor devices that exploit the avalanche effect are often in the 20-40 MV/m range, but vary greatly accordingly with the details of the device.

Once the necessary field strength has been achieved, all that is necessary to start the avalanche effect is a free electron, and since even in the best insulators a tiny number of free electrons are always present at room temperature, an avalanche will always occur. In devices that exploit the avalanche effect, the electric field is normally kept just below the threshold at which avalanche breakdown is possible, resulting in a current that is highly dependent on the generation of free electrons. In avalanche photodiodes, for example, incoming light is used to generate these free electrons.

As avalanche breakdown begins, free electrons are accelerated by the electric field to very high speeds. As these high-speed electrons move through the material they inevitably strike atoms. If their velocity is not sufficient for avalanche breakdown (because the electric field is not strong enough) they are absorbed by the atoms and the process halts. However, if their velocity is high enough, when they strike an atom, they knock an electron free from it, ionizing it. This is referred to as impact ionization for obvious reasons. Both the original electron and the one that has just been knocked free are then accelerated by the electric field and strike other atoms, in turn knocking additional electrons free. As this process continues, the number of free electrons moving through the material increases exponentially, often reaching a maximum in just picoseconds. The avalanche can result in the flow of very large currents, limited only by the external circuitry. When all electrons reach the anode, the process stops, unless of course holes are created also.

Applications

If the current is not externally limited, the process normally destroys the device where it has started, and in situations such as power line insulators, this can take the form of an explosive breakdown of the insulator. When avalanche current is externally limited, avalanche breakdown can successfully serve several purposes. In avalanche transistors and avalanche photodiodes, this effect is used to multiply normally tiny currents, thus increasing the gain of the devices: in avalanche photodiodes, current gains of over a million can be achieved. Also, the phenomenon is very fast, meaning that avalanche current quickly follows avalanche voltage variations or starting charge (number of free electrons available to start the process) variations, and this gives to avalanche transistors and avalanche photodiodes the capability of working in the microwave frequency range and in pulse circuits. In avalanche diodes, this effect is mainly used to construct over voltage protection circuits and voltage reference circuits: as a matter of fact, avalanche breakdown and Zener breakdown are jointly present in each avalanche diode, depending on breakdown voltage, which is the leading contributing process to the avalanche current.

7.6 DIODE EQUATION

The ideal diode equation or the diode law is the *I-V* characteristic of an ideal diode in either forward or reverse bias (or no bias). The equation is:

$$I = I_s\left(e^{V_D/\eta V_T} - 1\right),$$

where I is the diode current,

I_s is a scale factor called the saturation current,

V_D is the voltage across the diode,

V_T is the thermal voltage,

and η is the emission coefficient, also known as the ideality factor. The emission coefficient n varies from about 1 to 2 depending on the fabrication process and semiconductor material and in many cases is assumed to be approximately equal to 1 (thus the notation n is omitted).

The thermal voltage VT is approximately 25.85 mV at 300 K, a temperature close to "room temperature". At any temperature it is a known constant defined by:

$$V_T = \frac{kT}{q},$$

In the event the quasi-static approximation does not hold, that is, for very fast voltage changes occurring in times shorter than the transit time τF, the equations governing time-dependent transport in the device must be solved to find the charge in transit, for example the Boltzmann equation. That problem is a subject of continuing research under the topic of non-quasistat

where q is the magnitude of charge on an electron (the elementary charge),

k is Boltzmann's constant,

T is the absolute temperature of the *p-n* junction in kelvins

The ideal diode equation or the diode law is derived with the assumption that the only processes giving rise to current in the diode are drift (due to electrical field), diffusion, and thermal recombination-generation. It also assumes that the recombination-generation (*R-G*) current in the depletion region is insignificant. This means that the equation doesn't account for the processes involved in reverse breakdown and photon-assisted *R-G*. Additionally, it doesn't describe the "leveling off" of the *I-V* curve at high forward bias due to internal resistance.

Under reverse bias voltages the exponential in the diode equation is negligible, and the current is a constant (negative) reverse current value of *-IS*. The reverse breakdown region is not modeled by diode equation.

For even rather small forward bias voltages the exponential is very large because the thermal voltage is very small, so the subtracted '*1*' in the diode equation is negligible and the forward diode current is often approximated as

$$I = I_S e^{V_D/(\eta V_T)}$$

7.7 THE P-N JUNCTION CAPACITANCE

Any variation of the charge within a *p-n* diode with an applied voltage variation yields a capacitance which must be added to the circuit model of a *p-n* diode. The capacitance associated with the charge variation in the depletion layer is called the junction capacitance, while the capacitance associated with the excess carriers in the quasi-neutral region is called the diffusion capacitance.

Both types of capacitances are non-linear so that we will derive the small-signal capacitance in each case. We will find that the junction capacitance dominates for reverse-biased diodes, while the diffusion capacitance dominates in strongly forward-biased diodes. The total capacitance is the sum of both.

Expressions for the capacitances are obtained by calculating the change in charge for a change in applied voltage, or:

$$C = \frac{dQ}{dV}$$

Diffusion Capacitance

Diffusion capacitance is the capacitance due to transport of charge carriers between two terminals of a device, for example, the diffusion of carriers from anode to cathode in forward bias mode of a diode or from emitter to base (forward biased junction in active region) for a transistor. In a semiconductor device with a current flowing through it (for example, an ongoing transport of charge by diffusion) at a particular moment there is necessarily some charge in the process of transit through the device. If the applied voltage changes to a different value and the current changes to a different value, a different amount of charge will be in transit in the new circumstances. The change in the amount of transiting charge divided by the change in the voltage causing it is the diffusion capacitance. The adjective "diffusion" is used because the original use of this term was for junction diodes, where the charge transport was via the diffusion mechanism. See Fick's law.

To implement this notion quantitatively, at a particular moment in time let the voltage across the device be V. Now assume that the voltage changes with time slowly enough that at each moment the current is the same as the DC current that would flow at that voltage, say $I = I(V)$ (the quasistatic approximation). Suppose further that the time to cross the device is the forward transit time τF. In this case the amount of charge in transit through the device at this particular moment, denoted Q, is given by $Q = I(V)\tau F$.

Consequently, the corresponding diffusion capacitance: C_{diff} is

$$C_{diff} = \frac{dQ}{dV} = \frac{dI(V)}{dV} T_F$$

7.8 P-N JUNCTION DIODE

We create a *p-n* junction by joining together two pieces of semiconductor, one doped *n*-type, the other *p*-type. This causes a depletion zone to form around the junction (the join) between the two materials. This zone controls the behaviour of the diode.

Zero Bias

When a diode is connected in a Zero Bias condition no potential voltage is applied to the junction. A few holes (majority carriers) in the *P*-type material with enough energy to overcome the potential barrier will move across the junction against this barrier potential. This is known as the "Forward Current" and is referenced as I_f.

Likewise, holes generated in the *N*-type material (minority carriers), find this situation favourable and move across the junction in the opposite direction. This is known as the "Reverse Current" and is referenced as I_r.

Zero Biased Diode

An "Equilibrium" or balance will be established when the two hole currents are equal and both moving in opposite directions, so that the net result is zero current flowing in the circuit. When this occurs the junction is said to be in a state of "Dynamic Equilibrium".

Reverse Bias

When a diode is connected in Reverse Bias, a positive voltage is applied to the *N*-type material and negative voltage is applied to the *P*-type material. The positive voltage applied to the *N*-type material attracts electrons towards the positive electrode and away from the junction, while the holes in the *P*-type end are attracted away from the junction to the negative electrode. The result is that the depletion layer grows wider due to a lack of electrons and holes and presents a high impedance path, almost an insulator. The result is that a high potential barrier is created thus preventing current from flowing through the semi-conductor material.

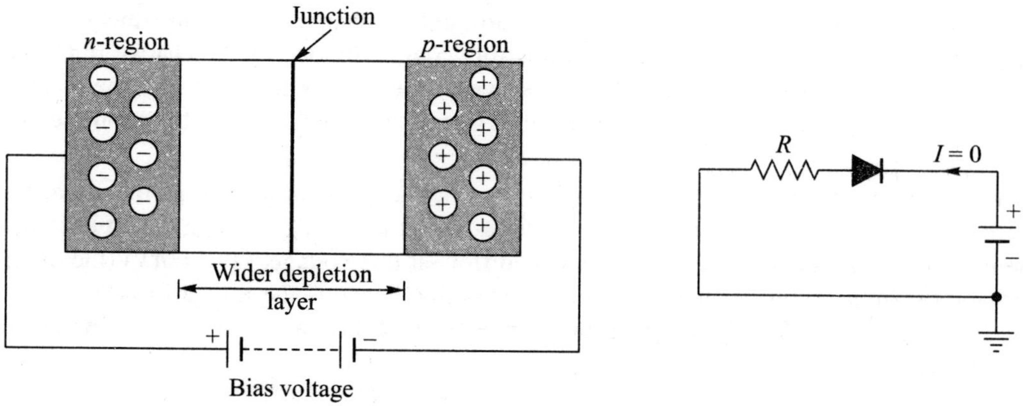

Fig. 7.15. Reverse Biased Diode showing the increase in width of the depletion layer.

This condition represents the high resistance direction in a *P-N* junction and practically no current flows through the diode with an increase in bias voltage. However, a very small leakage current does flow through the junction which can be measured in microamperes. One final point, if the reverse bias voltage is increased to a sufficiently high value it will cause the diode junction to overheat and fail due to the avalanche effect around the junction. This may cause the diode to become shorted and resulting in maximum circuit current to flow, Ohm's Law.

Sometimes this avalanche effect has practical applications in voltage stabilising circuits where a series limiting resistor is used with the diode to limit the reverse breakdown current to a set maximum value producing a fixed voltage output across the diode. These types of diodes are commonly known as Zener Diodes.

Forward Bias

When a diode is connected in forward bias a negative voltage is applied to the *N*-type material and a positive voltage is applied to the *P*-type material. The negative voltage pushes or repels electrons towards the junction allowing them to cross over and combine with the holes being pushed in the opposite direction towards the junction by the positive voltage. The result is that the depletion layer becomes very thin and narrow and represents a low impedance path thereby producing a very small potential barrier allowing high currents to flow. The point at which this takes place is represented on the static *I-V* characteristics curve above as the "knee" point.

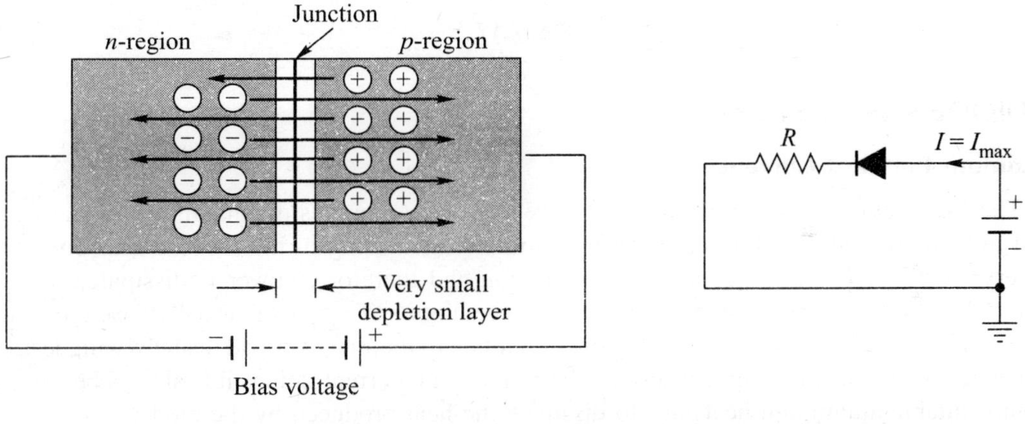

Fig. 7.16. Forward Biased Diode showing the reduction in width of the depletion layer.

This condition represents the low resistance direction in a *P-N* junction allowing very large currents to flow through the diode with only a small increase in bias voltage. The actual potential difference across the junction or diode is kept constant by the action of the depletion layer at about 0.3 V for germanium and about 0.6 V for silicon diodes. Since the diode can conduct "infinite" current above this knee point as it effectively becomes a short circuit, resistors are used in series with the device to limit current flow. Exceeding its maximum forward current specification causes the device to dissipate more power in the form of heat than it was designed for resulting in failure of the device.

7.9 V-I CHARACTERISTICS OF P-N JUNCTION DIODE

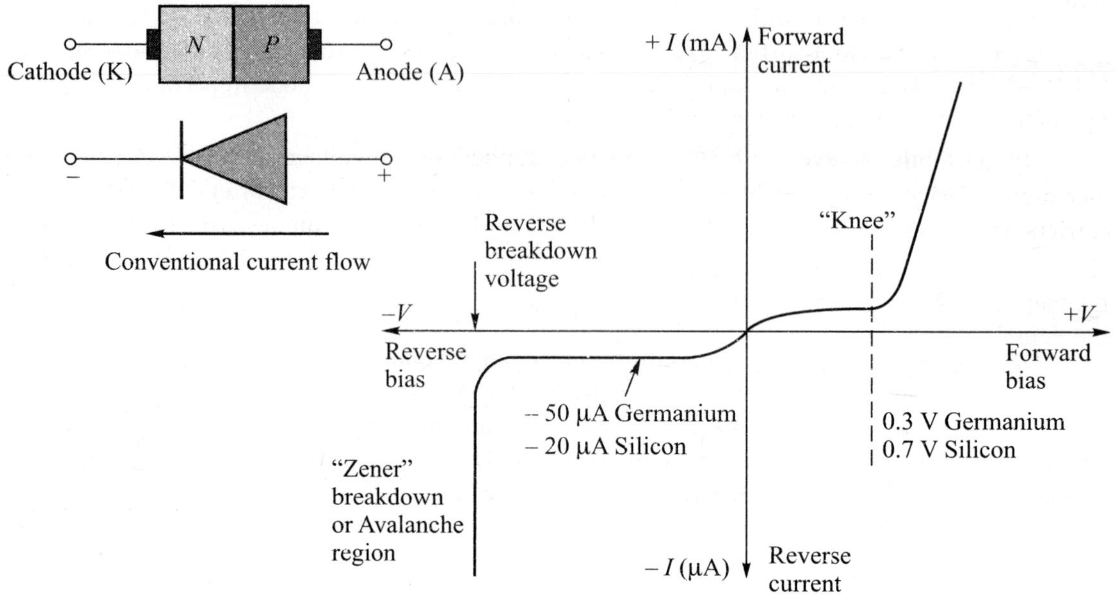

Fig. 7.17

7.10 DIODE PARAMETERS

1. Maximum Forward Current

The Maximum Forward Current ($I_{F(\max)}$) is as its name implies the maximum forward current allowed to flow through the device. When the diode is conducting in the forward bias condition, it has a very small "ON" resistance across the junction and therefore, power is dissipated across the junction (Ohm's Law) in the form of heat. Then, exceeding its ($I_{F(\max)}$) value will cause more heat to be generated across the junction and the diode will fail due to thermal overload, usually with destructive consequences. When operating diodes around their maximum current ratings it is always best to mount them onto suitable aluminium heatsinks to dissipate the heat produced by the diode.

2. Peak Inverse Voltage

The Peak Inverse Voltage (PIV) or Maximum Reverse Voltage ($V_{R(\max)}$), is the maximum allowable reverse operating voltage that can be applied to the diode without reverse breakdown occurring. This rating, therefore, is less than the "Avalanche Breakdown" level on the reverse bias characteristics. It is mainly used for rectifying diodes in a.c. rectifier circuits where the sinusoidal waveform changes from a positive to a negative value on each and every cycle.

3. Maximum Operating Temperature

The Maximum Operating Temperature actually relates to the Junction Temperature (T_J) of the device. It is the maximum temperature allowable before the structure of the device deteriorates and is expressed in units of degrees Centigrade per Watt. This value is linked closely to the maximum forward current of the device so that at this value the temperature of the junction is not exceeded. However, the maximum forward current will also depend upon the ambient temperature in which the device is operating so the maximum forward current is usually quoted for two or more ambient temperature values.

7.11 CURRENT THROUGH THE DIODE

The behaviour of a diode depends on its polarity in the circuit.

If the diode is reverse biased (positive potential on n-type material) the depletion region increases. The only charge carriers able to support a net current across the p-n junction are the minority carriers and hence the reverse current is very small.

A forward-biased diode (positive potential on p-type material) has a decreased depletion region; the majority carriers can diffuse across the junction. The voltage may become high enough to eliminate the depletion region entirely.

An approximation to the current in the p-n junction region is given by

$$I = I_0\left(e^{V/V_T} - 1\right)$$

where both I_0 and V_T are temperature dependent. This equation gives a reasonably accurate prediction of the current-voltage relationship of the p-n junction itself - especially the temperature variation - and can be improved somewhat by choosing I_0 and V_T empirically to fit a particular diode.

However, for a real diode, other factors are also important: in particular, edge effects around the border of the junction cause the actual reverse current to increase slightly with reverse voltage, and the finite conductivity of the doped semiconductor ultimately restricts the forward current to a linear increase with increasing applied voltage. A better current-voltage curve for the real diode is shown in the figure 7.18.

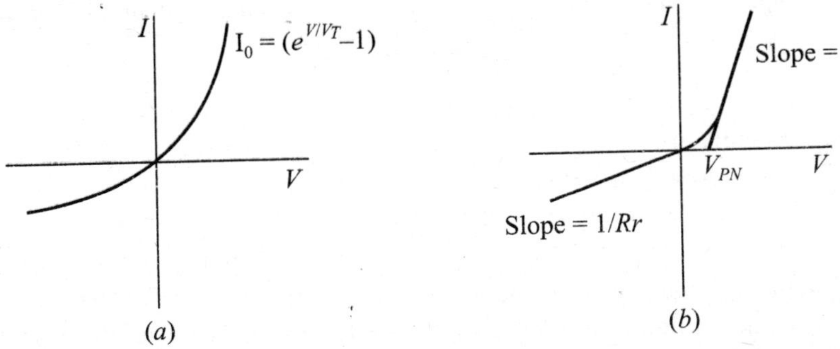

Fig. 7.18. Current versus voltage (a) in the p-n junction region and (b) for an actual p-n diode.

Various regions of the curve can be identified:

The linear region of forward-biasing, a non-linear transition region, a turn-on voltage (V_{pn}) and a reverse-biased region.

We can assign a dynamic resistance to the diode in each of the linear regions: R_f in the forward-biased region and R_r in the reverse-biased region. These resistances are defined as the inverse slope of the curve $\frac{1}{R} = \frac{\Delta I}{\Delta V}$.

The voltage V_{PN}, represents the effective voltage drop across a forward-biased *p-n* junction (the turn-on voltage). For a germanium diode, V_{PN} is approximately 0.3 V, while for a silicon diode it is close to 0.6 V.

7.12 APPLICATIONS

Half-wave Rectification

A rectifier is a circuit which converts Alternating Current (AC) into a Direct Current (DC) form and the simplest of all the rectifier circuits is that of the Half-Wave Rectifier. A half-wave rectifier circuit uses just one half of each complete sinewave or cycle of the a.c. supply in order to convert it to a d.c. supply. Then this type of circuit is called a "half-wave" rectifier because it passes only half of the incoming a.c. power signal as shown below.

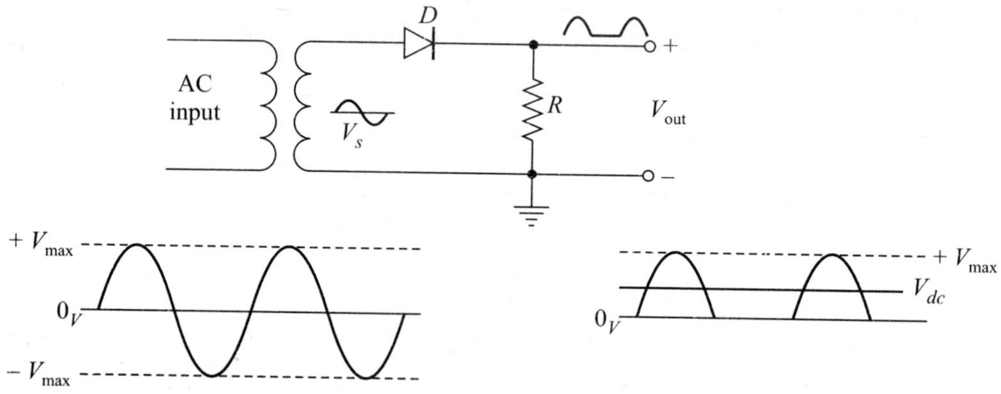

AC input waveform

Resultant output waveform

Fig. 7.19

During each "positive" half cycle of the a.c. sinewave, the diode is forward biased (anode is positive with respect to the cathode) and current flows through it. Since the d.c. load is resistive (resistor *R*), the current flowing in the load resistor is therefore proportional to the voltage (Ohm's Law), and the voltage across the load resistor is the same as the supply voltage V_s (minus V_f), that is the "d.c." voltage across the load is sinusoidal for the first half cycle only. Then $V_{out} = V_s$.

During each "negative" half cycle of the a.c. sinewave, the diode is reverse biased (anode is negative with respect to the cathode) and No current flows through it. Therefore, in the negative half cycle of the supply, no current flows in the load resistor as no voltage appears across it. Then $V_{out} = 0$.

The current on the d.c. side of the circuit flows in one direction only making the circuit unidirectional and the value of the d.c. voltage $V_{d.c.}$ across the load resistor is calculated as follows.

$$V_{d.c.} = \frac{V_{max}}{\pi} = 0.318V_{max} = 0.45V_S$$

where V_{max} is the maximum voltage value of the a.c. supply, and V_s is the r.m.s. value of the supply.

Full-wave Rectification

In the previous Power Diodes section we discussed ways of reducing the ripple or voltage variations on a direct DC voltage by connecting capacitors across the load resistance. While this method may be suitable for low power applications it is unsuitable to applications which need a "steady and smooth" d.c. voltage. One method to improve on this is to use every half-cycle of the input voltage instead of every other half-cycle. The circuit which allows us to do this is called a "Full-wave Rectifier".

In a full-wave rectifier circuit two diodes are now used, together with a transformer whose secondary winding is split equally into two and has a center tapped connection C. Each diode conducts in turn when its anode terminal is positive with respect to the center point C as shown below.

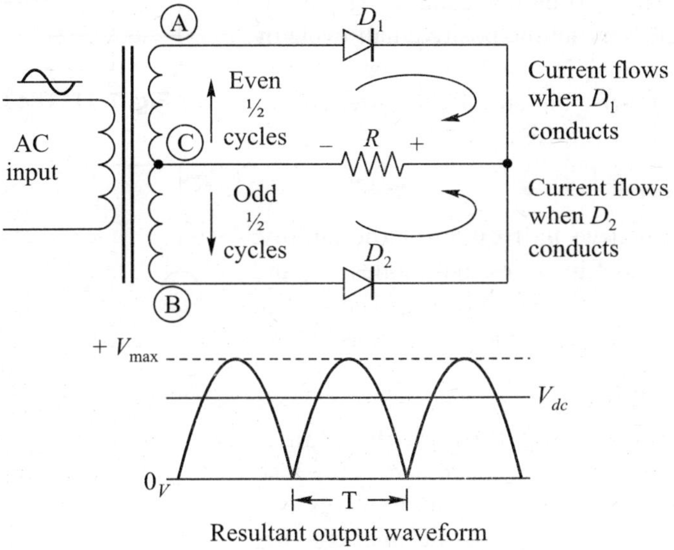

Fig. 7.20

The circuit consists of two Half-wave rectifiers connected to a single load resistance with each diode taking it in turn to supply current to the load. When point A is positive with respect to point B, diode $D1$ conducts in the forward direction as indicated by the arrows. When point B is positive (in the negative half of the cycle) with respect to point A, diode $D2$ conducts in the forward direction and the current flowing through resistor R is in the same direction for both circuits. As the output voltage across the resistor R is the sum of the two waveforms, this type of circuit is also known as a "bi-phase" circuit.

As the spaces between each half-wave developed by each diode are now being filled in by the other diode the average d.c. output voltage across the load resistor is now double that of the single half-wave rectifier circuit and is about 0.637Vmax of the peak voltage, assuming no losses.

$$V_{d.c.} = \frac{2V_{max}}{\pi} = 0.637\ V_{max} = 0.9V_S$$

The peak voltage of the output waveform is the same as before for the half-wave rectifier provided each half of the transformer windings have the same rms voltage value. To obtain a different d.c. voltage output different transformer ratios can be used, but one main disadvantage of this type of rectifier is that having a larger transformer for a given power output with two separate windings makes this type of circuit costly compared to a "Bridge Rectifier" circuit equivalent.

The Bridge Rectifier

Another type of circuit that produces full-wave rectification is that of the Bridge Rectifier. This type of single phase rectifier uses 4 individual rectifying diodes connected in a "bridged" configuration to produce the desired output but does not require a special center tapped transformer, thereby reducing its size and cost. The single secondary winding is connected to one side of the diode bridge network and the load to the other side as shown in Fig. 7.21.

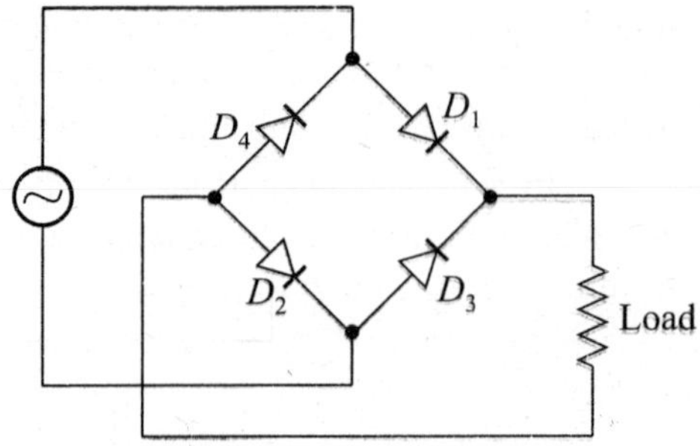

Fig. 7.21. The Diode Bridge Rectifier.

The 4 diodes labelled D_1 to D_4 are arranged in "series pairs" with only two diodes conducting current during each half cycle. During the positive half cycle of the supply, diodes D_1 and D_2 conduct in series while diodes D_3 and D_4 are reverse biased and the current flows through the load as shown in figure 7.22.

The Positive Half-cycle

During the negative half cycle of the supply, diodes D_3 and D_4 conduct in series, but diodes D_1 and D_2 switch of as they are now reverse biased. The current flowing through the load is the same direction as before.

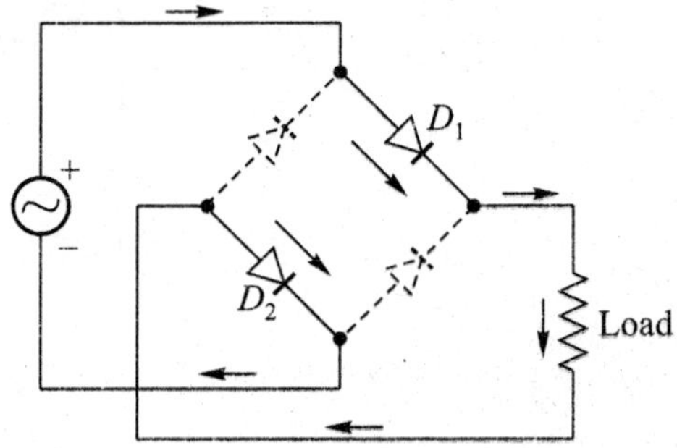

Fig. 7.22

The Negative Half-cycle

As the current flowing through the load is unidirectional, so the voltage developed across the load is also unidirectional the same as for the previous two diode full-wave rectifier, therefore the average d.c. voltage across the load is 0.637 V_{max} and the ripple frequency is now twice the supply frequency (*e.g.* 100 Hz for a 50 Hz supply).

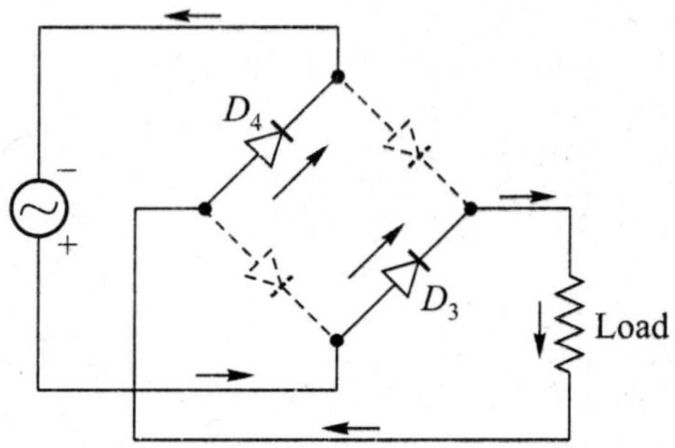

Fig. 7.23

The Smoothing Capacitor

We saw in the previous section that the single phase half-wave rectifier produces an output wave every half cycle and that it was not practical to use this type of circuit to produce a steady d.c. supply. The full-wave bridge rectifier however, gives us a greater mean d.c. value (0.637 V_{max}) with less superimposed ripple while the output waveform is twice that of the frequency of the input supply frequency. We can therefore increase its average d.c. output level even higher by connecting a suitable smoothing capacitor across the output of the bridge circuit as shown in Fig. 7.24.

The smoothing capacitor converts the full-wave rippled output of the rectifier into a smooth d.c. output voltage. Two important parameters to consider when choosing a suitable a capacitor are its Working Voltage, which must be higher than the no-load output value of the rectifier and its Capacitance Value, which determines the amount of ripple that will appear superimposed on top of the d.c. voltage. Too low a value and the capacitor has little effect. As a general rule of thumb, we are looking to have a ripple voltage of less than 100 mV peak to peak.

The main advantage of a full-wave bridge rectifier is that it has a smaller a.c. ripple value for a given load and a smaller reservoir or smoothing capacitor than an equivalent half-wave rectifier.

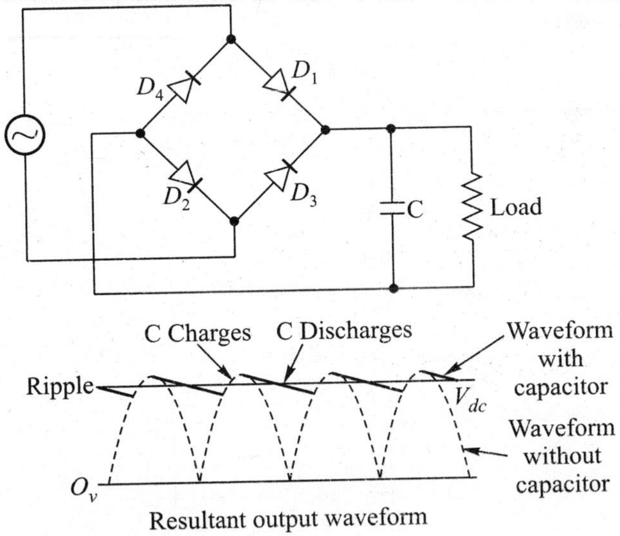

Fig. 7.24. Resultant output waveform.

The fundamental frequency of the ripple voltage is twice that of the a.c. supply frequency (100 Hz) where for the half-wave rectifier it is exactly equal to the supply frequency (50 Hz). The amount of ripple voltage that is superimposed on top of the d.c. supply voltage can be virtually eliminated by adding an improved π-filter (pi-filter) to the output terminals of the bridge rectifier. This type of low-pass filter consists of two smoothing capacitors, usually of the same value and a choke or inductance across them to introduce a high impedance path to the alternating ripple component. Another more practical and cheaper alternative is to use a 3-terminal voltage regulator IC, such as a LM7805 which can reduce the ripple by more than 70 dB while delivering over lamp of output current.

The Zener Diode

In the previous section we saw that a "reverse biased" diode passes very little current but will suffer breakdown or damage if the reverse voltage applied across it is made too high. However, the Zener Diode or "Breakdown Diode" as it is sometimes called, is basically the same as the standard junction diode but is specially made to have a low predetermined Reverse Breakdown Voltage, the "Zener Voltage" (Vz). In the forward direction it behaves just like a normal diode passing current, but when the reverse voltage applied to it exceeds the selected reverse breakdown voltage a process called Avalanche Breakdown occurs in the depletion layer and the current through the diode increases to the maximum circuit value, which is usually limited by a series resistor. The point at which current flows can be very accurately controlled (to less than 1% tolerance) in the doping stage of the diodes construction giving it a specific Zener Breakdown voltage (Vz) ranging from a few volts up to a few hundred volts.

Zener Diode V I-Characteristics

Zener Diodes are used in the "REVERSE" bias mode, *i.e.* the anode connects to the negative supply and from its I-V characteristics curve above, we can see that the zener diode has a region in its reverse bias characteristics of almost a constant voltage regardless of the current flowing through the diode. This voltage across the diode (it's Zener Voltage, Vz) remains nearly constant even with large

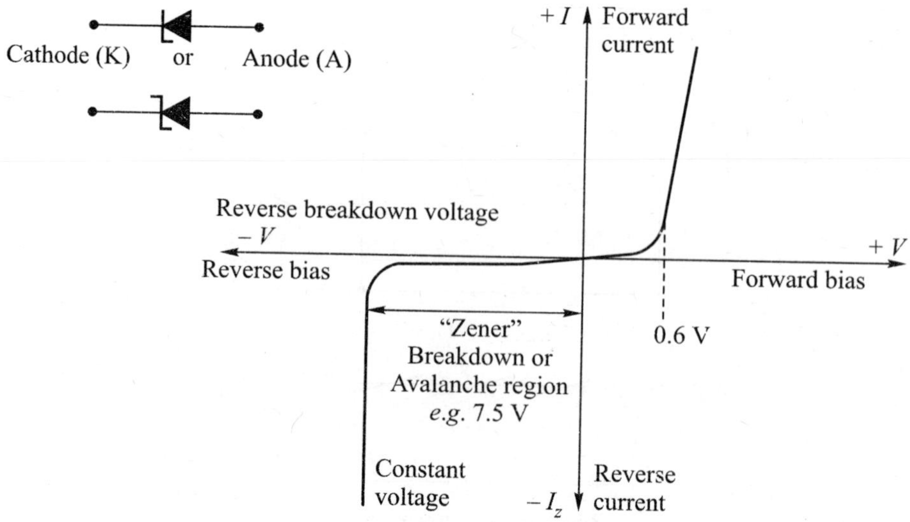

Fig. 7.25

changes in current through the diode caused by variations in the supply voltage or load. This ability to control itself can be used to great effect to regulate or stabilise a voltage source against supply or load variations. The diode will continue to regulate until the diode current falls below the minimum *Iz* value in the reverse breakdown region.

APPLICATION

The Zener Regulator

A zener diode can be used to produce a stabilised voltage output by passing a small current through it from a voltage via a suitable current limiting resistor. We remember from the previous sections that the d.c. output voltage from the half or full-wave rectifiers contains ripple superimposed onto the d.c. voltage and that as the load value changes so to the average output voltage. By connecting a simple zener stabiliser circuit as shown below across the output of the rectifier a stable reference voltage can be produced.

Zener Diode Stabiliser

The resistor, R_S is connected in series with the zener diode and the output from the voltage source, V_S is connected across the combination while the stabilised output voltage is taken from across the zener diode. The zener diode with its Cathode terminal connected to the positive terminal region of the d.c. supply it becomes reverse biased and will be operating in the breakdown. Resistor R_S is selected so to limit the maximum current flowing in the circuit. When no load resistance, R_L is connected to the circuit, no load current ($I_L = 0$), is drawn and all the circuit current passes through the zener diode which dissipates its maximum power. Care must be taken when selecting the appropriate value of resistance that the zeners maximum power rating is not exceeded under this "no-load" condition.

There is a minimum zener current for which the stabilisation of the voltage is effective and the zener current must stay above this value operating within its breakdown region at all times. The upper limit of current is of course dependant upon the power rating of the device.

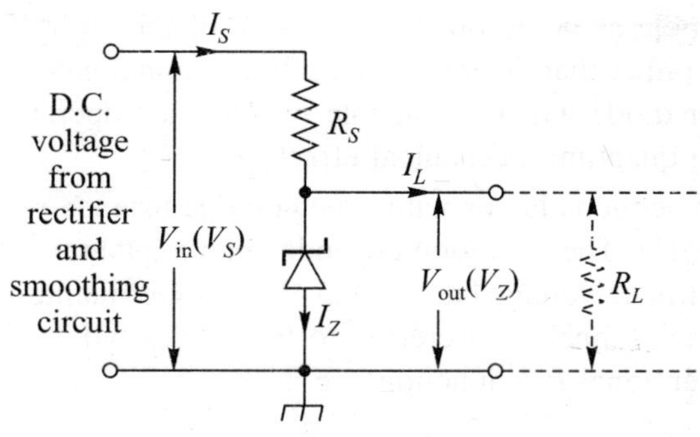

Fig. 7.26

Example

A 5.0 V stabilised power supply is required from a 12 V d.c. input source. The maximum power rating of the zener diode is 2W. Using the circuit above calculate:

(*a*) The maximum current flowing in the Zener Diode.

$$\text{Maximum Current} = \frac{\text{Watts}}{\text{Voltage}} = \frac{2\text{ W}}{5\text{ V}} = 400\text{ mA}$$

(*b*) The value of the series resistor RS

$$R_S = \frac{V_S - V_Z}{I_Z} = \frac{12 - 5}{400\text{ mA}} = 17.5\ \Omega$$

(*c*) The load current I_L if a load resistor of 1 kΩ is connected across the zener diode.

$$I_L = \frac{V_Z}{R_L} = \frac{5\text{ V}}{1000\ \Omega} = 5\text{ mA}$$

(*d*) The total supply current I_S

$$I_S = I_Z + I_L = 400\text{ mA} + 5\text{ mA} = 405\text{ mA}$$

Tunnel Diode

It was invented in 1958 by Leo Esaki, who in 1973 received the Nobel Prize in Physics for discovering the electron tunneling effect used in these diodes. These diodes have a heavily doped *p-n* junction only some 10 nm (100 Å) wide. The heavy doping results in a broken bandgap, where conduction band electron states on the *n*-side are more or less aligned with valence band hole states on the *p*-side. Tunnel diodes are usually made from germanium, but can also be made in gallium arsenide and silicon materials. The circuit symbol for the Tunnel diode is as shown in Fig. 7.27.

Fig. 7.27

In quantum mechanics, quantum tunnelling is a nanoscopic phenomenon in which a particle violates the principles of classical mechanics by penetrating a potential barrier or impedance higher than the kinetic energy of the particle. A barrier, in terms of quantum tunnelling, may be a form of energy state analogous to a "hill" or incline in classical mechanics, which classically suggests that passage through or over such a barrier would be impossible without sufficient energy.

On the quantum scale, objects exhibit wave-like behaviour; in quantum theory, quanta moving against a potential energy "hill" can be described by their wave-function, which represents the probability amplitude of finding that particle in a certain location at either side of the "hill". If this function describes the particle as being on the other side of the "hill", then there is the probability that it has moved through, rather than over it, and has thus "tunnelled". A tunnel diode or Esaki diode is a type of semiconductor diode which is capable of very fast operation, well into the microwave frequency region, by using quantum mechanical effects.

A tunnel diode is a semiconductor with a negative resistance region that results in very fast switching speeds, up to 5 GHz. The operation depends upon a quantum mechanic principle known as "tunneling" wherein the intrinsic voltage barrier (0.3 Volt for Germanium junctions) is reduced due to doping levels which enhance tunneling. Refering to the curves below, superimposing the tunneling characteristic upon a conventional *P-N* junction, we have:

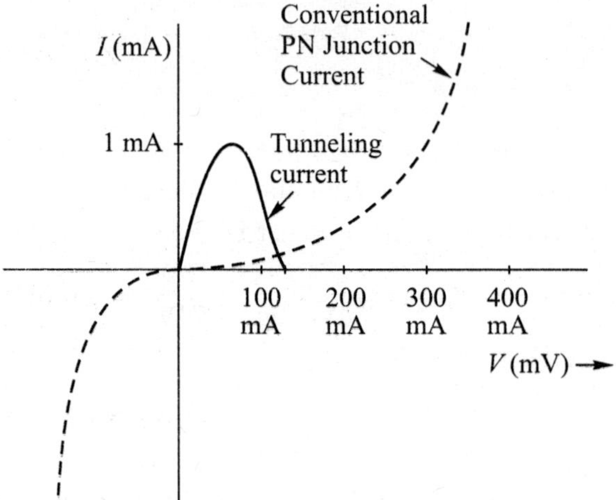

Fig. 7.28. Combination of tunneling current and conventional PN junction current.

Resulting in a composite characteristic which is the tunnel diode characteristic curve.

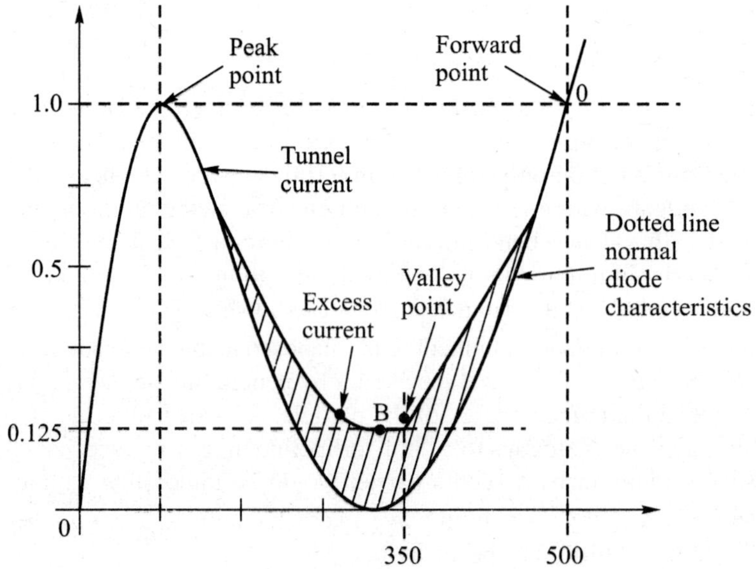

Fig. 7.29. Tunnel diode.

- Degenerate doping : $n > N_c$, $p > N_v \rightarrow E_f$ above E_c or below E_v
  —p^+ and n^+ junction
- Tunnel diode operation
  —Very high dopings → small depletion width and high electric field → tunneling with small forward or reverse biasing
  —Negative dynamic resistance (dV/dl) region : type *N* negative resistance, voltage controlled neg resistance
- Applications
  —Switching, oscillation, amplification
  —No time delay of diff or drift of tunneling → high speed circuits
  —Defect: low current operation

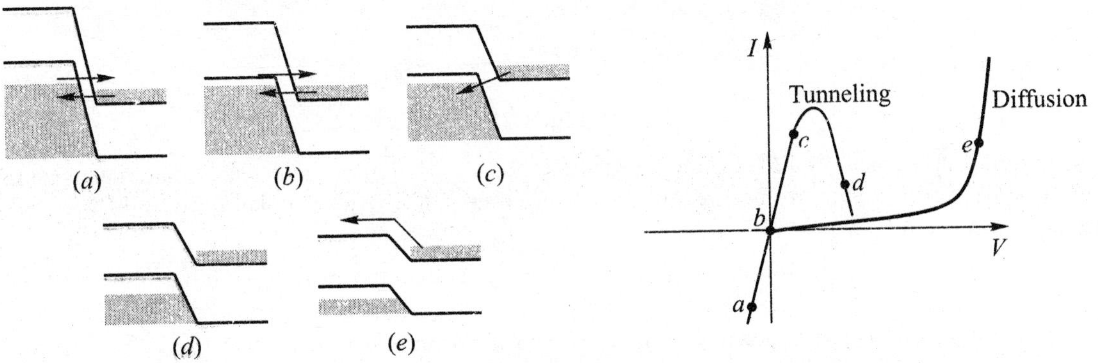

Fig. 7.30

The negative resistance region (between points A and B) is the important characteristic for the tunnel diode. In this region, as the voltage is increased, the current decreases; just the opposite of a conventional diode. The most important specifications for the tunnel diode are the Peak Voltage (V_p), Peak Current (I_p), Valley Voltage (V_v), and Valley Current (I_v).

The characteristic curve for a tunnel diode is illustrated in figure, Characteristic curve of a tunnel diode compared to that of a standard *PN* junction. The three most important aspects of this characteristic curve are (1) the forward current increase to a peak (*IP*) with a small applied forward bias, (2) the decreasing forward current with an increasing forward bias to a minimum valley current (*IV*), and (3) the normal increasing forward current with further increases in the bias voltage. The portion of the characteristic curve between *IP* and *IV* is the region of negative resistance.

USES

The negative resistance region is the most important and most widely used characteristic of the tunnel diode. A tunnel diode biased to operate in the negative resistance region can be used as either an oscillator or an amplifier in a wide range of frequencies and applications. Very high frequency applications using the tunnel diode are possible because the tunneling action occurs so rapidly that there is no transit time effect and therefore no signal distortion. Tunnel diodes are also used extensively in high-speed switching circuits because of the speed of the tunneling action. They can be used as oscillators, amplifiers, frequency converters and detectors.

Varactor Diode

The VARACTOR, or varicap is a diode that behaves like a variable capacitor, with the *PN* junction functioning like the dielectric and plates of a common capacitor. When a reverse voltage is applied to a *p-n* junction, the holes in the *p*-region are attracted to the anode and electrons in the *n*-region are attracted to the cathode terminal creating a region where there is little current. This depletion region, is essentially devoid of carriers and behaves as the dielectric of a capacitor. The depletion region increases as reverse voltage across it increases and since capacitance varies inversely as dielectric thickness, the junction capacitance will decrease as the voltage across the *p-n* junction increases. So by varying the reverse voltage across a *p-n* junction the junction capacitance can be varied. The varactor diode symbol is shown below with a diagram representation.

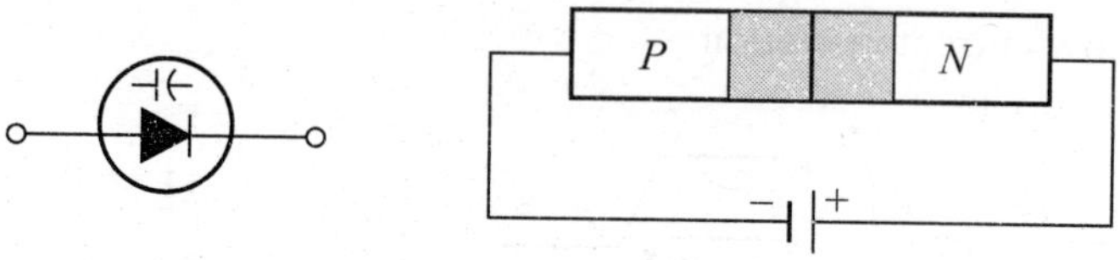

Fig. 7.31

Figure, which shows a *PN* junction, surrounding the junction of the *P* and *N* materials is a narrow region void of both positively and negatively charged current carriers as we know this area is called the depletion region.

The size of the depletion region in a varactor diode is directly related to the bias. Forward biasing makes the region smaller by repelling the current carriers toward the *PN* junction. If the applied voltage is large enough, the negative particles will cross the junction and join with the positive particles, as shown in figure 7.31. This forward biasing causes the depletion region to decrease, producing a low resistance at the *PN* junction and a large current flow across it. This is the condition for a forward-biased diode. On the other hand, if reverse-bias voltage is applied to the *PN* junction, the size of its depletion region increases as the charged particles on both sides move away from the junction. This condition, shown in figure 7.32, produces a high resistance between the terminals and allows little current flow (only in the microampere range). This is the operating condition for the varactor diode, which is nothing more than a special *PN* junction.

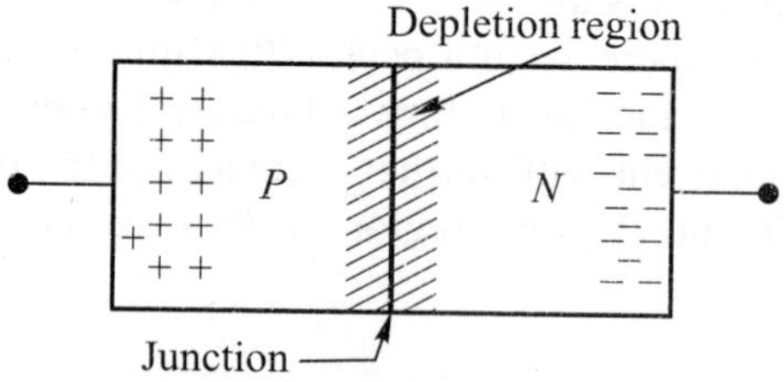

Fig. 7.32

As the figure shows, the insulation gap formed by reverse biasing of the varactor is comparable to the layer of dielectric material between the plates of a common capacitor. Furthermore, the formula used to calculate capacitance

$$C = \frac{AK}{d}$$

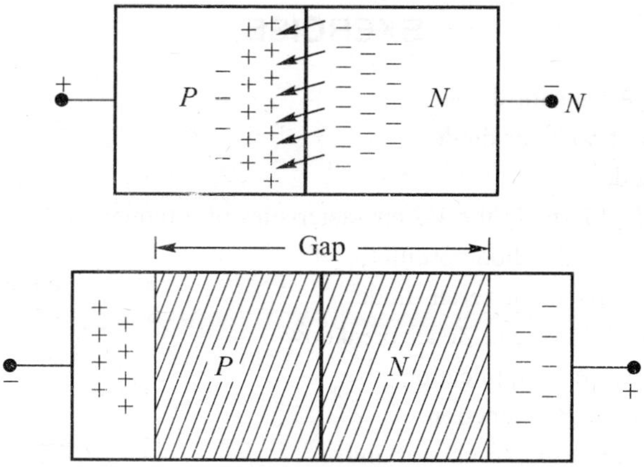

Fig. 7.33

Where A = plate area K = a constant value d = distance between plates can be applied to both the varactor and the capacitor. In this case, the size of the insulation gap of the varactor, or depletion region, is substituted for the distance between the plates of the capacitor. By varying the reverse-bias voltage applied to the varactor, the width of the "gap" may be varied. An increase in reverse bias increases the width of the gap (d) which reduces the capacitance (C) of the PN junction. Therefore, the capacitance of the varactor is inversely proportional to the applied reverse bias. The typical varactor voltage-capacitance curve is shown in below.

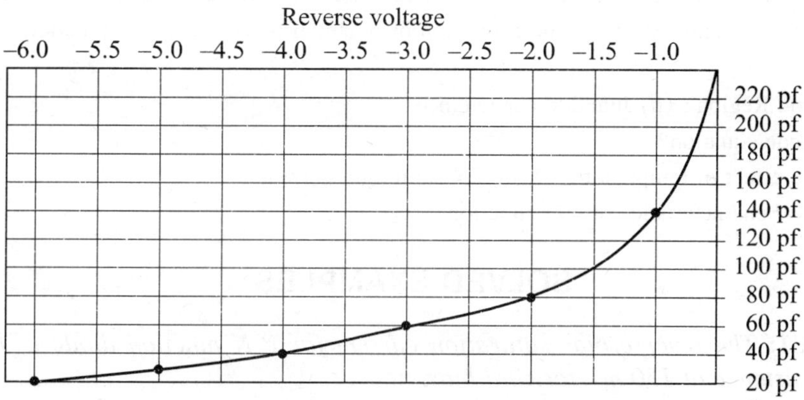

Fig. 7.34. Varactor voltage-capacitance curve.

The nonlinear increase in capacitance as the reverse voltage is decreased. This nonlinearity allows the varactor to be used also as a harmonic generator. In general, varactors are used to replace the old style variable capacitor tuning. They are used in tuning circuits of more sophisticated communication equipment and in other circuits where variable capacitance is required. One advantage of the varactor is that it allows a dc voltage to be used to tune a circuit for simple remote control or automatic tuning functions. One such application of the varactor is as a variable tuning capacitor in a receiver or transmitter tank circuit.

EXERCISE

1. Explain the working of varactor diode.
2. Give the applications of varactor diode.
3. What is a varactor diode ?
4. What is a tunnel diode ? Explain the *V-I* characteristics of a tunnel diode.
5. Explain the working of tunnel diode oscillator.
6. Explain the *V-I* characteristics of zerier diode.
7. Explain Avalanche breakdown and Zener breakdown.
8. What is zener diode? Give its applications.
9. Describe the formation of *P-N* junction.
10. What is *P-N* junction diode? Give its working.
11. Give the applications of *P-N* junction diode.
12. Differentiate *P-N* junction diode and zener diode.
13. Explain how zener diode maintains constant voltage across load.
14. Write a short notes on the following:
    (*i*) Breakdown voltage
    (*ii*) Knee voltage
    (*iii*) Limitations in the operating conditions of *P-N* junction.
15. Draw and explain the *V-I* characteristics of a *P-N* junction.
16. Discuss the behaviour of a *P-N* junction under forward and reverse biasing.
17. What is a *P-N* junction? Explain the formation of potential barrier in a *P-N* junction.
18. Describe
    (*i*) Depletion Region ; (*ii*) Junction capacitance
19. What is diode equation?
20. What is the effect of temperature of reverse saturation current?
21. Draw the *V-I* characteristics of *P-N* junction.

SOLVED EXAMPLES

Example 1: *The reverse bias saturation current for P-N junction diode is 1 mA at 300K. Determine its resistance at 150 mV forward bias.*

Solution:
$$r_{ac} = \frac{dv}{dt} = \frac{1}{(I_O / kT)\exp(V / kT)}$$

At temperature 300 K,

$$KT = (8.62 \times 10^{-5})(300) = 25.86 \times 10^{-3}\text{V}$$

$$r_{ac} = \frac{1}{(1\mu\text{A} / 25.86 \times 10^{-3}\text{V})\exp(0.15 / 25.86 \times 10^{-3}\text{V})}$$

$$r_{ac} = 77.65 \text{ ohm}$$

Example 2: *Find the static and dynamic resistance of a P-N junction diode, if the temperature is 27°C and I_0=1 μA for an applied forward bias of 0.2 V.*

Solution:
$$I = I_0\left[\exp\left[\frac{ev}{kT}\right] - 1\right]$$

$$I = 1 \times 10^{-6}\left[\exp\left[\frac{1.6\times10^{-19}\times0.2}{1.38\times10^{-23}\times300}\right]-1\right]$$

$$I = 2.27 \times 10^{-3}\text{ A}$$

Static resistance $$r_{dc} = \frac{V}{I} = \frac{0.2}{2.27\times10^{-3}} = 88\ \Omega$$

Dynamic resistance $$r_{ac} = \frac{0.026}{2.27\times10^{-3}} = \frac{26}{227} = 11.4\ \Omega$$

EXERCISES

1. An ideal silicon *P-N* diode has reverse saturation current of 2 μA at temperature of 140°C. At this temperature of 140°, find the dynamic resistance of the diode for 0.2 volts bias in (*a*) forward direction; (*b*) reverse direction. [**Ans.** 2.146 kΩ, 0.5906 μΩ]
2. A silicon *P-N* diode has reverse saturation current of 10 nA at 27°C. At 147°C, calculate the dynamic resistance of the diode for forward bias of 0.1 volt. [**Ans.** 444 Ω]

CHAPTER 8

Bipolar Junction Transistor (BJT)

8.1 INTRODUCTION

The invention of the transistor by American physicists John Bardeen, Walter H. Brattain, and William Shockley, later jointly awarded a Nobel Prize, was announced by the Bell Telephone Laboratories in 1948. It was also independently developed nearly simultaneously by Herbert Mataré and Heinrich Welker, German physicists working at Westinghouse Laboratory in Paris. Since then many types have been designed. Transistors are very durable, are very small, have a high resistance to physical shock, and are very inexpensive. At one time, only discrete devices existed; they were usually sealed in ceramic, with a wire extending from each segment to the outside, where it could be connected to an electric circuit. Although discrete transistors are still used significantly, the vast majority of transistors are built as parts of integrated circuits. Transistors are used in virtually all electronic devices, including radio receivers, computers, and space vehicles and guided missiles.

The word Transistor is a combination of the words Transfer of resistor used to describe their mode of operation back in their early days of development. There are two basic types of transistors, NPN and PNP, which basically describes the physical arrangement of the P-type and N-type semiconductor materials from which they are made. The principle of operation of the two transistor types NPN and PNP, is exactly the same the only difference being in the biasing and the polarity of the power supply for each type.

The transistor is an arrangement of semiconductor materials that share common physical boundaries. Materials most commonly used are silicon, gallium-arsenide, and germanium, into which impurities have been introduced by a process called "doping". In *n*-type semiconductors the impurities or dopants result in an excess of electrons, or negative charges, in *p*-type semiconductors the dopants lead to a deficiency of electrons and therefore an excess of positive charge carriers or "holes."

8.2 BIPOLAR JUNCTION TRANSISTOR

The bipolar junction transistor (BJT) was the first type of transistor to be mass-produced. Bipolar transistors are so named because they conduct by using both majority and minority carriers. Although a small part of the transistor current is due to the flow of minority carriers, most of the transistor current is due to the flow of majority carriers and so BJTs are classified as majority carriers devices. The three terminals of the BJT are named emitter, base and collector. Two *p-n* junctions exist inside a BJT, the base/emitter junction and base/collector junction. "The BJT is useful in amplifiers because the currents at the emitter and collector are controllable by the relatively small base current." In an NPN transistor operating in the active region, the emitter-base junction is forward biased, and electrons are injected into the base region. Because the base is narrow, most of these electrons will diffuse into the reverse-biased base-collector junction and be swept into the collector, perhaps one-hundredth of the electrons will recombine in the base, which is the dominant mechanism in the base current. By controlling the number of electrons that can leave the base, the number of electrons entering the collector can be controlled.

8.3 STRUCTURE

A BJT consists of three differently doped semiconductor regions, the emitter region, the base region and the collector region. These regions are, respectively, *p* type, *n* type and *p* type in a PNP, and *n* type, *p* type and *n* type in a NPN transistor. Each semiconductor region is connected to a terminal, appropriately labeled: emitter (*E*), base (*B*) and collector (*C*).

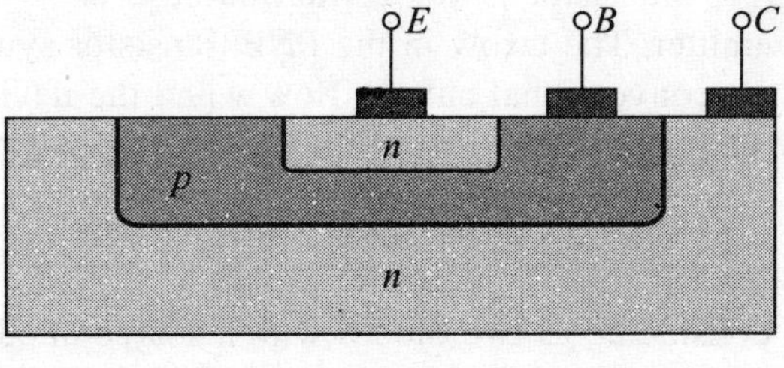

Fig. 8.1. Simplified cross section of a planar NPN bipolar junction transistor.

The base is physically located between the emitter and the collector and is made from lightly doped, high resistivity material. The collector surrounds the emitter region, making it almost impossible for the electrons injected into the base region to escape being collected, thus making the resulting value of α very close to unity, and so, giving the transistor a large β (to be discussed later). A cross section view of a BJT indicates that the collector–base junction has a much larger area than the emitter–base junction.

Small changes in the voltage applied across the base–emitter terminals causes the current that flows between the emitter and the collector to change significantly. This effect can be used to amplify the input voltage or current. BJTs can be thought of as voltage-controlled current sources, but are more simply characterized as current-controlled current sources, or current amplifiers, due to the low impedance at the base.

Early transistors were made from germanium but most modern BJTs are made from silicon. A significant minority are also now made from gallium arsenide, especially for very high speed applications.

8.4 NPN

NPN is one of the two types of bipolar transistors, in which the letters "*N*" and "*P*" refer to the majority charge carriers inside the different regions of the transistor. Most bipolar transistors used today are NPN, because electron mobility is higher than hole mobility in semiconductors, allowing greater currents and faster

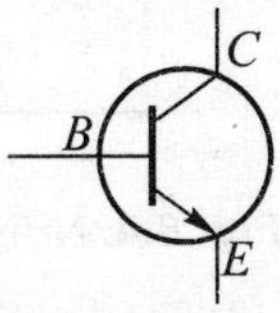

Fig. 8.2. The symbol of an NPN Bipolar Junction Transistor.

operation. NPN transistors consist of a layer of *P*-doped semiconductor "base" between two *N*-doped layers. A small current entering the base in common-emitter mode is amplified in the collector output. In other terms, an NPN transistor is "on" when its base is pulled **high** relative to the emitter. The arrow in the NPN transistor symbol is on the emitter leg and points in the direction of the conventional current flow when the device is in forward active mode. One mnemonic device for identifying the symbol for the NPN transistor is "**n**ot **p**ointing i**n**".

8.5 PNP

The other type of BJT is the PNP with the letters "*P*" and "*N*" referring to the majority charge carriers inside the different regions of the transistor.

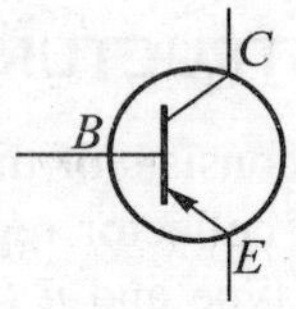

Fig. 8.3. The symbol of a PNP Bipolar Junction Transistor.

PNP transistors consist of a layer of *N*-doped semiconductor between two layers of *P*-doped material. A small current leaving the base in common-emitter mode is amplified in the collector output. In other terms, a PNP transistor is "on" when its base is pulled **low** relative to the emitter. The arrow in the PNP transistor symbol is on the emitter leg and points in the direction of the conventional current flow when the device is in forward active mode. One mnemonic device for identifying the symbol for the PNP transistor is "**p**oints i**n** **p**roudly".[5]

8.6 WORKING

An NPN transistor can be considered as two diodes with a shared anode region. In typical operation, the emitter–base junction is forward biased and the base–collector junction is reverse biased. In a NPN transistor, for example, when a positive voltage is applied to the base–emitter junction, the equilibrium between thermally generated carriers and the repelling electric field of the depletion region becomes unbalanced, allowing thermally excited electrons to inject into the base region. These electrons wander (or "diffuse") through the base from the region of high concentration near the emitter towards the region of low concentration near the collector. The electrons in the base are called minority carriers because the base is doped *p*-type which would make holes the majority carrier in the base.

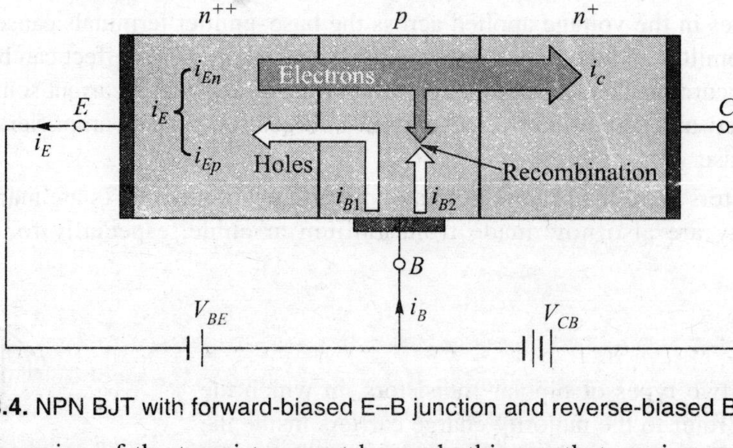

Fig. 8.4. NPN BJT with forward-biased E–B junction and reverse-biased B–C junction.

The base region of the transistor must be made thin, so that carriers can diffuse across it in much less time than the semiconductor's minority carrier lifetime, to minimize the percentage of carriers that recombine before reaching the collector–base junction. To ensure this, the thickness of the base is much less than the diffusion length of the electrons. The collector–base junction is reverse-biased, so little electron injection occurs from the collector to the base, but electrons that diffuse through the base towards the collector are swept into the collector by the electric field in the depletion region of the collector–base junction.

8.7 VOLTAGE, CURRENT, AND CHARGE CONTROL

The collector–emitter current can be viewed as being controlled by the base–emitter current (current control), or by the base–emitter voltage (voltage control). These views are related by the current–voltage relation of the base–emitter junction, which is just the usual exponential current–voltage curve of a *p-n* junction (diode).

The physical explanation for collector current is the amount of minority-carrier charge in the base region. Detailed models of transistor action account for the distribution of this charge explicitly to explain transistor behavior more exactly. The charge-control view easily handles photo-transistors, where minority carriers in the base region are created by the absorption of photons, and handles the dynamics of turn-off, or recovery time, which depends on charge in the base region recombining. However, since base charge is not a signal that is visible at the terminals, the current- and voltage-control views are usually used in circuit design and analysis.

8.8 TRANSISTOR 'ALPHA' AND 'BETA'

The proportion of electrons able to cross the base and reach the collector is a measure of the BJT efficiency. The heavy doping of the emitter region and light doping of the base region cause many more electrons to be injected from the emitter into the base than holes to be injected from the base into the emitter. The common-emitter current gain is represented by β_F or h_{fe}. It is approximately the ratio of the DC collector current to the DC base current in forward-active retion, and is typically greater than 100. Another important parameter is the common-base current gain, α_F. The common-base current gain is approximately the gain of current from emitter to collector in the forward-active region. This ratio usually has a value close to unity; between 0.98 and 0.998. Alpha and beta are more precisely related by the following identities (NPN transistor):

$$\alpha_T = \frac{I_C}{I_E}$$

$$\beta_F = \frac{I_C}{I_B}$$

$$\beta_F = \frac{\alpha_T}{1-\alpha_T} \Leftrightarrow \alpha_T = \frac{\beta_F}{\beta_F + 1}$$

8.9 TRANSISTOR CONNECTIONS

The Common Base Configuration

As its name suggests, in the Common Base or Grounded Base configuration, the BASE connection is common to both the input signal AND the output signal with the input signal being applied between the base and the emitter terminals. The corresponding output signal is taken from between the base and the collector terminals as shown with the base terminal grounded or connected to a fixed reference voltage point. The input current flowing into the emitter is quite large as its the sum of both the base current and collector current respectively therefore, the collector current output is less than the emitter current input resulting in a Current Gain for this type of circuit of less than "1", or in other words it "Attenuates" the signal.

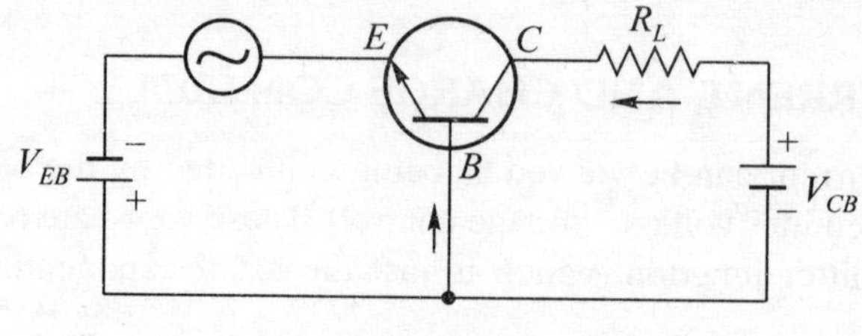

Fig. 8.5

This type of amplifier configuration is a non-inverting voltage amplifier circuit, in that the signal voltages V_{in} and V_{out} are In-Phase. This type of arrangement is not very common due to its unusually high vol.age gain characteristics. Its Output characteristics represent that of a forward biased diode while the Input characteristics represent that of an illuminated photo-diode. Also this type of configuration has a high ratio of Output to Input resistance or more importantly "Load" resistance (R_L) to "Input" resistance (R_{in}) giving it a value of "Resistance Gain". Then the Voltage Gain for a common base can therefore be given as:

Common Base Voltage Gain

$$A_V = \frac{I_C \times R_L}{I_e \times R_{IN}} = \alpha \times \frac{R_L}{R_{IN}}$$

The Common Base circuit is generally only used in single stage amplifier circuits such as microphone pre-amplifier or RF radio amplifiers due to its very good high frequency response.

The Common Emitter Configuration

In the **Common Emitter** or Grounded Emitter configuration, the input signal is applied between the base, while the output is taken from between the collector and the emitter as shown. This type of configuration is the most commonly used circuit for transistor based amplifiers and which represents the "normal" method of connection. The common emitter amplifier configuration produces the highest voltage, current and pcwer gain of all the three transistor configurations. This is mainly because the input impedance is LOW as its connected to a forward-biased junction, while the output impedance is HIGH as it is taken frum a reverse-biased junction.

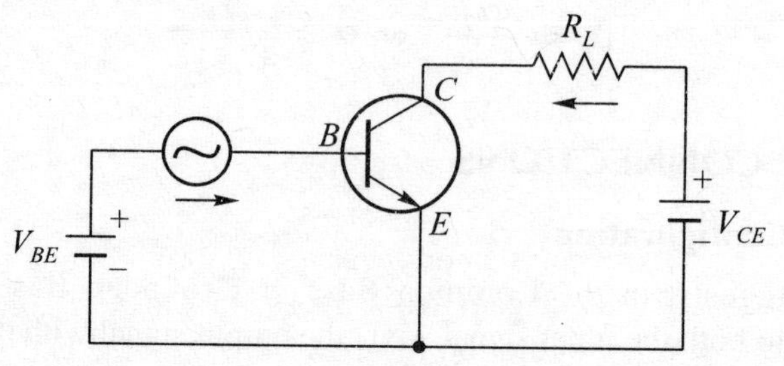

Fig. 8.6

In this type of configuration, the current flowing out of the transistor must be equal to the currents flowing into the transistor as the emitter current is given as $I_e = I_c + I_b$. Also, as the load resistance (R_L) is connected in series with the collector, the Current gain of the Common Emitter Transistor Amplifier is quite large as it is the ratio of I_c/I_b and is given the symbol of Beta, (β). Since the relationship between these three currents is determined by the transistor itself, any small change in the base current will result in a large change in the collector current. Then, small changes in base current will thus control the current in the Emitter/Collector circuit.

By combining the expressions for both α and β the mathematical relationship between these parameters and therefore the current gain of the amplifier can be given as:

$$I_E = I_C + I_B$$

$$\alpha = \frac{I_C}{I_E} \quad \text{and} \quad \beta = \frac{I_C}{I_B}$$

$$\alpha = \frac{\beta}{\beta + 1} \qquad \beta = \frac{\alpha}{1 - \alpha}$$

Where: "I_c" is the current flowing into the collector terminal, "I_b" is the current flowing into the base terminal and "I_e" is the current flowing out of the emitter terminal.

Then to summarize, this type of transistor configuration has a greater input impedance, Current gain and Power gain than that of the common base configuration but its Voltage gain is much lower. The common emitter is an inverting amplifier circuit resulting in the output signal being 180° out of phase with the input voltage signal.

The Common Collector Configuration

In the **Common Collector** or Grounded Collector configuration, the collector is now common and the input signal is connected to the base, while the output is taken from the emitter load as shown. This type of configuration is commonly known as a **Voltage Follower** or **Emitter Follower** circuit. The emitter follower configuration is very useful for impedance matching applications because of the very high input impedance, in the region of hundreds of thousands of Ohms, and it has relatively low output impedance.

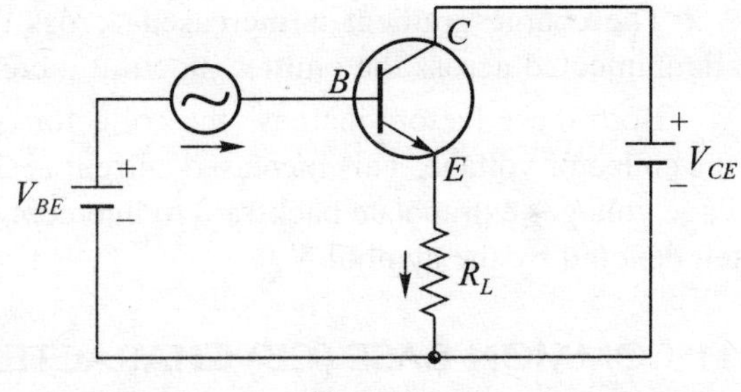

Fig. 8.7

The common emitter configuration has a current gain equal to the β value of the transistor itself. In the common collector configuration the load resistance is situated in series with the emitter so its current is equal to that of the emitter current. As the emitter current is the combination of the collector AND base currents combined the load resistance in this type of amplifier configuration also has both the collector current and the input current of the base flowing through it. Then the current gain of the circuit is given as:

This type of transistor configuration is a non-inverting amplifier circuit in that the signal voltages of V_{in} and V_{out} are In-Phase. It has a voltage gain that is always less than "1" (unity). The load resistance of the common collector amplifier configuration receives both the base and collector currents giving a large current gain (as with the Common Emitter configuration) therefore, providing good current amplification with very little voltage gain.

8.10 EARLY EFFECT

The **Early effect** is the variation in the width of the base in a BJT due to a variation in the applied base-to-collector voltage, named after its discoverer James M. Early. A greater reverse bias across the collector-base junction, for example, increases the collector-base depletion width, decreasing the width of the charge neutral portion of the base.

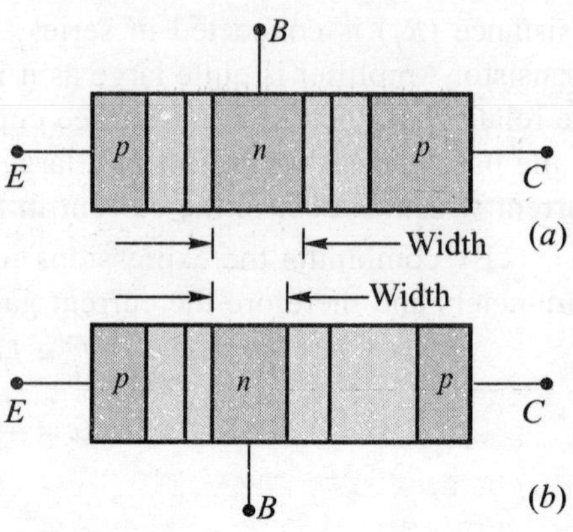

Fig. 8.8

From figure top: *pnp* base width for low collector-base reverse bias; Bottom: narrower *pnp* base width for large collector-base reverse bias. Light colors are depleted regions.

In the figure the neutral base width is dark blue, and the depleted base regions are light blue. The neutral emitter and collector regions are dark red and the depleted regions pink. Under increased collector-base reverse bias, the lower panel of Figure 8.8(*a*) shows a widening of the depletion region in the base and the associated narrowing of the neutral base region.

The collector depletion region also increases under reverse bias, more than does that of the base, because the collector is less heavily doped. The principle governing these two widths is charge neutrality. The emitter-base junction is unchanged because the emitter-base voltage is the same.

Base-narrowing has two consequences that affect the current:

There is a lesser chance for recombination within the "smaller" base region.

The charge gradient is increased across the base, and consequently, the current of minority carriers injected across the emitter junction increases.

Both these factors increase the collector or "output" current of the transistor with an increase in the collector voltage. This increased current is shown in Figure 8.8(*b*). Tangents to the characteristics at large voltages extrapolate backward to intercept the voltage axis at a voltage called the **Early voltage,** often denoted by the symbol $\mathbf{V_A}$.

8.11 COMMON-BASE (CB) CHARACTERISTICS

(1) Input Characteristics: The curves showing the variation of the Emitter current I_E with Emitter voltage V_{EB} at constant collector voltage V_{CB} are called the Emitter or input characteristics of a transistor.

Fig. (8.9) shows the circuit arrangement of a *CB* circuit for obtaining the input characteristics of a *P-N-P* transistor. Emitter base junction is biased in the forward direction by connecting the Emitter to the positive terminal of a variable battery V_{BB} whose negative terminal is connected to the base. The Emitter to base voltage V_{EB} is read by a voltmeter (*V*) and the Emitter current I_E by a milliammeter (mA). The collector-base circuit is biased in the reverse direction by connecting the collector to the negative terminal of another variable battery V_{CC} whose positive terminal is connected to the base. The collector-to-base voltage V_{CB} is read by another voltmeter and the collector current I_C by another milliammeter.

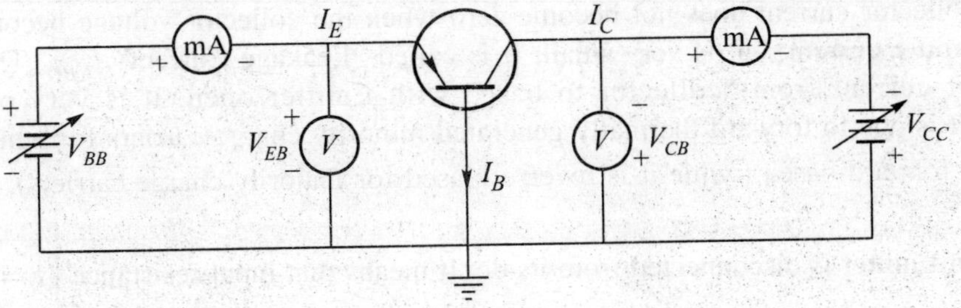

Fig. 8.9

To obtain the input characteristics, collector-to-base voltage V_{CB} is kept constant at a suitable value. The Emitter-to-base voltage V_{EB} is now varied in small steps and the Emitter current I_E is noted every time. A curve is then plotted between V_{EB} and I_E as shown in Fig. (8.10). A number of these characteristic curves are obtained for various collector-to-base voltage V_{CB}. It may be seen that these curves resemble very much to the *P-N* junction diode characteristics which is forward biased and shown in by the curve marked V_{CB} open.

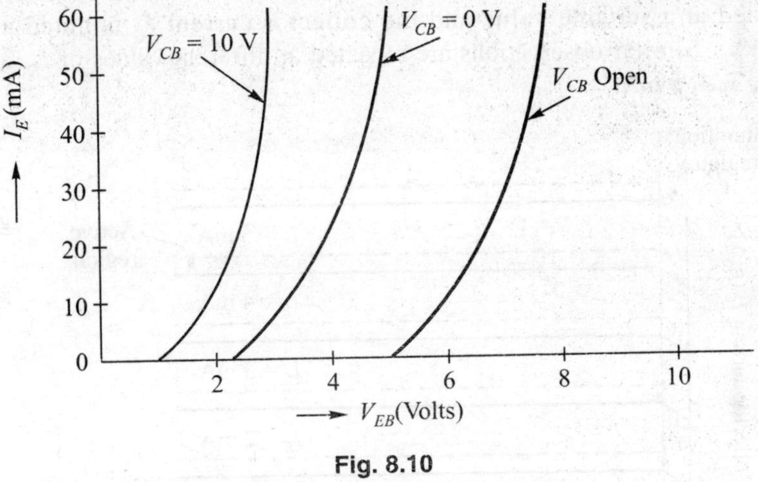

Fig. 8.10

A close look at these curves reveals the following facts:

(*i*) The Emitter current I_E increases rapidly with increasing values of Emitter-to-base voltage V_{EB} and constant increase in the collector current I_C for low voltages. Thus entire variation in collector current takes place at very low values of collector voltage.

(*ii*) There exists a well-defined knee above which the collector current I_C is almost independent of the collector voltage V_{CB} and depends only upon Emitter Current I_E.

(*iii*) A very large change in the collector-base voltage V_{CB} produces only a tiny change in the collector current. It means that the output resistance $\left[r_O = \dfrac{\Delta V_{CB}}{\Delta I_C} \right]$ is very high.

(*iv*) The collector current I_C is always a little less than the corresponding Emitter current due to a small percentage of charge carriers being lost in the base due to electronnale combination.

(*v*) The collector current does not become zero when the collector voltage becomes zero for any value of Emitter current, but is very small. It is called "Leakage Current" I_{CBO}. The subscripts *CBO* stands for current from "collector to base" with Emitter open. It is very often simply written as I_{CO}. It is due to forward thermally generated minority charge carriers for which collector base junction is forward based (while it is reverse biased for majority charge carriers). This current flows even when Emitter is disconnected from its dc. It means that input resistance $\left[r_i = \left[\frac{\Delta V_{EB}}{\Delta I_E} \right]_{V_{CB}} \right]$ is very small.

(*vi*) As the collector-to-base voltage V_{CB} is made more negative, more holes are attracted by the collector and hence Emitter current I_E rises more rapidly.

(2) Output Characteristics: Curves showing the variation of the collector current (I_C) with the collector voltage $V_{CB\,(sat)}$ constant Emitter current are called the collector or output characteristics of a transistor.

To obtain the output characteristics, circuit of Fig. 8.11 is again used. Here the Emitter current I_E is adjusted at a suitable value and the collector current I_C is noted as a function of collector voltage V_{CB}. Similar observations are repeated at different values of I_E, as shown in Fig. (8.11) for I_E = 1, 2, 3, 4, 5 mA.

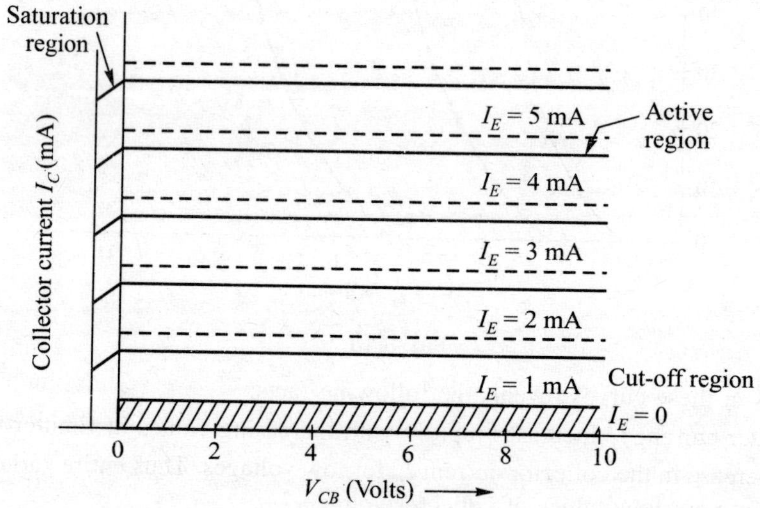

Fig. 8.11

A close look at these curves reveals the following facts:

(*i*) As the collector voltage V_{CB} is made more negative (collector is reverse biased) starting from zero, there is a rapid supply as shown in Fig. 8.12. It flows in the same direction as the collector current of majority carriers. It is extremely temperature dependent and doubles for every 10°C rise in temperature for Ge and 6°C for Si. Its value at room temperature varies from 2 μA to 5 μA for Ge and 0.1 μA to 1 μA for Si transistor.

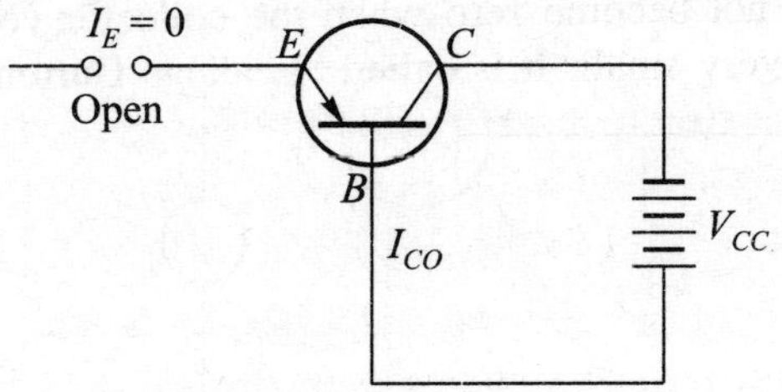

Fig. 8.12

The family of characteristics is divided into three regions.

(1) Active Region: Although values of V_{CB} are negative, V_{CB} is generally plotted along positive *x*-axis.

For the first curve when $I_E = 0$ *i.e.*, when input terminals are open, the collector current $I_C = I_{CO}$ which is very small. Now when input circuit is made on and I_E flows, then αI_E flows towards collector region and collector current becomes $(\alpha I_E + I_{CO})$ which is independent of collector voltage. Hence curves appear to be practically flat.

(2) Saturation Region: The Region to the left of ordinate $V_{CB} = 0$ and above $I_E = 0$ characteristics is called saturation region. Here both Emitter and collector junctions are forward biased. Thus forward biasing of collector is responsible for large change in collector current I_C for a small change in V_{CB} and I_C increases exponentially with voltage.

(3) Cut-off Region: The characteristics for $I_E = 0$ is similar to other characteristics but passes through origin. It does not coincide with *x*-axis (voltage axis) but there is a small gap between the characteristics and voltage axis. The region below $I_E = 0$ characteristics, for which both Emitter and collector junctions are reverse biased, is called "cut-off region".

8.12 COMMON EMITTER CHARACTERISTICS

(1) Input Characteristics: The curves showing the variation of the base current I_B with base-Emitter voltage V_{BE} at constant collector-Emitter voltage V_{CE} are called the "input characteristics" of a transistor.

The input characteristics of a *P-N-P CE* transistor can be obtained by the circuit arrangement shown in Fig. 8.13. The Emitter-base junction is biased in the forward direction by connecting the negative terminal of the variable battery V_{BB} to the base and positive terminal to the Emitter (*P*-type semiconductor). The base-Emitter voltage V_{BE} is read by a voltmeter and the base current I_B by a microammeter (μA). The collector junction is reverse biased by junction connecting the collector (*P*-type semiconductor) to the negative terminal of another variable battery V_{CC} whose positive terminal is connected to Emitter.

The collector-to-Emitter voltage V_{CE} is read by another voltmeter and the collector current I_C by a milliammeter (mA).

The input characteristics can be obtained by keeping V_{CE} fixed and noting the base current I_B as a function of input potential V_{BE}.

The procedure is repeated for another value of V_{CE}. Typical input characteristics are shown in Fig. 8.14 for $V_{CE} = 0, -10$ and -20 voltages.

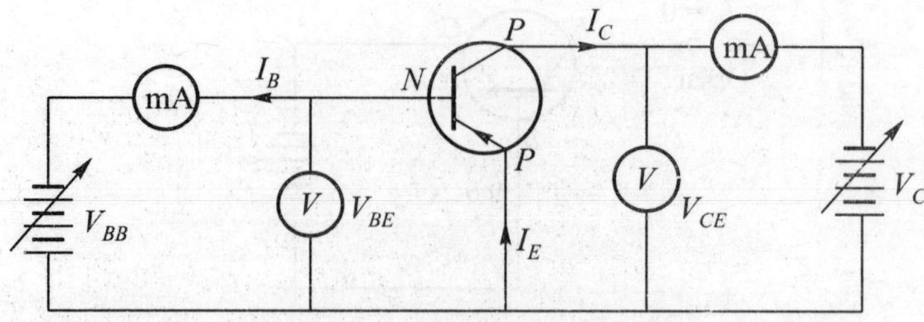

Fig. 8.13

A study of these curves reveals the following facts:

(*i*) The characteristics resemble to those of a forward biased junction diode because the base-Emitter section of transistor is a junction diode and it is forward biased.

(*ii*) The base current increases non-linearly with increase in base voltage.

(*iii*) Base current I_B increases less rapidly with V_{BE} as compared to *CB* configuration, thus, input resistance $\left[r_i = \left[\frac{\Delta V_{BE}}{\Delta I_B} \right]_{V_{CE}} \right]$ of a *CE* circuit is higher compared to that in *CB* configuration.

(*iv*) Input characteristics are only slightly dependent upon the collector-to-Emitter voltage V_{CE}.

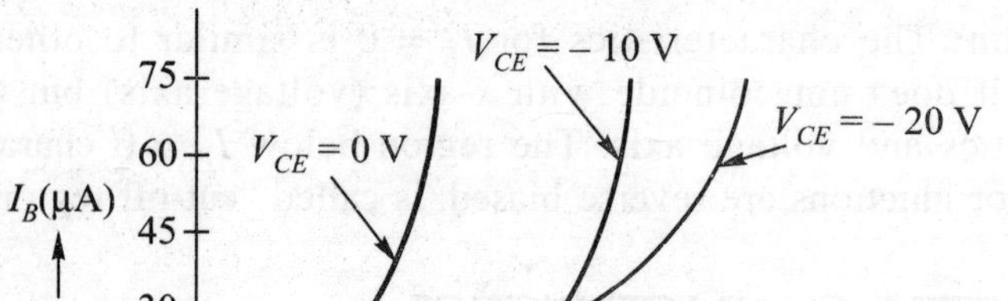

Fig. 8.14

(2) Output Characteristics: The curves showing the variation of the collector current I_C with collector-Emitter voltage V_{CE} at constant base current I_B are called output characteristics of a transistor.

The output characteristics of a *CE* transistor circuit can be drawn with the help of circuit arrangement shown in Fig. 8.15. For it, the collector current I_C is measured by varying the collector voltage V_{CE} for a constant value of base current I_B. Observations are repeated for different settings of base current I_B and a family of curves is obtained as shown in Fig. 8.15.

A study of these curves reveals the following facts:

(*i*) The collector current I_C varies rapidly with V_{CE} for very small voltage (say upto V_{CE} = 2 volts). After this collector current becomes constant and is decided entirely by base current I_B. It then becomes independent of V_{CE}.

(*ii*) Collector voltage has only a minute effect on the collector current for low values of base current. But as the base current rises, the effect of collector voltage on the collector current also increases.

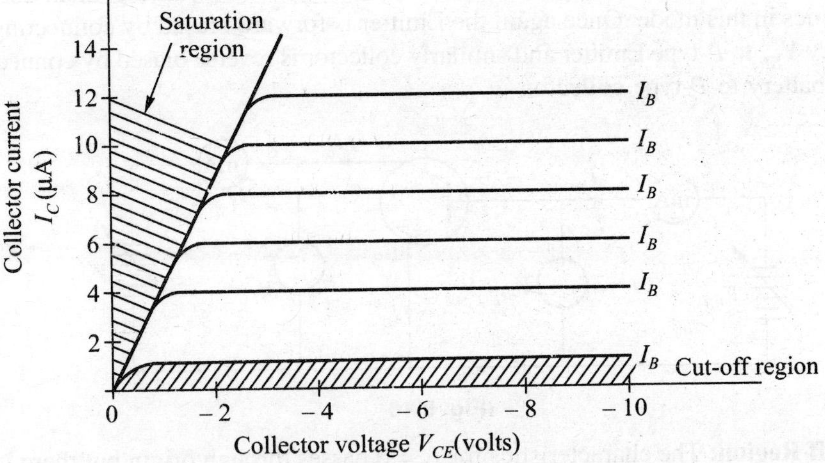

Fig. 8.15

(*iii*) The collector current I_C is not zero when base current is zero. This is due to the inherent intrinsic condition in semiconductors, *i.e.* due to minority charge carriers.

(*iv*) Since the input current I_B is measured in microampere and output current I_C is measured in milliampere, the common Emitter configuration exhibits a current amplification.

$$B = \left(\frac{\Delta I_B}{\Delta I_B}\right)_{V_{CE}}$$

Thus, small input current I_B produces a large output current I_C or in other words, the collector current is B times greater than the base current.

Just like common base configuration, the family of curves is divided into three regions.

(1) Active Region: When Emitter junction is forward biased and collector junction is reverse biased, then in Fig. 8.15 the area above $I_B = 0$ and to the right of ordinate V_{CE} = a few tenth of a volt, is known as active region. For a given value of I_B, Emitter injects certain charge carriers into base region. When collector-Emitter voltage V_{CE} is small, the collector is unable to collect all charge carries. Therefore, when V_{CE} is increased, collector current I_C first increases rapidly and then rate of increase is quite small.

(2) Saturation Region: Referring to Fig. 8.15, the region very close to the zero voltage axis where all the curves, merge and fall rapidly towards origin, is known as saturation region. In this region, the collector junction as well as the Emitter junction is forward biased.

$$\therefore \quad I_C = \alpha I_E - I_{CEO}$$

Also

$$\begin{aligned} I_B &= I_E - I_C \\ &= I_E - (\alpha I_E - I_{CEO}) \\ &= (1 - \alpha)I_E - I_{CEO} \end{aligned}$$

8.13 COMMON COLLECTOR CHARACTERISTICS

In this arrangement input is applied b/w base and collector while output is taken from the Emitter. Here, collector is common to both the input and ouptut circuits.

Fig. 8.16 shows the common collector *P-N-P* transistor circuit arrangement for drawing the static characteristics in this mode. Once again the Emitter is forward biased by connecting the positive terminal of battery V_{EE} to *P*-type Emitter and similarly collector is reverse biased by connecting negative terminal of this battery to *P*-type collector.

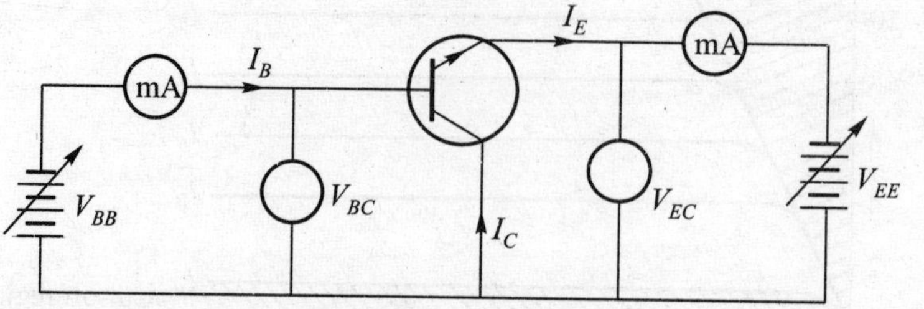

Fig. 8.16

(3) Cut-off Region: The characteristics for $I_B = 0$ passes through origin but there is appreciable gap between *X*-axis and the curve for $I_B = 0$. Thus, appreciable collector current and the area under $I_B = 0$ and *X*-axis is known as cut-off region.

Since I_C is not zero when $I_B = 0$, there is a leakage current from collector to Emitter. It is called I_{CEO}. The sub-scripts CEO standing for "Collector to Emitter" with base open. It has been shown in Fig. 8.17 for a common Emitter circuit of NPN transistor whose base lead is open.

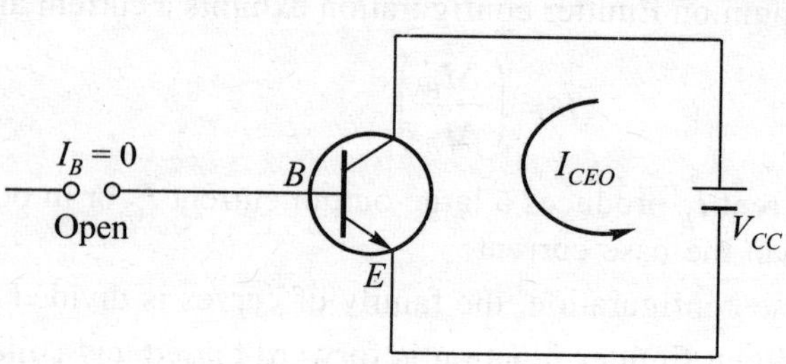

Fig. 8.17

Taking leakage current into account,

$$I_C = \beta I_B + I_{CEO} = \beta I_B + (1 + \beta)\, I_{CO}$$

Thus
$$I_{CEO} = (1 + \beta) I_{CO} = (1 + \beta) I_{CBO}$$

$$= \frac{1}{1-\alpha} I_{CO}$$

and
$$I_C = \beta I_B + \frac{I_{CO}}{1-\alpha}$$

∴
$$I_C = \left(\frac{\alpha}{1-\alpha}\right) I_B + \frac{I_{CO}}{1-\alpha}$$

But
$$\beta I_B = \alpha I_E$$

(1) Input Characteristics: The curves showing the variation of base current I_B with collector base voltage V_{CB} at constant collector emitter voltage V_{CE} is called the input characteristics.

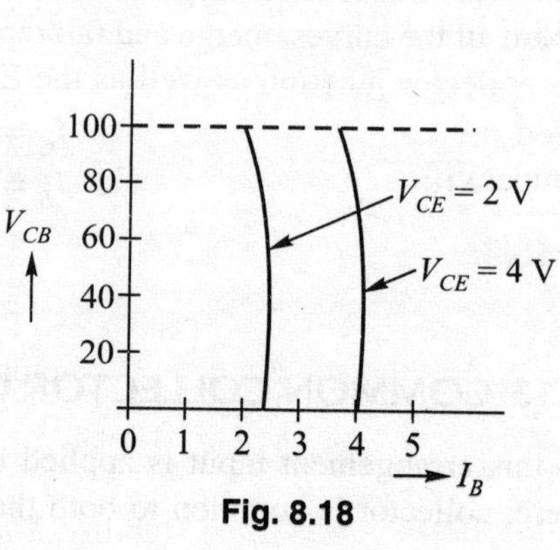

Fig. 8.18

Fig. 8.18 gives a plot of I_B versus V_{CB} for different values of V_{CE}. These characteristics are quite different from those for *CB* or *CE* circuits because the input voltage V_{CB} is large determined by the value of *CE* voltage. It may be seen that with an increase in V_{CB}, V_{BE} gets decreased, thereby decreasing I_B.

(2) Output Characteristics: The curves showing the variation of the emitter current I_E with collector emitter voltage V_{CE} at constant base current I_B are called the output characteristics of the transistor.

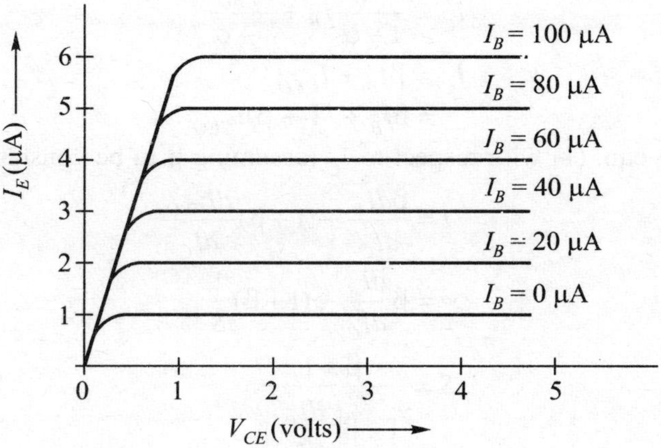

Fig. 8.19

Fig. 8.19 gives the plot of I_E versus V_{CE} for several fixed values of I_B. It may be seen that these characteristics are practically identical to that of the *CE* circuit because $I_C \simeq I_E$.

Thermal Runaway: It is not sufficient to establish the operating point, but it has to be maintained. Because the collector current of a transistor is sensitive to temperature and depend on amplification factor life of a transistor. As a result of this the operating point (*Q*-point) changes when the temperature or a transistor is changed.

According to the equation

$I_C = \beta I_b + (1 + \beta) I_{CEO}$, at a constant V_{BE}, the base current increases with temperature and the leakage current I_{CEO} gets doubled for even 10°C rise in temperature. Thus I_C is strongly depend on temperature. This causes a sort of instability of the operating point. The increase in I_C increases in the temperature of the collector junction, this in turn increases I_{CEO} leads to further increase in I_C. This is cumulative and within no time the transistor may get burnt. This stage is called as Thermal Runaway and it must be avoided.

The self-destruction of a transistor due to proper unbias is called as thermal runaway. So we need to stabilize the operating point. The process of making the operating point independent of temperature or transistor variations is called bias stabilization.

Stability Factor

The stability of an operating point can be measured by a factor called stability factor. Mathematically stability factor can be defined as

$$S = \frac{dI_C}{dI_E}$$

For Good thermal stability, *S* should be small.

Let us consider

$$I_C = \alpha I_E + I_{CBO}$$

[I_{CBO} : collector-base current with emitter open]

$$I_C - I_{CBO} = \alpha I_E$$

$$\frac{I_C}{\alpha} - \frac{I_{CBO}}{\alpha} = I_E = I_B + I_C$$

When
$$I_C = \frac{\alpha}{1-\alpha} I_B + \frac{I_{CBO}}{1-\alpha}$$

$\therefore$
$$I_C = \beta I_B + I_{CEO}$$
$$= \beta I_B + (1+\beta) I_{CEO}$$

Differentiating eqn. (1) with respect to I_C (assuming β to be constant) we get,

$$I = \frac{\beta dI_B}{dI_C} + (1+\beta)\frac{dI_{CBO}}{dI_C}$$

$$= \beta \frac{dI_B}{dI_C} + (1+\beta)\frac{1}{S}$$

$$S = \frac{\beta+1}{1-\beta\left(\dfrac{dI_B}{dI_C}\right)}$$

The values of $\left(\dfrac{dI_B}{dI_C}\right)$ depend upon the type of biasing arrangement. The stability factor S pertains to variation due to changes in I_{CEO}.

8.14 BIASING

Biasing in electronics is the method of establishing predetermined voltages and/or currents at various points of a circuit to set an appropriate operating point.

The operating point of a device, also known as bias point or quiescent point (or simply Q-point), is the DC voltage and/or current which, when applied to a device, causes it to operate in a certain desired fashion. The term is normally used in connection with devices such as transistors and diodes which are used in amplification or rectification.

8.15 BIPOLAR TRANSISTOR BIASING

For bipolar junction transistors the bias point is chosen to keep the transistor operating in the active mode, using a variety of circuit techniques, establishing the Q-point DC voltage and current. A small signal is then applied on top of the Q-point bias voltage, thereby either modulating or switching the current, depending on the purpose of the circuit.

The quiescent point of operation is typically near the middle of DC load line. The process of obtaining certain DC collector current at a certain DC collector voltage by setting up operating point is called biasing.

After establishing the operating point, when input signal is applied, the output signal should not move the transistor either to saturation or to cut-off. However, this unwanted shift might occur due to various reasons outlined below:

1. Parameters of transistor depend on temperature. As it increases, leakage current due to minority charge carriers (I_{CBO}) increases. As I_{CBO} increases, I_{CEO} also increases, causing increase in collector current I_C. This produces heat at the collector junction. This process repeat, and finally Q-point may shift into saturation region. Sometimes the excess heat produced at the junction may even burn the transistor. This is known as **thermal runaway.**

2. When a transistor is replaced by another of the same type, the Q-point may shift, due to change in parameters of transistor such as *current gain* (β) which changes from unit to unit.

To avoid a shift of Q-point, bias-stabilization is necessary. Various biasing circuits can be used for this purpose.

Small signal bipolar transistor amplifiers must be properly biased to operate correctly. In circuits made with individual devices (discrete circuits), biasing networks consisting of discrete components such as resistors and capacitors are commonly employed. Much more elaborate biasing arrangements are used in integrated circuits, for example, bandgap voltage references and current mirror.

Requirements Upon Biasing Circuit

The operating point of a device, also known as bias point or quiescent point (or simply Q-point), is the DC voltage and/or current which, when applied to a device, causes it to operate in a certain desired fashion.

1. For analog circuit operation, the Q-point is placed so the transistor stays in active mode (does not shift to operation in the saturation region or cut-off region) when input is applied. For digital operation, the Q-point is placed so the transistor does the contrary - switches from "on" to "off state. Often, Q-point is established near the center of active region of transistor characteristic to allow similar signal swings in positive and negative directions.

2. Q-point should be stable. In particular, it should be insensitive to variations in transistor parameters (for example, should not shift if transistor is replaced by another of the same type), variations in temperature, variations in power supply voltage and so forth.

The circuit must be practical: easily implemented and cost-effective.

Note on Temperature Dependence

At constant current, the voltage across the emitter-base junction V_{BE} of a bipolar transistor *decreases* about 2 mV for each 1°C rise in temperature.[1] Oppositely, if V_{BE} is held constant and the temperature rises, I_C increases, also increasing the power consumed in the transistor, tending to further increase its temperature. Unless steps are taken to control this positive feedback of increased temperature ⟶ increased current ⟶ increased temperature, **thermal runaway** ensues.[2] An electrical approach to avoid thermal runaway is to use negative feedback, as described in conjunction with some of the circuits below.

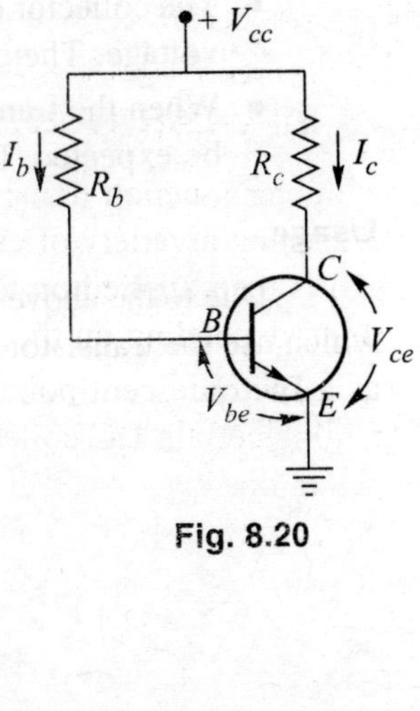

Fig. 8.20

8.16 TYPES OF BIAS CIRCUIT

The following discussion treats five common biasing circuits used with bipolar transistors:

1. Fixed bias
2. Collector-to-base bias
3. Fixed bias with emitter resistor
4. Voltage divider bias
5. Emitter bias
6. Fixed bias (base bias)

8.17 FIXED BIAS (BASE BIAS)

This form of biasing is also called *base bias.* In the example image on the right, the single power source (for example, a battery) is used for both collector and base of transistor, although separate batteries can also be used.

In the given circuit,

$$V_{CC} = I_B R_B + V_{be}$$

Therefore,

$$I_B = (V_{CC} - V_{be})/R_B$$

For a given transistor, V_{be} does not vary significantly during use. As V_{CC} is of fixed value, on selection of R_B, the base current I_B is fixed. Therefore this type is called *fixed bias* type of circuit.

Also for given circuit,

$$V_{CC} = I_C R_C + V_{ce}$$

Therefore,

$$V_{ce} = V_{CC} - I_C R_C$$

From this equation we can obtain V_{ce}. Since $I_C = \beta I_B$, we can obtain I_C as well. In this manner, operating point given as (V_{CE}, I_C) can be set for given transistor.

Merits

- It is simple to shift the operating point anywhere in the active region by merely changing the base resistor (R_B).
- Very few number of components are required.

Demerits

- The collector current does not remain constant with variation in temperature or power supply voltage. Therefore the operating point is unstable.
- When the transistor is replaced with another one, considerable change in the value of β can be expected. Due to this change the operating point will shift.

Usage

Due to the above inherent drawbacks, fixed bias is rarely used in linear circuits, *i.e.* those circuits which use the transistor as a current source. Instead it is often used in circuits where transistor is used as a switch.

8.18 COLLECTOR-TO-BASE BIAS

In this form of biasing, the base resistor R_B is connected to the collector instead of connecting it to the battery V_{CC}. That means this circuit employs negative feedback to stabilize the operating point.

From Kirchhoff's voltage law, the voltage across the base resistor is

$$V_{Rb} = V_{CC} - (I_C + I_b)R_C - V_{be}.$$

From Ohm's law, the base current is

$$I_b = V_{Rb}/R_b.$$

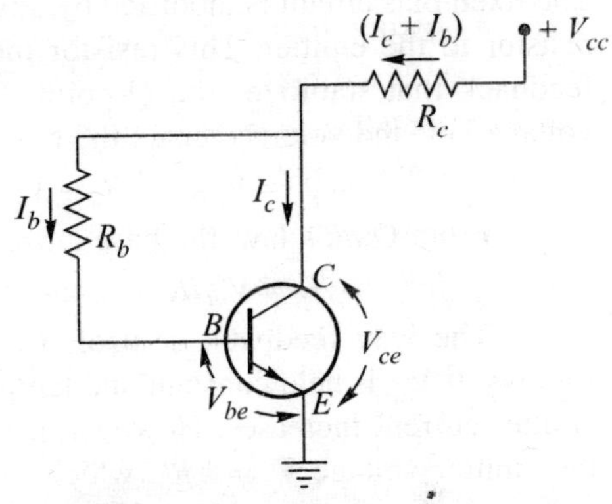

Fig. 8.21. Collector-to-base bias.

The way feedback controls the bias point is as follows. If V_{be} is held constant and temperature increases, collector current increases. However, a larger I_C causes the voltage drop across resistor R_C to increase, which in turn reduces the voltage V_{Rb} across the base resistor. A lower base-resistor voltage drop reduces the base current, which results in less collector current, so increase in collector current with temperature is opposed, and operating point is kept stable.

For the given circuit,

$$I_B = (V_{CC} - V_{be})/(R_B + \beta R_C).$$

Merits

- Circuit stabilizes the operating point against variations in temperature and β (*i.e.* replacement of transistor)

Demerits

- In this circuit, to keep I_C independent of β the following condition must be met:

$$I_C = \beta I_B = \frac{\beta(V_{CC} - V_{be})}{R_B + \beta R_C} \approx \frac{(V_{CC} - V_{be})}{R_C}$$

which is approximately the case if

$$\beta R_C >> R_B.$$

- As β-value is fixed for a given transistor, this relation can be satisfied either by keeping R_C fairly large, or making R_B very low.
  - If R_C is of large value, high V_{CC} is necessary. This increases cost as well as precautions necessary while handling.
  - If R_B is low, the reverse bias of the collector-base is small, which limits the range of collector voltage swing that leaves the transistor in active mode.
- The resistor R_B causes an ac feedback, reducing the voltage gain of the amplifier. This undesirable effect is a trade-off for greater *Q*-point stability.

Usage

The feedback also decreases the input impedance of the amplifier as seen from the base, which can be advantageous. Due to the gain reduction from feedback, this biasing form is used only when the trade-off for stability is warranted.

8.19 FIXED BIAS WITH EMITTER RESISTOR

The fixed bias circuit is modified by attaching an external resistor to the emitter. This resistor introduces negative feedback that stabilizes the Q-point. From Kirchhoff's voltage law, the voltage across the base resistor is

$$V_{Rb} = V_{CC} - I_e R_e - V_{be}.$$

From Ohm's law, the base current is

$$I_b = V_{Rb}/R_b.$$

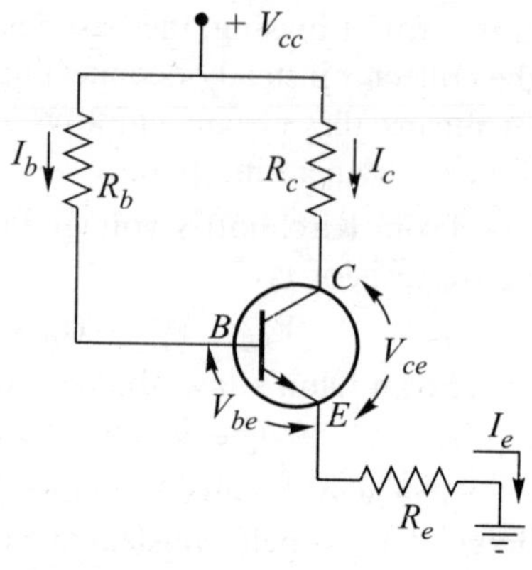

Fig. 8.22. Fixed bias with emitter resistor.

The way feedback controls the bias point is as follows. If V_{be} is held constant and temperature increases, emitter current increases. However, a larger I_e increases the emitter voltage $V_e = I_e R_e$, which in turn reduces the voltage V_{Rb} across the base resistor. A lower base-resistor voltage drop reduces the base current, which results in less collector current because $I_c = \beta\, I_B$. Collector current and emitter current are related by $I_c = \alpha I_e$ with $\alpha = 1$, so increase in emitter current with temperature is opposed, and operating point is kept stable.

Similarly, if the transistor is replaced by another, there may be a change in I_C (corresponding to change in β-value, for example). By similar process as above, the change is negated and operating point kept stable.

For the given circuit,

$$I_B = (V_{CC} - V_{be})/(R_B + (\beta + 1)R_E).$$

Merits

The circuit has the tendency to stabilize operating point against changes in temperature and β-value.

Demerits

In this circuit, to keep I_C independent of β the following condition must be met:

$$I_C = \beta I_B = \frac{\beta(V_{CC} - V_{be})}{R_B + (\beta + 1)R_E} \approx \frac{(V_{CC} - V_{be})}{R_E}$$

which is approximately the case if

$$(\beta + 1)R_E >> R_B.$$

- As β-value is fixed for a given transistor, this relation can be satisfied either by keeping R_E very large, or making R_B very low.
  - If R_E is of large value, high V_{CC} is necessary. This increases cost as well as precautions necessary while handling.
  - If R_B is low, a separate low voltage supply should be used in the base circuit. Using two supplies of different voltages is impractical.
- In addition to the above, R_E causes ac feedback which reduces the voltage gain of the amplifier.

Usage

The feedback also increases the input impedance of the amplifier as seen from the base, which can be advantageous. Due to the above disadvantages, this type of biasing circuit is used only with careful consideration of the trade-offs involved.

8.20 VOLTAGE DIVIDER BIAS

The voltage divider is formed using external resistors R_1 and R_2. The voltage across R_2 forward biases the emitter junction. By proper selection of resistors R_1 and R_2, the operating point of the transistor can be made independent of β. In this circuit, the voltage divider holds the base voltage fixed independent of base current provided the divider current is large compared to the base current. However, even with a fixed base voltage, collector current varies with temperature (for example) so an emitter resistor is added to stabilize the Q-point, similar to the above circuits with emitter resistor.

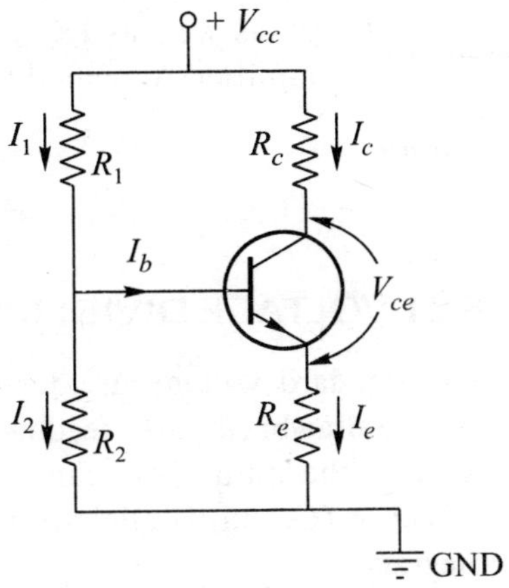

Fig. 8.23. Voltage divider bias.

In this circuit the base voltage is given by:

$$V_B = \text{voltage across } R_2 = V_{ce}\frac{R_2}{(R_1+R_2)} - I_B\frac{R_1R_2}{(R_1+R_2)}$$

$$\approx V_{CC}\frac{R_2}{(R_1+R_2)} \quad \text{provided } I_b << I_2 = V_B/R_2$$

Also $V_B = V_{be} + I_E R_E$

For the given circuit,

$$I_B = \frac{\dfrac{V_{CC}}{1+R_1/R_2} - V_{be}}{(\beta+1)R_E + R_1 \parallel R_2}$$

Merits

- Unlike above circuits, only one dc supply is necessary.
- Operating point is almost independent of β variation.
- Operating point stabilized against shift in temperature.

Demerits

In this circuit, to keep I_C independent of β the following condition must be met:

$$I_C = \beta I_B = \beta\frac{\dfrac{V_{CC}}{1+R_1/R_2} - V_{be}}{(\beta+1)R_E + R_1 \parallel R_2} \approx \frac{\dfrac{V_{CC}}{1+R_1/R_2} - V_{be}}{R_E}$$

which is approximately the case if

$$(\beta + 1)R_E >> R_1 \parallel R_2$$

where $R_1 \parallel R_2$ denotes the equivalent resistance of R_1 and R_2 connected in parallel.

- As β-value is fixed for a given transistor, this relation can be satisfied either by keeping R_E fairly large, or making $R_1 \parallel R_2$ very low.

- If R_E is of large value, high V_{CC} is necessary. This increases cost as well as precautions necessary while handling.
- If $R_1 \parallel R_2$ is low, either R_1 is low, or R_2, is low, or both are low. A low R_1 raises V_B closer to V_C, reducing the available swing in collector voltage, and limiting how large R_C can be made without driving the transistor out of active mode. A low R_2 lowers V_{be}, reducing the allowed collector current. Lowering both resistor values draws more current from the power supply and lowers the input resistance of the amplifier as seen from the base.
- AC as well as DC feedback is caused by R_E, which reduces the AC voltage gain of the amplifier. A method to avoid AC feedback while retaining DC feedback is discussed below.

Usage

The circuit's stability and merits as above make it widely used for linear circuits.

8.21 VOLTAGE DIVIDER WITH AC BYPASS CAPACITOR

The standard voltage divider circuit discussed above faces a drawback - AC feedback caused by resistor R_E reduces the gain. This can be avoided by placing a capacitor (C_E) in parallel with R_E, as shown in circuit diagram.

This capacitor is usually chosen to have a low enough reactance at the signal frequencies of interest such that R_E is essentially shorted at AC, thus grounding the emitter. Feedback is therefore only present at DC to stabilize the operating point. Of course, any AC advantages of feedback are lost.

Of course, this idea can be used to shunt only a portion of R_E, thereby retaining some AC feedback.

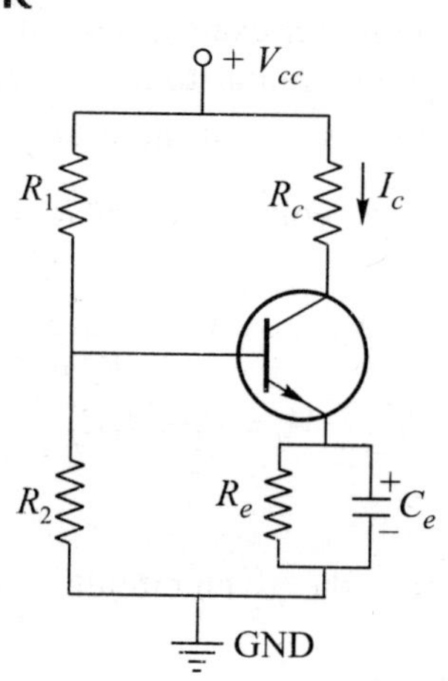

Fig. 8.24. Voltage divider with capacitor

8.22 EMITTER BIAS

When a split supply (dual power supply) is available, this biasing circuit is the most effective. The negative supply V_{EE} is used to forward-bias the emitter junction through R_E. The positive supply V_{CC} is used to reverse-bias the collector junction. Only three resistors are necessary.

We know that,

$$V_B - V_E = V_{be}$$

If R_B is small enough, base voltage will be approximately zero. Therefore emitter current is,

$$I_E = (V_{EE} - V_{be})/R_E$$

The operating point is independent of β if $R_E >> R_B/\beta$.

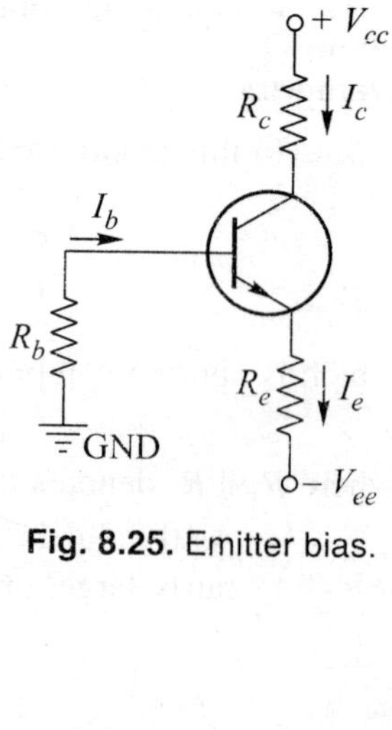

Fig. 8.25. Emitter bias.

Merit

Good stability of operating point similar to voltage divider bias.

Demerit

This type can only be used when a split (dual) power supply is available.

8.23 LOAD LINE

A **load line** is used in graphic analysis of circuits, representing the constraint other parts of the circuit place on a non-linear device, like a diode or transistor. A load line represents the response of a resistor which shares a current with the device in question. Since both currents are the same, the operating point of the circuit will be at the intersection of the curve with the load line.[1]

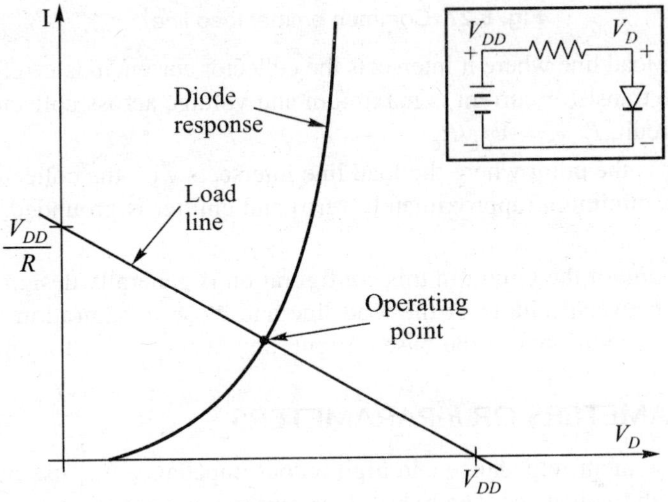

Fig. 8.26. Diode load line.

In the simple case of a diode shown, there is a single voltage across the diode and a single current through it. The load line represents the current in the resistor. When $V_D = V_{DD}$, there will be no voltage across the resistor, so the current will be 0. If $V_D = 0$, the current will be at its maximum.

In a BJT circuit, the BJT has a different current-voltage ($I_C - V_{CE}$) characteristic depending on the Base current. Placing a series of these curves on the graph shows how the base current will effect the operating point of the circuit.

It should be noted that the load line is used for dc analysis, and has no bearing on small-signal analysis once an operating point is identified.

8.24 LOAD LINE (COMMON EMITTER CONFIGURATION)

The given load line diagram is for the Common emitter configuration.

The load line diagram illustrates all possible values of collector current (I_C) and the collector voltage (V_{CE} in this case) for a given load resistor (R_C).

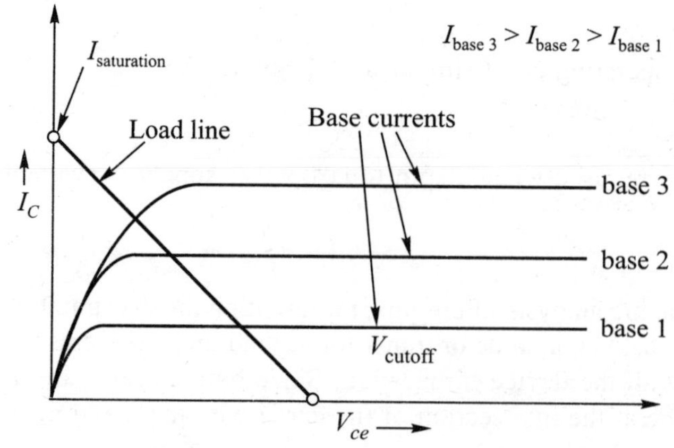

Fig. 8.27. Common emitter load line.

The point on the load line where it intersects the collector current axis is referred to as *saturation point.* At this point, the transistor current is maximum and voltage across collector is minimum, for a given load. For this circuit, $I_{C\text{-SAT}} = V_{CC}/R_C$.

The *cutoff point* is the point where the load line intersects with the collector voltage axis. Here the transistor current is minimum (approximately zero) and emitter is grounded. Hence $V_{CE\text{-CUTOFF}} = V_{cc}$.

The *operating point* of the circuit in this configuration is generally designed to be in the *active region,* approximately between middle of the load line and close to saturation point. In this region, the collector current is proportional to the base current, and hence useful for amplifier applications.

8.25 HYBRID PARAMETERS OR *h*-PARAMETERS

As the transistor has low input impedance and high output impedance, the use of *Z* and *Y* parameters becomes awkward at high frequencies. The hybrid *h* parameters appears most useful for circuit analysis because they form a combination of impedance and admittance parameters, and are selected to suit the low input and high output impedances of the transistor.

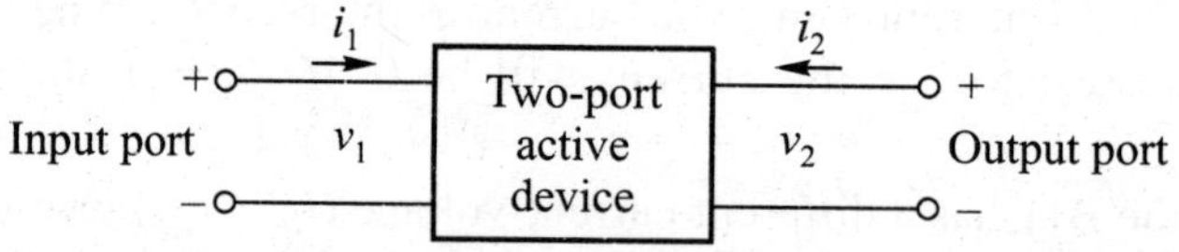

Fig. 8.28. A Two-Port Network.

Consider a linear circuit shown in figure. In this circuit input voltage and current are shown by v_1 and i_i while output voltage and current are shown by v_2 and i_2. Considering the standard conventions, the currents i_1 and i_2 are assumed to flow into the box. The input and output voltages v_1 and v_2 are assumed positive from the upper to the lower terminals. To analyse a particular circuit if the voltages are of opposite polarity or currents flow out of the box, they are treated as negative quantities.

In this circuit v_2 and i_1 are chosen as independent variables while v_1 and i_2 are taken as dependent variables. The different parameters have different dimensions and hence they are known as hybrid parameters, $v_1 = f(i_1, v_2)$, and $i_2 = f(i_1, v_2)$.

Taking the total differentials: $dv_1 = \left(\frac{\partial v_1}{\partial i_1}\right)_{v_2} di_1 + \left(\frac{\partial v_1}{\partial v_2}\right)_{i_1} dv_2$ ...(1)

and $$di_2 = \left(\frac{\partial i_2}{\partial i_1}\right)_{v_2} di_1 + \left(\frac{\partial i_2}{\partial v_2}\right)_{i_1} dv_2 \quad ...(2)$$

Again, we have $$v_1 = h_{11}i_1 + h_{12}v_2 \quad ...(3)$$
$$i_2 = h_{21}i_1 + h_{22}v_2 \quad ...(4)$$

or $$\begin{bmatrix} v_1 \\ i_2 \end{bmatrix} = \begin{bmatrix} h_{11} & h_{12} \\ h_{21} & h_{22} \end{bmatrix} \begin{bmatrix} i_1 \\ v_2 \end{bmatrix} \quad ...(5)$$

In equations (3) and (4), h's are fixed constants for a given circuit and are called h parameters. If we short-circuit the output terminals of figure shown above, then $v_2 = 0$. From equations (3) and (4) we get

$$v_1 = h_{11}i_1 + h_{12} \times 0 \quad ...(6)$$

and $$i_2 = h_{21}i_1 + h_{22} \times 0 \quad ...(7)$$

$\therefore$ $h_{11} = h_i = v_1/i_1$ for $v_2 = 0$ *i.e.* output shorted.

and $h_{21} = h_f = i_2/i_1$ for $v_2 = 0$ *i.e.* output shorted.

Here h_{11} is the ratio of voltage and current (impedance) and hence h_{11} is called as input impedance with output shorted. Similarly, h_{21} is the ratio of output and input currents *i.e.*, it will be dimensionless and called as current gain with output shorted.

Now consider that input terminals are open *i.e.* $i_1 = 0$. Now

$$v_1 = h_{11} \times 0 + h_{12}v_2 \quad ...(8)$$

and $$i_2 = h_{21} \times 0 + h_{22}\, v_2 \quad ...(9)$$

$\therefore$ $h_{12} = h_r = \frac{v_1}{v_2}$ for $i_1 = 0$ *i.e.* input open.

So that h parameters are:

$h_{11} = (v_1/i_1)$ = short circuit input impedance, $v_2 = 0$
$h_{12} = (v_1/v_2)$ = open circuit reverse voltage ratio, $i_1 = 0$
$h_{21} = (i_2/i_1)$ = short circuit forward current ratio, $v_2 = 0$
$h_{22} = (i_2/v_2)$ = open circuit output admittance, $i_1 = 0$
$h_{22} = h_0 = i_2/v_2$ for $i_1 = 0$ *i.e.*, input open.

Here h_{12} is dimensionless and is called voltage feedback ratio with input open. Similarly, h_{22} has the dimension of admittance and is called output admittance with input open.

The subscripts are modified in different configurations. For example, in common base following hybrid parameters are used: h_{ib}, h_{rb}, h_{fb} and h_{ob}. Similarly, h parameters for common emitter configuration are h_{ie}, h_{re}, h_{fe} and h_{oe}.

8.26 *h*-PARAMETER EQUIVALENT CIRCUIT

h-Parameter equivalent circuit is shown in Fig. 8.29. This circuit is derived from the following two equations

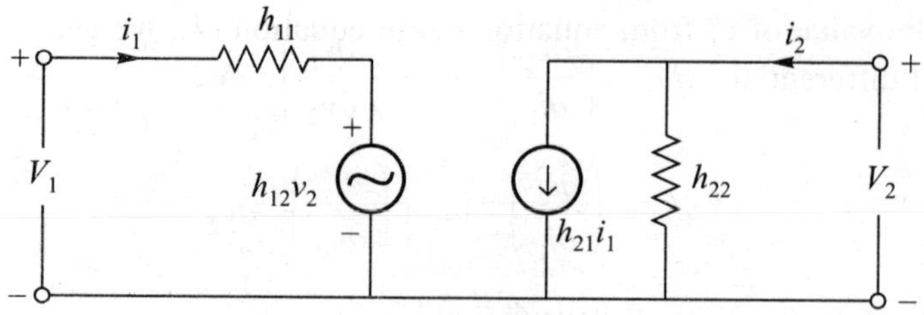

Fig. 8.29

$$v_1 = h_{11}\, i_1 + h_{12}v_2 \qquad ...(1)$$

$$i_2 = h_{21}i_1 + h_{22}v_2 \qquad ...(2)$$

The input circuit is derived with the help of equation (1). It appears as a resistance h_{11} in series with a voltage generator $h_{12}v_2$. The output circuit is derived with the help of equation (2). It involves a current generator $h_{21}i_1$ and a shunt resistance h_{22}.

The circuit is called as hybrid equivalent because its input portion is a Thevenin equivalent or voltage generator with series resistance while output circuit is Norton equivalent *i.e.*, current generator with shunt resistance.

In the equivalent circuit, the effect of output upon input is represented by equivalent generator $h_{12}v_2$ and its value depends upon output voltage. Similarly, the effect of input upon output is represented by current generator $h_{21}i_1$ and its value depends upon input current. So the equivalent circuit is extremely useful.

8.27 PERFORMANCE OF LINEAR CIRCUIT USING *h*-PARAMETERS

Consider a linear circuit with a load resistance R_L across the output terminals as shown in figure.

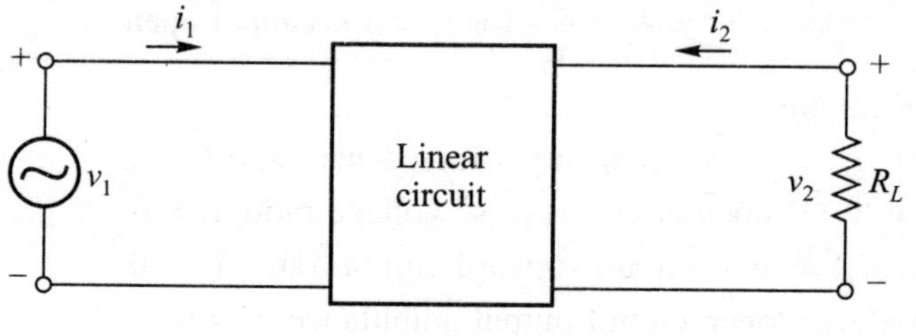

Fig. 8.30

In terms of *h* parameters, we have $v_1 = h_{11}i_1 + h_{12}v_2$ ...(1)

and $i_2 = h_{21}i_1 + h_{22}v_2$ ...(2)

From figure, $i_2 = -\dfrac{v_2}{R_L}$ ...(3)

Here the negative sign is used because the load current is opposite to the direction of i_2.

(1) Input Impedance

The input impedence Z_{in} of the circuit is the ratio of input voltage to input current, *i.e.*,

$$Z_{in} = \frac{v_1}{i_1} \qquad ...(4)$$

Substituting the value of v_1 from equation (1) in equation (4), we get

$$Z_{in} = \frac{h_{11}i_1 + h_{12}v_2}{i_1} = h_{11} + \frac{h_{12}v_2}{i_1} \quad ...(5)$$

Substituting the value of i_2 from equation (3) in equation (2), we get

$$-\frac{v_2}{R_L} = h_{21}i_1 + h_{22}\,v_2.$$

Solving we get

$$\frac{v_2}{i_1} = \frac{-h_{21}}{\left\{h_{22} + \left(\frac{1}{R_L}\right)\right\}} \quad ...(6)$$

Substituting the value of $\frac{v_2}{i_1}$ from equation (6) in equation (5), we get

$$Z_{in} = h_{11} - \frac{h_{12}h_{21}}{h_{22} + \left(\frac{1}{R_L}\right)} \quad ...(7)$$

(2) Current Gain

The current gain A_i of the circuit is given by $A_i = \frac{i_2}{i_1}$

From equations (2) and (3) $i_2 = h_{21}i_i + h_{22}(-i_2R_L)$ or $i_2(1 + h_{22}R_L) = h_{21}i_1$

or
$$\frac{i_2}{i_1} = \frac{h_{21}}{1 + h_{22}R_L}$$

$\therefore$
$$A_i = \frac{h_{21}}{1 + h_{22}R_L} \quad ...(8)$$

(3) Voltage Gain

The voltage gain A_v of the circuit is given by $A_v = \frac{v_2}{v_1}$

Now
$$A_v = \frac{v_2}{v_1} = \frac{v_2}{(i_1 Z_{in})}$$

From equation (6),
$$\frac{v_2}{i_1} = \frac{-h_{21}}{\left\{h_{22} + \left(\frac{1}{R_L}\right)\right\}}$$

$\therefore$
$$A_v = \frac{-h_{21}}{Z_{in}\left\{h_{22} + \left(\frac{1}{R_L}\right)\right\}} \quad ...(9)$$

(4) Output Impedance

The output impedance is denoted by Z_{out} and is defined as the ratio of output voltage to the output current.

$$Z_{\text{out}} = \frac{v_2}{i_2}$$

By using equations (1) and (2), we derive as following:

$$Z_{\text{out}} = \frac{1}{\left(h_{22} - \dfrac{h_{21}^2 R_L}{1 + h_{22} R_L}\right)}$$

The reciprocal of output impedance is known as output admittance.

SOLVED EXAMPLES

Example 1: *In a common base connection, $I_E = 1$ mA, $I_c = 0.95$ mA. Calculate the value of I_B.*

Solution: Using the relation

$$I_E = I_B + I_C$$
$$1 = I_B + 0.95$$
$$I_B = 1 - 0.95$$
$$= 0.05 \text{ mA.}$$

Example 2. *In a common base connection $I_C = 0.95$ mA and $I_B = 0.05$ mA. Find the value of α.*

Solution: We know

$$I_E = I_B + I_C$$
$$= 0.05 + 0.95$$
$$= 1 \text{ mA}$$

$\therefore$ Current amplification factor $\alpha = \dfrac{I_C}{I_E} = \dfrac{0.95}{1} = 0.95.$

Example 3: *In a common base connection, current amplification factor is 0.9. If the emitter current is 1 mA, determine the value of base current.*

Solution: Here $\alpha = 0.9$ $I_E = 1$ mA

$$\alpha = \frac{I_C}{I_E}$$
$$I_C = \alpha I_E = 0.9 \times 1$$
$$= 0.9 \text{ mA}$$
$$I_E = I_B + I_C$$

Base current

$$I_B = I_E - I_C$$
$$= 1 - 0.9 = 0.1 \text{ mA.}$$

Example 4: *In a common base connection, the emitter current is 1 mA. If the emitter circuit is open the collector current is 50 μA. Find the total collector current given that $\alpha = 0.92$.*

Solution: Here $I_E = 1$ mA, $\alpha = 0.92$, $I_{CBO} = 50$ μA

$\therefore$ Total collector current,

$$I_C = \alpha I_E + I_{CBO}$$
$$= 0.92 \times 1 + 50 \times 10^{-3}$$
$$= 0.92 + 0.05 = 0.97 \text{ mA.}$$

Example 5: *A common base transistor amplifier has an input resistance of 20 Ω and output resistance of 100 kΩ. The collector load is 1 kΩ if a signal of 500 mV is applied between emitter and base, find the voltage amplification. Assume α_{ac} to be nearly one.*

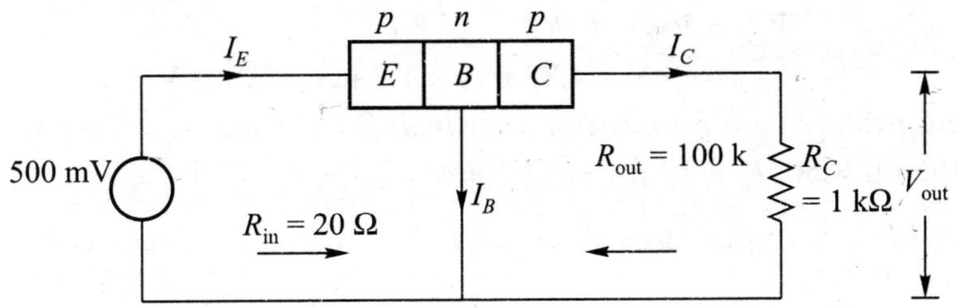

Fig. 8.31

Solution: Fig. 8.31 shows the condition of the problem. Note that output resistance is very high as compared to input resistance. This is not surprising because input junction (base to emitter) of the transistor is forward biased while the output junction (base to collector) is reverse biased.

Input current, $$I_E = \frac{\text{Signal}}{R_{in}} = \frac{500 \text{ mV}}{20\ \Omega}$$

Since α_{ac} is nearly 1, output current, $I_C = I_E = 25$ mA

Output voltage, $$V_{out} = I_C R_C = 25 \text{ mA} \times 1 \text{ k}\Omega = 25 \text{ V}$$

Voltage amplification, $$A_v = \frac{V_{out}}{\text{Signal}} = \frac{25 \text{ V}}{500 \text{ mV}} = 50$$

Example 6: *In a common base connection $\alpha = 0.95$. The voltage drop across 2 kΩ resistance which is connected in the collector is 2 V. Find the base current.*

Solution: Fig. 8.32 shows the required common base connection. The voltage drop across R_C = (= 2 kΩ) is 2 V.

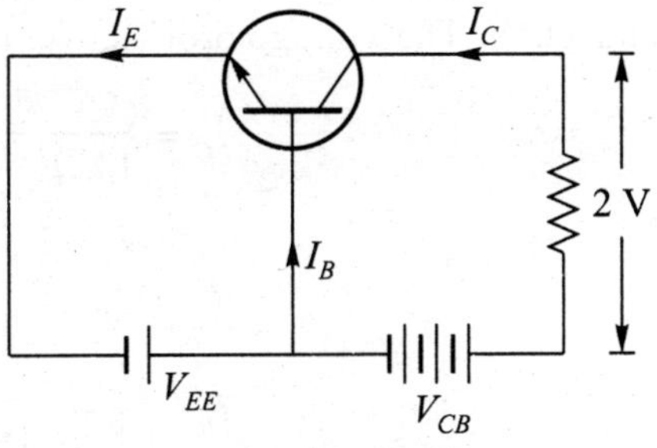

Fig. 8.32

$\therefore$ $$I_C = 2 \text{ V}/2 \text{ k}\Omega = 1 \text{ mA}$$

$$\alpha = I_C/I_E$$

$\therefore$ $$I_E = \frac{I_C}{\alpha} = \frac{1}{0.95} = 1.05 \text{ mA}$$

Using the relation $I_E = I_B + I_C$

$$I_B = I_E - I_C = 1.05 - 1 = 0.05 \text{ mA.}$$

Example 7: *For the common base circuit shown in Fig. 8.33 transistor to be of silicon.*

Solution: Since the transistor is of silicon, V_{BE} = 0.7 V. App. side loop we get

$$V_{EE} = I_E R_E + V_{BE}$$

or $$I_E = \frac{V_{EE} - V_{BE}}{R_E} = \frac{8 \text{ V} - 0.7 \text{ V}}{1.5 \text{ k}\Omega} = 4.87 \text{ mA}$$

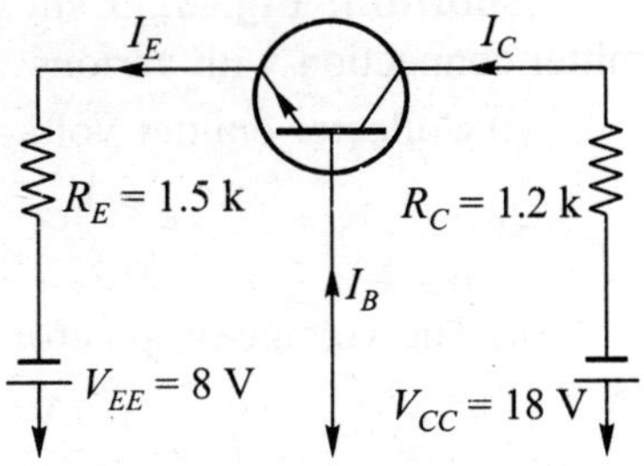

Fig. 8.33

$\therefore$ $$I_C \simeq I_E = 4.87 \text{ mA}$$

Applying Kirchhoff's voltage law to the collector-side loop we have

$$V_{CC} = I_C R_C + V_{CB}$$

$$\therefore \quad V_{CB} = V_{CC} - I_C R_C$$
$$= 18 \text{ V} - 4.87 \text{ mA} \times 1.2 \text{ k}\Omega = 12.16 \text{ V}.$$

Example 8: *Calculate I_E in a transistor for which $\beta = 50$, and $I_B = 20\ \mu A$.*

Solution: Here $\beta = 50$, $I_B = 20\ \mu\text{A} = 0.02$ mA

$$\beta = \frac{I_C}{I_B}$$
$$I_C = \beta I_B$$
$$= 50 \times 0.02 = 1 \text{ mA}$$

Using the relation $I_E = I_B + I_C$

$$= 0.02 + 1 = 1.02 \text{ mA}.$$

Example 9: *Find the value of β if (i) $\alpha = 0.9$, (ii) $\alpha = 0.98$, (iii) $\alpha = 0.99$.*

Solution:
$$\beta = \frac{\alpha}{1-\alpha} = \frac{0.9}{1-0.9} = 9$$
$$\beta = \frac{\alpha}{1-\alpha} = \frac{0.98}{1-0.98} = 49$$
$$\beta = \frac{\alpha}{1-\alpha} = \frac{0.99}{1-0.99} = 99.$$

Example 10: *For a transistor $\beta = 45$ and voltage drop across 1 kΩ which is connected in the collector circuit is 1 Volt. Find the base current for common emitter connection.*

Solution: Fig. 8.34 shows the required common emitter connection. The voltage drop across R_C (= 1 kΩ) is 1 Volt.

$$\therefore \quad I_C = \frac{1 \text{ V}}{1 \text{ k}\Omega} = 1 \text{ mA}$$
$$\beta = \frac{I_C}{I_B}$$
$$I_B = \frac{I_C}{\beta} = \frac{1}{45} = 0.022 \text{ mA}.$$

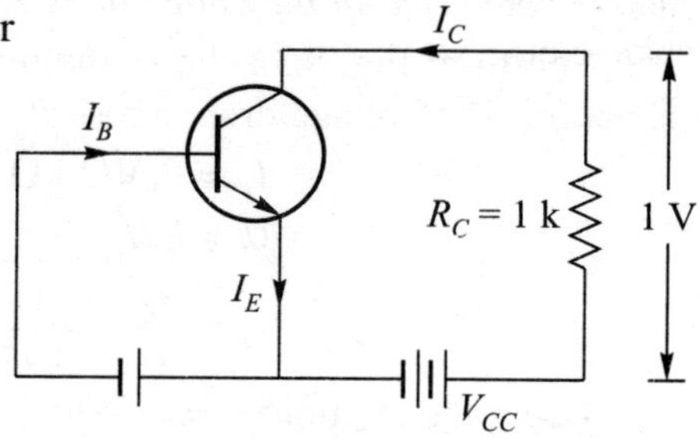

Fig. 8.34

Example 11: *A transistor is connected in common emitter (CE) configuration in which collector supply is 8V and the voltage drop across resistance R_C connected in the collector circuit is 0.5 V. The voltage of R_C= 800 Ω if $\alpha = 0.96$ determine (i) collector emitter voltage, (ii) base current.*

Solution: Fig. 8.35 shows the required common emitter connection with various values,

(*i*) Collector emitter voltage

$$V_{CE} = V_{CC} - 0.5$$
$$= 8 - 0.5 = 7.5 \text{ V}$$

(*ii*) The voltage drop across R_C (= 800 Ω) is 0.5 V

$$\therefore \quad I_C = \frac{0.5 \text{ V}}{800\ \Omega} = \frac{5}{8} \text{ mA} = 0.625 \text{ mA}$$

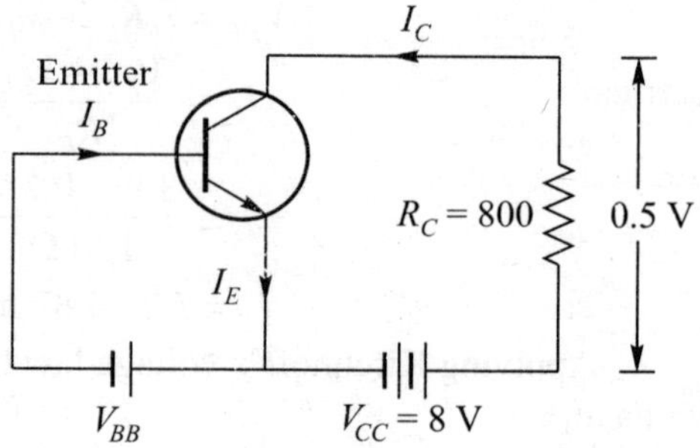

Fig. 8.35

Now $$\beta = \frac{\alpha}{1-\alpha} = \frac{0.96}{1-0.96} = 24$$

$\therefore$ Base Current $$I_B = \frac{I_C}{\beta} = \frac{0.625}{24} = 0.026 \text{ mA.}$$

Example 12: *Find the α rating of the transistor shown in Fig. 8.36. Hence determine the value of I_C using both α and β rating of the transistor.*

Solution: Fig. 8.36 shows the condition of the problem

$$\alpha = \frac{\beta}{1+\beta} = \frac{49}{1+49} = 0.98$$

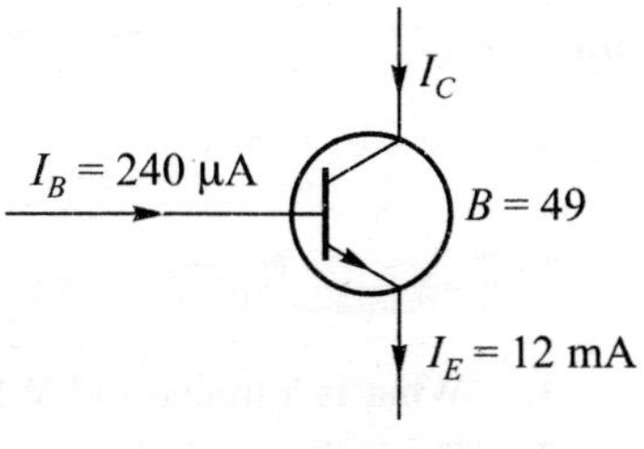

Fig. 8.36

The value of I_C can be found by using either α or β rating as under:

$$I_C = \alpha I_E = 0.98(12 \text{ mA}) = 11.76 \text{ mA}$$
$$I_C = \beta I_B = 49(240\ \mu\text{A}) = 11.76 \text{ mA.}$$

Example 13: *An n-p-n transistor at room temperature has its emitter disconnected. A voltage of 5V is applied between collector and base with collector positive, a current of 0.2 μA flows when the base is disconnected and the same voltage is applied between collector and emitter the current is found to be 20 μA. Find α, I_E and I_B when collector current is 1 mA.*

Solution: When the emitter circuit is open the collector base junction is reverse biased. A small leakage current I_{CBO} flows due to minority carriers.

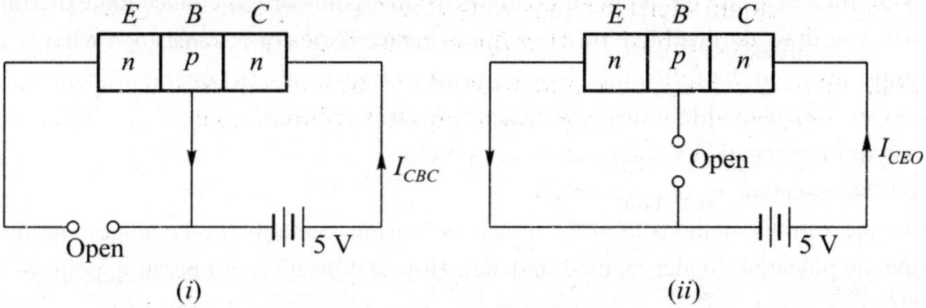

Fig. 8.37

When base is open [See Fig. 8.37 (*ii*)], a small leakage current I_{CEO} flows due to minority carriers.

$\therefore$ $$I_{CEO} = 20\ \mu\text{A}$$

We know $$I_{CEO} = \frac{I_{CBO}}{1-\alpha}$$

or $$20 = \frac{0.2}{1-\alpha}$$

$\therefore$ $\alpha = 0.99$

Now $I_C = \alpha I_E + I_{CBO}$

Hence $I_C = 1 \text{ mA} = 1000 \text{ μA}$

$\alpha = 0.99;$

$I_{CBO} = 0.2 \text{ μA}$

$1000 = 0.99 \times I_E + 0.2$

or $$I_E = \frac{1000 - 0.2}{0.99} = 1010 \text{ μA}$$

and $I_B = I_E - I_C = 1010 - 1000 = 10 \text{ μA}.$

EXERCISES

1. What is a transistor? Why is it so called?
2. Draw the symbol of npn and pnp transistor and specify the leads.
3. Show by means of a diagram how you normally connect external batteries
   (*i*) pnp transistor, (*ii*) npn transistor.
4. Describe the transistor action in detail.
5. Name the three possible transistor connections.
6. Define α. Show that it is always less than unity.
7. Draw the input and output characteristics of *CB* connection. What do you infer from these characteristics?
8. Define β show that $\beta = \frac{\alpha}{1-\alpha}$.
9. How will you determine the input and output characteristics of *CE* connection experimentally?
10. How will you draw d.c load line on the output characteristics of a transistor? What is its importance?
11. Write short notes on the following (*i*) advantages of transistors; (*ii*) operating point; (*iii*) dc load line.
12. What do you understand by transistor biasing? What is its need?
13. What do you understand by stabilisation of operating point?
14. Mention the essentials of biasing circuit.
15. Describe the various methods used for transistor biasing. State their advantages and disadvantages.
16. Describe the potential divider method in detail. How stabilisation of operating point is achieved by this method?
17. Mention the steps that are taken to design the transistor biasing and stabilisation circuits.
18. Write short notes on the following:
    (*i*) Operating point; (*ii*) Stabilisation of operating point.

UNIT–IV

CHAPTER 9

Field Effect Transistor (FET)

9.1 INTRODUCTION

In the bipolar transistor we saw that the output collector current is determined by the amount of current flowing into the base terminal and therefore it is a current operated device. Field Effect Transistors, or simply FET's however, use the voltage that is applied to their input terminal to control the output current, since their operation relies on the electric field (hence the name field effect) generated by the input voltage, in other words, they are a voltage operated devices. Field effect transistors possess very similar properties to those of the bipolar transistor *i.e.* high efficiency, instant operation, robust and cheap, and may be used in most circuit applications that use bipolar junction transistors.

FET's are a semiconductor equivalent of the thermionic valve but have one major advantage over bipolar junction transistors, their input impedance is extremely high and sensitive which means that they can be easily damaged by static electricity. There are two main types of field effect transistor, the Junction-gate FET (JFET) and the Insulated-gate FET (IGFET), which is more commonly known as the standard MOSFET (Metal Oxide Semiconductor Field Effect Transistor).

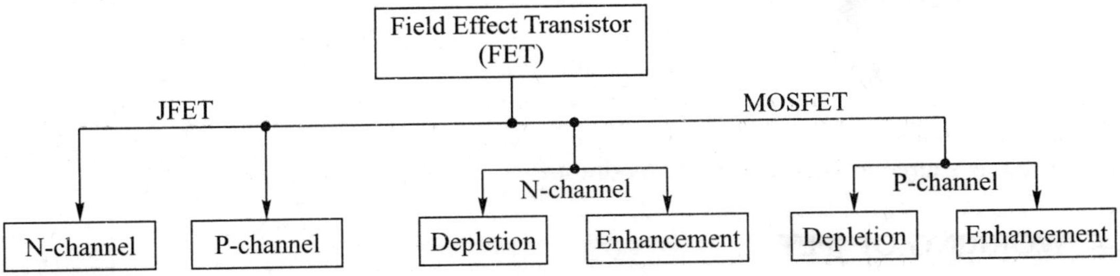

Fig. 9.1

9.2 JUNCTION FIELD EFFECT TRANSISTOR (JFET)

A JFET is a three terminal semiconductor device in which current conduction is by one type of carrier *i.e.* electrons or holes. The JFET was developed about the same time as the transistor but it came into general use only in late 1960's. In a JFET, the current conduction is either by electrons or holes and is controlled by means of electric field between the gate electrode and the conducting channel of the device. The JFET has high input impedance and low noise level.

9.3 CONSTRUCTION OF JFET

We know that a bipolar junction transistor is constructed using two *pn* junction in the main current path between the emitter and the collector. The FET has no junction but instead has a narrow channel of *n*-type or *p*-type silicon with electrical connections at either end called the Drain and the Source respectively. Both *P*-channel and *N*-channel FET's are available. Within the channel there is a third connection which is called the Gate and this is also a *P* or *N*-type material forming a *pn* junction and thereby giving the device its name.

The cross section of a JFET and its corresponding circuit symbols are given below.

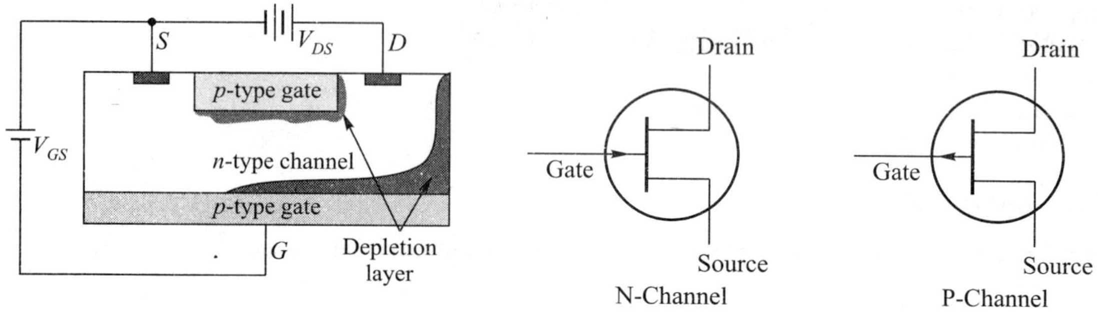

Fig. 9.2

The cross sectional diagram above shows an *N*-type semiconductor channel with a *P*-type region called the gate diffused into the *N*-type channel forming a reverse biased *pn* junction and this junction which forms the depletion layer around the gate area. This depletion layer restricts the current flow through the channel by reducing its effective width and thus increasing the overall resistance of the channel. With a small external voltage (V_{DS}) applied between the drain and the source a current (I_D) will flow through the channel slightly restricted by the small depletion layer. If the negative voltage (V_{GS}) is applied to the gate the depletion layer increases reducing the overall effective area of the channel and thus reducing the current flowing through it, a sort of "squeezing" effect. As the gate voltage (V_G) is made more negative, the width of the channel decreases until no more current flows between the drain and the source and the FET is said to be pinched-off. In the pinch-off region V_{GS} controls the channel current, V_{DS} has little or no effect and the FET acts like a voltage dependant resistor (*VDR*). The drain and source terminals are interchangeable *i.e.* either end can be used as source and the other and as drain.

9.4 WORKING OF JFET

Above figure shows how the *JFET* operates in a zero gate bias condition. Five volts are applied across the JFET so that current flows through the bar from source to drain, as indicated by the arrow through a channel between the depletion layers. The size of these layers determines the width of the channel and hence the current conduction through the bar. The gate terminal is tied to ground. This

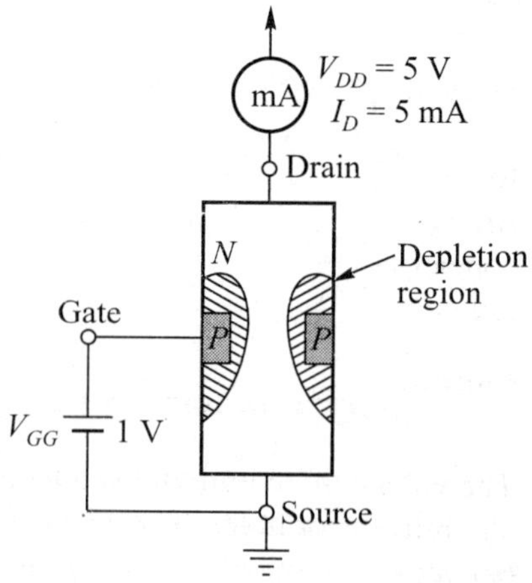

Fig. 9.3. JFET operation with zero gate bias.

is a zero gate bias condition. In this condition, a typical bar represents a resistance of about 500 ohms. A milliammeter, connected in series with the drain lead and dc power, indicates the amount of current flow. With a drain supply (V_{DD}) of 5 volts, the milliammeter gives a drain current (I_D) reading of 10 milliamperes. The voltage and current subscript letters (V_{DD}, I_D) used for an FET correspond to the elements of the FET just as they do for the elements of transistors.

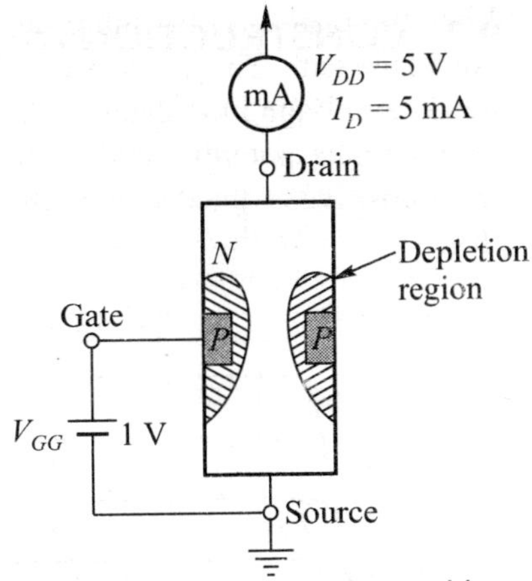

Fig. 9.4. JFET with reverse bias.

In figure shown above, a small reverse-bias voltage is applied to the gate of the JFET. A gate-source voltage (V_{GS}) of negative 1 volt applied to the *P*-type gate material causes the junction between the *P*- and *N*-type material to become reverse biased. Just as it did in the varactor diode, a reverse-bias condition causes a "depletion region" to form around the *PN* junction of the JFET. Because this region has a reduced number of current carriers, the effect of reverse biasing is to reduce the effective cross-sectional area of the "channel". This reduction in area increases the source-to-drain resistance of the device and decreases current flow. On the other hand, if the reverse voltage on the gate is decreased the width of the depletion layers also decreases. This increases the width of the conducting channel and hence source to drain current. It is clear that current from source to drain can be controlled by the application of potential on the gate. For this reason, the device is called as FET. It may be noted that a *p*-channel JFET operates in the same manner as the *n*-channel JFET except that channel current carriers will be holes instead of electrons and the polarities of V_{GS} and V_{DS} are reversed.

Comparison with Other Transistors

The JFET gate presents a small current load which is the reverse leakage of the gate-to-channel junction. The MOSFET has the advantage of extremely low gate current (measured in picoamperes) because of the insulating oxide between the gate and the channel. However, compared to the base current of a bipolar junction transistor the JFET gate current is much lower, and the JFET has higher transconductance than the MOSFET. Therefore JFETs are used to advantage in some low-noise, high input-impedance op-amps and sometimes used in switching applications.

The JFET had been predicted as early as 1925 by Julius Lilienfeld, and the theory of operation of the device was sufficiently well known by the mid 1930s for a patent to be issued for it. However, technology at the time was not sufficiently advanced to produce doped crystals with enough precision for the effect to be seen until many years later. In 1947, researchers John Bardeen, Walter Houser Brattain and William Shockley were attempting to construct a JFET when they discovered the point-contact transistor. The first practical JFETs were thus constructed many years after the first bipolar junction transistors, in spite of it having been conceived much earlier.

Output Characteristics of JFET

The output characteristic voltage-current curves of a typical junction FET are shown below. The voltage V_{GS} applied to the gate controls the current flowing between the drain and source terminals. V_{GS} refers to the voltage applied between the gate and the source while V_{DS} refers to the voltage applied between the drain and the source. Because an JFET is a voltage controlled device, no current flows into the gate then the source current (I_S) flowing out of the device equals the drain current flowing into it and therefore ($I_D = I_S$).

The control element for the JFET comes from depletion of charge carriers from the *n*-channel.

When the Gate is made more negative, it depletes the majority carriers from a larger depletion zone around the gate. This reduces the current flow for a given value of Source-to-Drain voltage. Modulating the Gate Voltage modulates the current flow through the device. The control of the drain current by a negative gate potential makes the EFT useful as a switch and it is essential that the gate voltage is never positive as the channel current will flow to the gate and not the drain resulting in damage to the FET.

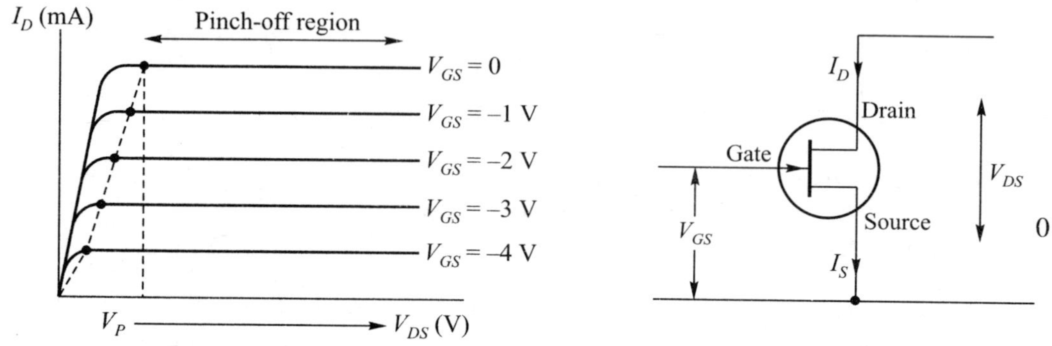

Fig. 9.5

9.5 TRANSFER CHARACTERISTICS OF FET

The transfer characteristic for the JFET is useful for visualizing the gain from the device and identifying the region of linearity. The gain is proportional to the slope of the transfer curve. The current value I_{DSS} represents the value will be part of the data supplied by the manufacturer. The gate voltage at which the current reaches zero is called the "pinch voltage", V_P. Note that the dashed line representing the gain in the linear region of operation strikes the zero current line at about half the pinch voltage.

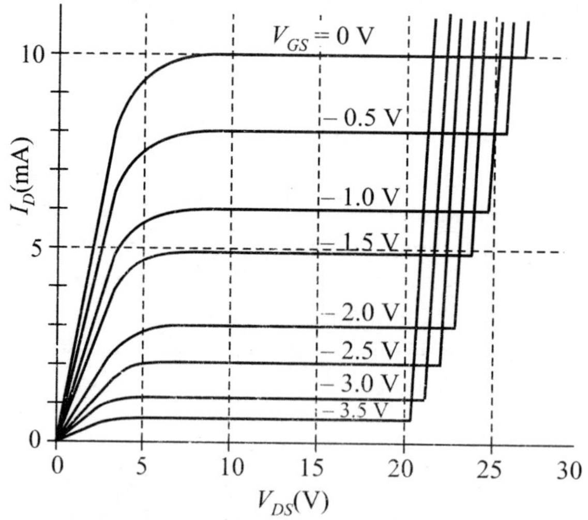

Common source JFET characteristic curves

Fig. 9.6

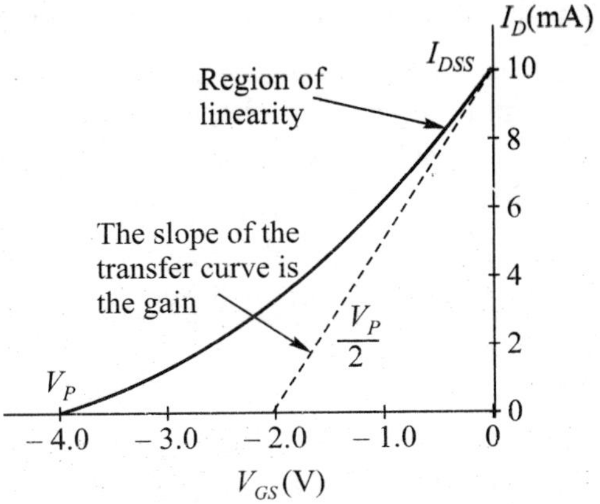

Transfer curve for JFET
(after Diefenderfer & Holton)

Fig. 9.7

9.6 DETERMINATION OF FET PARAMETERS

In the analysis of a JFET circuit, the following important terms are often used:

(1) Short-circuited drain current (I_{DSS}): It is the drain current with source short circuited to gate [*i.e.* $V_{GS} = 0$] and drain voltage (V_{DS}) equal to pinch off voltage. It is sometimes called zero bias current.

(2) Pinch off Voltage (V_P): It is minimum drain-source voltage at which the drain current essentially becomes constant.

(3) Gate-sourced cut-off voltage ($V_{GS(Off)}$): It is the gate-source voltage where the channel is completely cut-off and drain current becomes zero.

We note that gate-source cut-off voltage [$V_{GS(Off)}$] on the transfer characteristic is equal to pinch off voltage V_P on the drain characteristic *i.e.*,

$$V_P = |\, V_{GS(Off)} \,|$$

A rather complex mathematical analysis yields the following expression for drain current:

$$I_D = I_{DSS}\left[1 - \frac{V_{GS}}{V_{GS\,(Off)}}\right]^2$$

where I_D = drain current at given V_{GS}

I_{DSS} = short-gate drain current at V_{GS}

V_{GS} = gate-source voltage

$V_{GS(Off)}$ = gate-source cut-off voltage

The parameters of a JFET are **(1) a.c. drain resistance (r_d)**, **(2) Transconductance (g_m)**, **(3) Amplification factor (μ)**.

(1) A.C. drain resistance (μ): It is the ratio of change in drain-source voltage (ΔV_{DS}) to the change in drain current (ΔI_D) at constant gate-source voltage *i.e.*,

A.C. drain resistance, $r_d = \Delta V_{DS}/\Delta I_D$ at constant V_{GS}

Drain resistance of a JFET has a large value, ranging from 10 kΩ to 1 MΩ.

(2) Transconductance (g_m): It is the ratio of change in drain current (ΔI_D) to the change in gate-source voltage (ΔV_{GS}) at constant drain-source voltage *i.e.*,

Transconductance, $g_m = \Delta I_D / \Delta V_{GS}$ at constant V_{DS}

The transconductance of a JFET is usually expressed either in mA/volt or micromho.

(3) Amplification factor (μ): It is the ratio of change in drain-source voltage (ΔV_{DS}) to the change in gate-source voltage (ΔV_{GS}) at constant drain current *i.e.*

Amplification factor = $\Delta V_{DS} / \Delta V_{GS}$ at constant I_D.

Amplification factor of a JFET indicates how much more control the gate voltage has over drain current than has the drain voltage.

$$\mu = r_d \times g_m$$

9.7 APPLICATION OF FET AS VOLTAGE VARIABLE RESISTOR

In most linear applications of field effect transistors the device is operated in the constant-current portion of it's output characteristics. Now, we consider FET transistor operation in the region before pinch off, where V_{DS} is small. In this region the FET is useful as a voltage controlled resistor *i.e.*, the drain to source resistance is controlled by the bias voltage V_{GS}. In such applications the FET is also referred as a voltage variable resistor [VVR] or voltage dependant resistor [VDR].

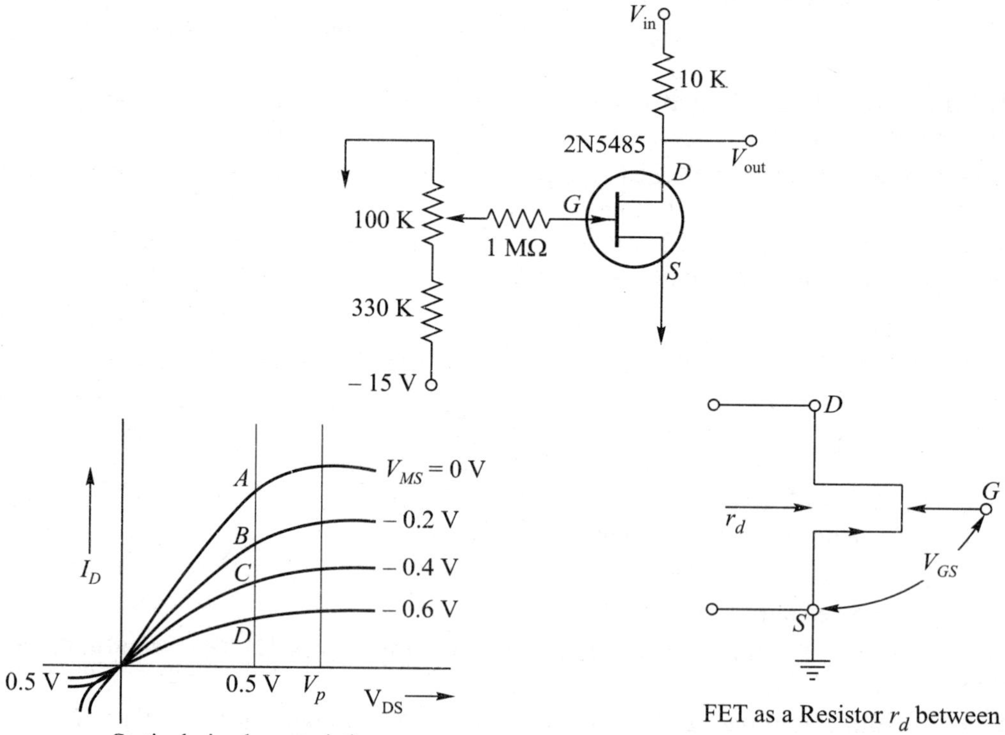

Static drain characteristics

FET as a Resistor r_d between drain and source

Fig. 9.8

In JFET the drain to source conductance $g_m = I_D/V_{DS}$, for all values of V_{DS}, which may expressed as

$$g_m = g_{mo}\left[1 - \left(\frac{V_{GS}}{V_P}\right)^{1/2}\right]$$

where g_{mo} is the value of the drain conductance when the bias voltage V_{GS} is zero.

The variation of the r_d with V_{GS} can be closely approximated by the empirical expression.

$$r_d = r_o/1 - KV_{GS}$$

where r_d = drain resistance at zero gate bias

K = a constant, dependant on FET type

V_{GS} = gate to source voltage.

Thus small signal FET drain resistance r_d varies with applied gate voltage V_{GS} and FET acts like a variable passive resistor. FET finds wide applications where VVR property is useful. For example, the VVR can be used in Automatic Gain Control (AGC) circuit of a multistage amplifier. The FET can be used as a voltage controlled resistor which can be used in audio applications. To ensure that distortion is minimized simple steps are taken to linearise the resistance. Applications include oscillators (where both frequency of operation and the feedback can be controlled), filters, modulators.

9.8 METAL OXIDE SEMICONDUCTOR FIELD EFFECT TRANSISTOR (MOSFET)

Introduction

The Metal Oxide Semiconductor Field Effect Transistor, MOSFET is another type of field effect transistor which ha a metal oxide gate [usually silicon dioxide (glass)], which is electrically insulated from the main semiconductor *N* or *P*-channel. The MOSFET is also called an IGFET (insulated gate field effect transistor) and is the most commonly used Field Effect Transistor. Both *P*-channel and *N*-channel MOSFET's are available in two basic forms, the Enhancement type and the Depletion type. The input impedance of a MOSFET is much more than that of a JFET because of very small gate leakage current. Therefore, all the equations apply equally well to the MOSFET and JFET in amplifier connections.

Construction and Working of Enhancement MOSFET

An *N* channel MOSFET is shown in figure. There are two highly doped [N^+] regions (one source and the other drain) on a highly doped *P*-type (substrate). A layer of insulating silicon dioxide is placed over it. Now over the insulating SiO_2 layer a gate is formed. Three different terminals for source, gate, and drain are taken out as shown in figure. The insulating layer between gate and channels provides a very high input impedance. When a positive voltage is applied at the gate, it induces negative charges in insulating layer and correspondingly negative charges in semiconductor. The negative charge region in semiconductor increases as the positive voltage at the gate increases. This negative charge in semiconductor helps in conducting between source and drain. In this way the drain current is enhanced by positive gate voltage.

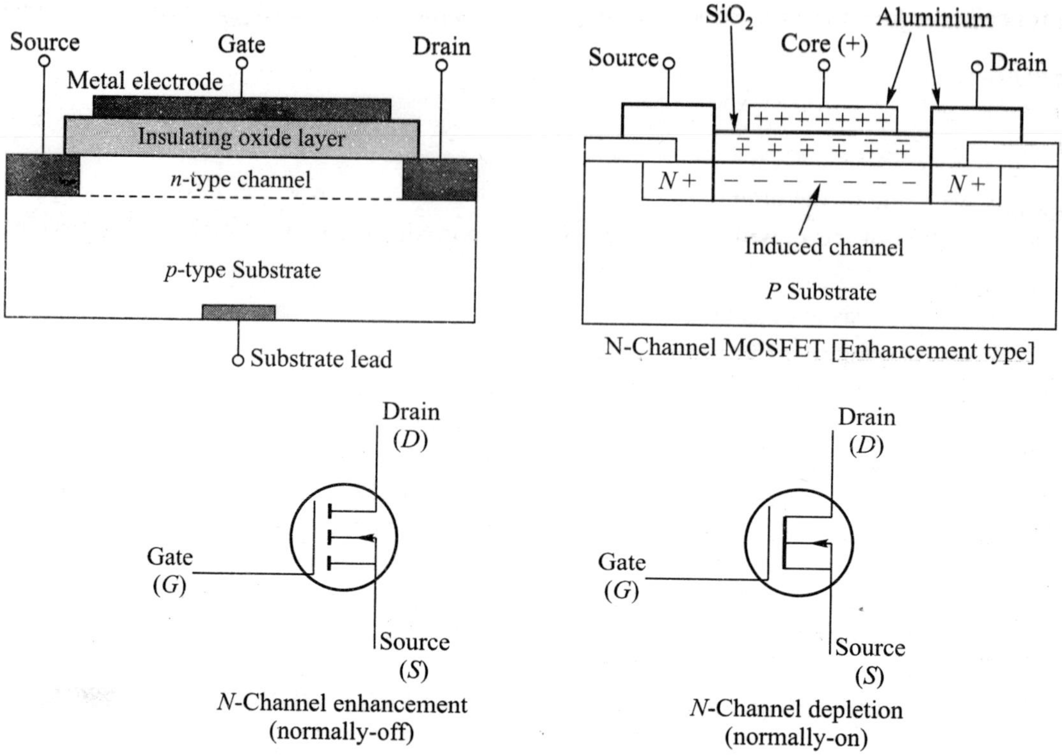

Fig. 9.9

Construction and Working of Depletion MOSFET

An N-channel depletion MOSFET is shown in figure. The difference between depletion MOSFET and enhanced MOSFET is that in depletion MOSFET an N-type layer is also introduced between source and drain. When a negative voltage is applied at the gate, positive charges are induced in the channel. We know that electrons are the majority carriers in N-type material but the positive charges reduce the conductivity of this channel by forming a depletion layer. In this way the drain current can be reduced by applying a negative voltage at the gate. However, depletion of MOSFET can be used in enhancement mode by applying positive grid voltage.

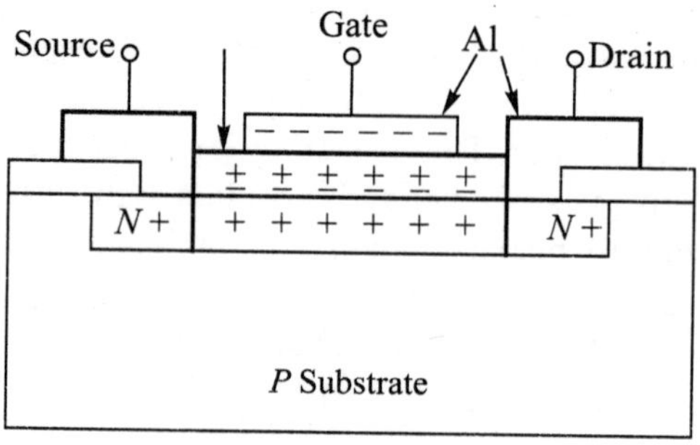

Depletion type MOSFET

Fig. 9.10

9.9 OUTPUT AND TRANSFER CHARACTERISTICS OF MOSFET

(1) Enhancement Type MOSFET: The drain static characteristics and transfer characteristics of *N*-channel enhancement type MOSFET are shown in below figure. It is noted that the drain current for V_{GS} greater than zero is very small and as V_{GS} is made more positive the current increases slowly at the first and then at a much more rapid rate. The gate source voltage at which the channel is formed to let through the flow of drain current of predefined value [say 10 μA] is called gated-source threshold value. The gate source threshold value is denoted by V_{GST} or simply V_T as shown in figure. The current $I_{D(ON)}$ represents the maximum permissible current on the drain characteristics. These values [V_{GST} and $I_{D(ON)}$] are sometimes specified by the manufacturer for particular MOSFET.

(2) Depletion MOSFET: It is obvious from the characteristics of Depletion MOSFET, that for $V_{GS} = 0$, the drain current is not zero but has appreciable value. The figure shows the gate source cutoff voltage $V_{GS(Off)}$ at which the drain current I_D reduces to some specified negligible value at a recommended drain voltage V_{DS}. This voltage $V_{GS(Off)}$ corresponds to a pinch off voltage V_P of JFET. The operation and the characteristics of a *P*-channel MOSFET are similar to *N*-channel MOSFET except that the signs of currents and voltages are reversed.

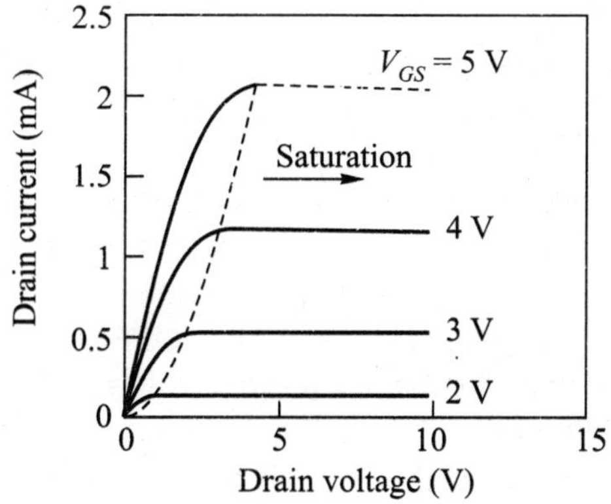

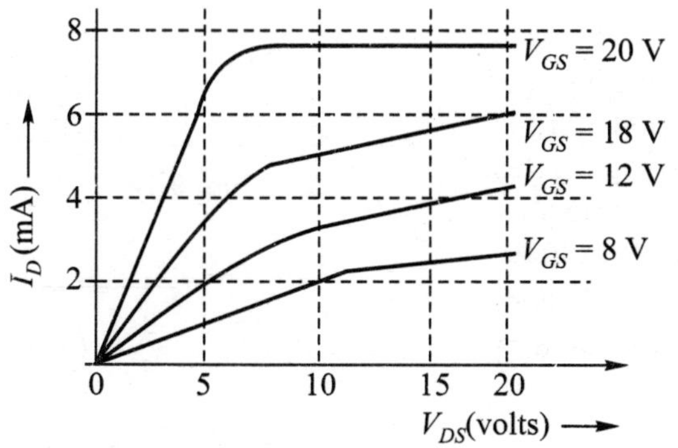

The drain characteristics of *N*-channel enhancement type MOSFET

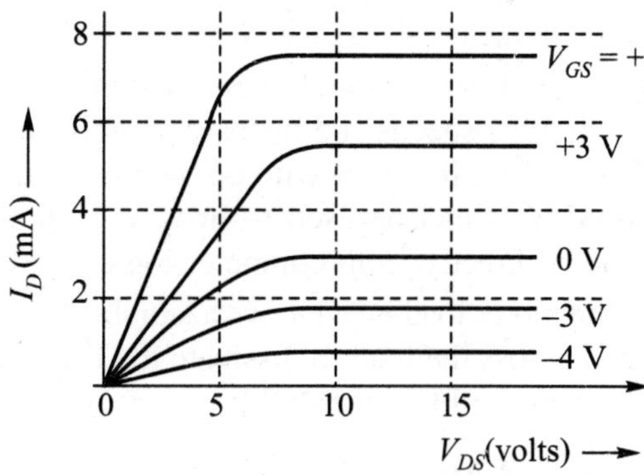

Drain characteristics of depletion MOSFET

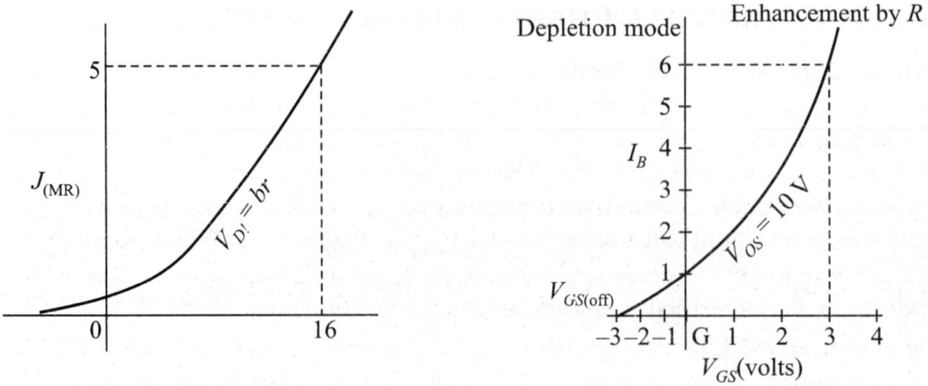

Transfer characteristics of enhancement MOSFET

Transfer characteristics of N-channel dual mode MOSFET

Fig. 9.11

9.10 MOSFET AS A SWITCH

By applying a suitable voltage to the gate of an FET the resistance of the drain to source channel can be varied from an "ON-resistance" of less than 1 Ω, effectively a short circuit, to an "OFF-resistance" of many hundreds of kΩ's, effectively an open circuit. This ability to turn "ON" and "OFF" allows the device to be used as a very efficient switch with switching speeds much faster than standard bipolar junction transistors. An example of a MOSFET switch is given below.

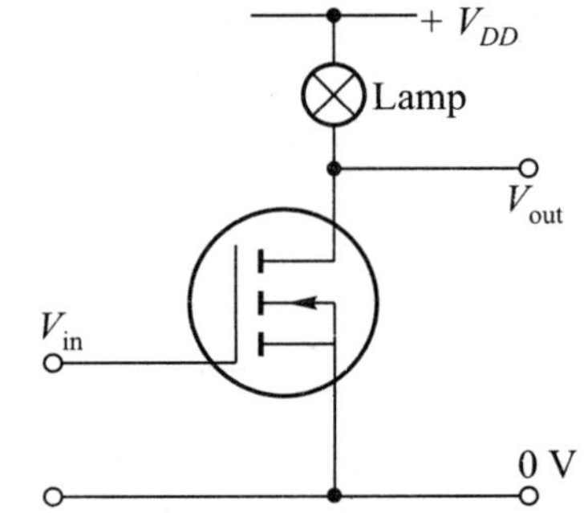

Fig. 9.12

In this arrangement an enhancement mode N-channel MOSFET is being used to switch a simple lamp "ON" and "OFF". The gate input voltage V_{GS} is taken to an appropriate positive voltage level to turn the device and the lamp fully "ON", (V_{GS} = +ve) or a negative voltage level to turn the device fully "OFF" (V_{GS} = –ve).

The power dissipated in the FET depends upon the current flowing through the device and also the ON-resistance of the device.

Because of the very high input gate impedance of FET's, its very fast switching speeds and the ease at which they can be driven makes them ideal to interface with op-amps or standard logic gates. Power MOSFET's can be used to drive d.c. motors or brushless stepper motors directly from computer logic or PWM controllers.

As the motor load is inductive a simple "free-wheeling" diode is connected across the load to dissipate the back emf generated by the motor when the FET turns OFF.

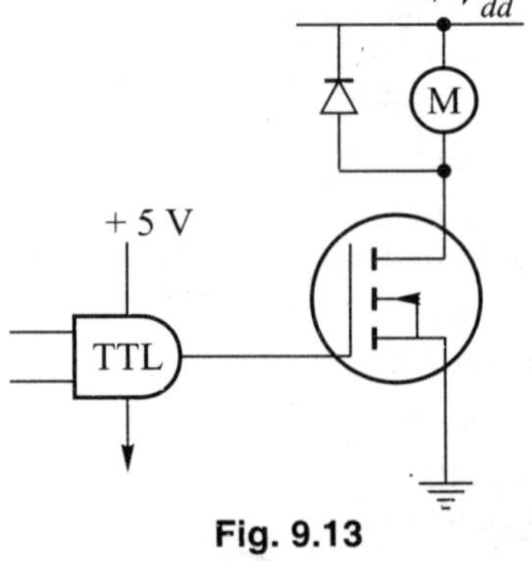

Fig. 9.13

Advantages and Applications of MOSFET

The insulated gate in MOSFETs results in an input resistance several orders of magnitude higher than that of the JFET, typical input resistances being of the order of 10 to 15 ohms. Another advantage of the MOSFETs is that the interelectrode capacitances are independent of the bias voltage and these capacitances are smaller than in the case of JFET. The MOSFETs may be used in any of the circuits covered for JFET, with substrate connected to the source. Because of their input high resistances, the enhancement type MOS devices have been used as micro resistors in integrated microcircuits. It is easier to fabricate MOSFET.

9.11 DIFFERENCES BETWEEN FET AND BJT

Field Effect Transistor (FET)	Bipolar Junction Transistor (BJT)
1. Low voltage gain	High voltage gain
2. High current gain	Low current gain
3. High input impedance	Low input impedance
4. High output impedance	Low output impedance
5. Low noise generation	Medium noise generation
6. Fast switching time	Medium switching time
7. Requires an input to turn it "OFF"	Requires zero input to turn it "OFF"
8. Voltage controlled device	Current controlled device
9. Exhibits the properties of a resistor	
10. More expensive than bipolar	Less expensive
11. Difficult to bias	Easy to bias

Advantages of FET over BJT

1. A JFET has a smaller size, longer life, and high efficiency.
2. A JFET has a very high power gain and this eliminates the necessity of using driver stages.
3. A JFET has a negative temperature coefficient of resistance and this avoids the risk of thermal runaway.
4. The operation of a JFET depends upon the bulk material current carriers that do not cross junctions.
5. JFET has high input impedance over BJT.

Summary of FET

Field effect transistors, FET's are voltage operated devices and can be divided into junction-gate devices called JFET's and insulated-gate devices called IGFET's or more commonly MOSFET's. Insulated gate devices can also be sub-divided into enhancement types and depletion types. All forms are available in both *N*-channel and *P*-channel versions.

FET's have very high input impedances so very little or no current (MOSFET types) flows into the input terminal and they can be used as voltage controlled resistances and electronic switches. They can be used as ideal switches due to their high OFF resistance, low ON resistance and fast switching speed. The input impedance of the MOSFET is even higher than that of the JFET due to the insulating layer and therefore static electricity can easily damage MOSFET's so care needs to be taken when handling these devices.

Field effect transistors can be used to replace normal unipolar junction transistors in electronic circuits and a comparison between FET's and transistors stating both their advantages and their disadvantages is given below.

SOLVED EXAMPLES

Example 1: *Following readings were obtained experimentally for a FET*

V_{GS}	*0 volt*	*0 volt*	*0.5 volt*
V_{DS}	*5 volt*	*12 volt*	*12 volt*
I_D	*8 mA*	*9.2 volt*	*8.6 volt*

Determine the parameters of the FET.

Solution. (*i*) With V_{GS} constant at 0 volt

$$\Delta V_{DS} = 12 - 5 = 7 \text{ volt}$$

$$\Delta I_D = 98.2 - 8 = 10.2 \text{ mA}$$

∴ AC Drain resistance $$r_d = \frac{\Delta V_{DS}}{\Delta I_D} = \frac{5 \text{ volt}}{1.2 \text{ mA}} = 4.166 \text{ k}\Omega$$

(*ii*) With V_{DS} constant at 10 volt

$$\Delta V_{GS} = 0.5 - 0 = 0.5 \text{ volt}$$

$$\Delta I_D = 9.2 - 8.6 = 0.6 \text{ mA}$$

∴ Transconductance $$g_m = \frac{\Delta I_D}{\Delta V_{GS}} = \frac{0.6 \text{ mA}}{0.3 \text{ volt}} = 3 \text{ mA/volt} = 3 \text{ ms}$$

Amplification factor $$\mu = g_m \cdot r_d = (3 \times 10^{-3}) \times (4.166 \times 10^3)\text{A/r}$$

$$\mu = 12.499$$

Example 2: *For an N-channel JFET, I_{DES} = 8.7 mA, V_1 = –3 V, V_{GS} = – 1 volt. Find the values I_D, g_{mo}, I_m.*

Solution: $$I_D = I_{DSS}\left[1 - \frac{V_{GS}}{V_P}\right]^2 = 8.7\left[1 - \left(\frac{-1}{-3}\right)\right]^2 = 3.866 \text{ mA}$$

$$g_{mo} = -\frac{2I_{DSS}}{2I_{DSS}} = -\frac{2 \times 8.7}{-3} = 5.8 \text{ ms}$$

$$g_m = g_{mo}\left[1 - \frac{V_{GS}}{V_P}\right] = 5.8\left[1 - \left(\frac{-1}{-3}\right)\right] = 5.8 \times \frac{2}{3} = 3.866 \text{ mA.}$$

Example 3: *An N-channel JFET has a pinch-off voltage of – 4.5 V and I_{DSS} = 9 mA. At what value of V_{GS} in the pinch off region will I_D equal to 3 mA?*

Solution: From Shockley equation

$$I_D = I_{DSS}\left[1 - \frac{V_{GS}}{V_P}\right]^2$$

$$3 = 9\left[1 - \frac{V_{GS}}{-4.5}\right]^2$$

$$\frac{3}{9} = \left[1 + \frac{V_{GS}}{4.5}\right]^2$$

$$1 + \frac{V_{GS}}{4.5} = \frac{1.732}{3} = 0.577$$

$$\frac{V_{GS}}{4.5} = 0.577 - 1 = 0.423$$

$$V_{GS} = -4.5 \times 0.423 = -1.9 \text{ volt}$$

Example 4: *In a FET when value of gate voltage is changed from (– 3.0 volt) to (– 2.9 volt), the drain current increases for 1 mA to 1.2 mA. Find mutual conductance of transistor.*

Solution: Mutual conductance $(g_m) = \frac{\Delta I_D}{\Delta V_{GS}} = \frac{(1.2 - 1) \text{ mA}}{-2.9 - (-3) \text{ volt}} = \frac{0.2 \text{ mA}}{0.1 \text{ volt}} = 2$ ms.

Example 5: *An N-channel FET has $a = 4 \times 10^{-4}$ cm and $N_D = 0.8 \times 10^{15}$ electron/cm^3. Find (a) the pinch off voltage V_P and (b) the channel half width for $V_{GS} = \frac{1}{3} V_P$ and $I_D = 0$.*

Given that the relative dielectric constant of silicon is 12

$$q = 1.6 \times 10^{-19} \text{ coulomb}$$

and $$\epsilon_0 = 1/(36\pi \times 10^9) F/m$$

Solution: (*i*) Dielectric constant of silicon is

$$E = 12 \epsilon_0 = \frac{12}{36\pi \times 10^9}$$

$$N_D = 0.8 \times 10^{15} \text{ electrons/Cm}^3 = 0.8 \times 10^{21} \text{ electrons/m}^3.$$

$$q = 1.6 \times 10^{-19} \text{ coulomb}$$

$$a = 4 \times 10^{-4} \text{ cm} = 4 \times 10^{-6} \text{ meter}$$

Hence pinch off voltage is

$$V_P = \frac{qN_d}{2\epsilon} a^2 = \frac{1.6 \times 10^{-9} \times 0.8 \times 10^{21} \times (4 \times 10^{-6})^2 \times 36\pi \times 10^9}{2 \times 12} \text{ volt} = 9.65 \text{ V}$$

(*ii*) $$V_{GS} = \frac{1}{3} V_P$$

$$V_{GS} = \left[1 - \frac{b}{a}\right]^2 V_P$$

$$b = a\left[1 - \sqrt{\frac{V_{GS}}{V_P}}\right] = 4 \times 10^{-6}\left[1 - \sqrt{\frac{1}{3}}\right] \text{ meter}$$

$$b = 1.704 \times 10^{-4} \text{ cm}$$

Example 6: *An N-channel silicon FET has $a = 4 \times 10^{-4}$ cm and channel resistivity of 10 Ω – cm. Compute the pinch off voltage V_P, for $I_{DSS} = 10$ μA compute g_{mo} and for $V_{GS} = -4$ volts, compute I_{DS} and g_m.*

Solution: (1) $$\sigma = \frac{1}{P} = n\mu_n q = N_D \mu_n q$$

$$qN_D = \frac{1}{p\mu_n} = \frac{1}{10 \times 1300} = 7.7 \times 10^{-5}$$

Hence $$|V_P| = \frac{qN_D a^2}{2\epsilon} = \frac{7.7 \times 10^{-5} \times 36\pi \times 10^{11} \times (4 \times 10^{-4})^2}{2 \times 12} = 5.8 \text{ volts}$$

$$V_P = 5.8 \text{ volts}$$

(2) $V_{GS} = -4$ volts.

Hence $$I_{DS} = I_{DSS}\left[1 - \frac{V_{GS}}{V_P}\right]^2 = 100 \times 10^{-6}\left[1 - \frac{4}{5.8}\right]^2 A = 9.6\ \mu A$$

(3) $$g_{mo} = \frac{-2I_{DSS}}{V_P}\left[\frac{-2\times 100\times 10^{-6}}{5.8}\right]$$

$$g_{mo} = \frac{-2I_{DSS}}{V_P}\left[1 - \frac{V_{GS}}{V_P}\right] = \frac{-2\times 100\times 10^{-6}}{5.8} = 34.5 \times 10^{-6}\ S$$

(4) $$g_m = -\frac{-2I_{GS}}{V_P}\left[1 - \frac{V_{GS}}{V_P}\right] = -\frac{2\times 100\times 10^{-6}}{5.8}\left[1 - \frac{4}{5.8}\right] S = 10.7\ \mu S$$

Example 7: *A JFET has following parameters: I_{DSS} = 32 mA; $V_{GS(Off)}$ = – 8 V, V_{DGS} = – 4.5 V. Find the value of drain current.*

Solution: $$I_D = I_{DSS}\left[1 - \frac{V_{GS}}{V_{GS(Off)}}\right]^2 = 32\left[1 - \left(\frac{-45}{-8}\right)\right]^2 = 6.12\ mA$$

Example 8: *A JFET has a drain current of 5 mA. If I_{DSS} = 10 mA and $V_{GS(Off)}$ = – 6 V, find the value of V_{GS}, V_P.*

Solution: $$I_D = I_{DSS}\left[1 - \frac{V_{GS}}{V_{GS(Off)}}\right]^2$$

$$5 = 10\left[1 + \frac{V_{GS}}{6}\right]^2$$

$$1 + \frac{V_{GS}}{6} = \sqrt{\frac{5}{10}} = 0.707$$

$$V_{GS} = -1.76\ V$$

$$V_P = -V_{GS(Off)} = 6\ V.$$

EXERCISES

1. Sketch the transfer curve for a P-channel JFET with I_{DSS} = 4 mA, and V_P = 3 V.
2. For a JFET, I_{DSS} = 9 mA, V_P = – 3.5 V. Determine I_D when V_{GS} = 0, V_{GS} = – 2V. [**Ans.** 9 mA, 1.65 mA]
3. A 1 V increase in gate voltage changes the drain current 10 mA in a FET. Calculate g_m value. [**Ans.** 10 ms]
4. When the reverse gate voltage of 15 V is applied to a FET, the gate current is 10^{-3} μA. Find the resistance between gate and source. [**Ans.** 15000 Ω]
5. A JFET circuit has the following values: V_{GS} = – 2V, V_{DS} = 10 V, I_D = 6 μA. Calculate r_d, μ, g_m of given JFET. [**Ans.** 1.66 mΩ, 3 ms, 5]
6. A gate current of 10^{-3} μA is obtained with a reverse gate voltage of 10 V in a FET. Determine gate to source resistance. $\left[\textbf{Ans.}\ 10000\ m\Omega = \frac{V_{GS}}{I_G} = R_{GS}\right]$

QUESTIONS

A. Short Answer Questions:

1. Describe the operation of FET and draw it's output characteristics.
2. Compare JFET and BJT.
3. Give the basic structure of *P*-channel JFET.
4. Discuss the advantages and disadvantages of JFET over BJT.
5. Explain the characteristics of MOSFET.
6. State the structural differences of JFET and MOSFET.
7. Define the pinch off voltage of JFET.
8. Explain the construction and working of enhancement type of MOSFET.
9. Discuss the FET parameters.
10. Draw the circuit symbol of JFET and MOSFET.
11. Explain, how FET works as a voltage variable resistor?
12. Describe the function of MOSFET as switch.
13. Discuss the transfer characteristics of a MOSFET.

B. Long Answer Questions:

1. Explain the construction, working and *V-I* characteristics of JFET.
2. Explain the characteristics of MOSFET. Write applications of FET and MOSFET and their advantages over BJT.
3. What are MOSFETs? Describe depletion MOSFET. Compare them with JFET.
4. Describe the construction and working enhancement MOSFET. Explain it's static and transfer characteristics of it. Explain it's function as a switch.
5. Explain the output characteristics and transfer characteristics of JFET. Explain the action of FET as VVR.
6. Describe the construction and working of enhancement type and depletion MOSFET.
7. Explain the derivation of FET parameters.

CHAPTER 10

Unijunction Transistor (UJT)

10.1 INTRODUCTION

The UNIJUNCTION TRANSISTOR (UJT), originally called a double-based diode, is a three-terminal, solid-state device that has several advantages over conventional transistors. It is very stable over a wide range of temperatures and allows a reduction of components when used in place of conventional transistors. The UJT circuit has fewer components. Reducing the number of components reduces the cost, size and probability of failure. This device has a unique characteristic that when it is triggered, the emitter current increases regeneratively until it is limited by emitter power supply. Due to this characteristic, the UJT can be employed in a variety of application like switching, pulse generator, saw-tooth generator etc.

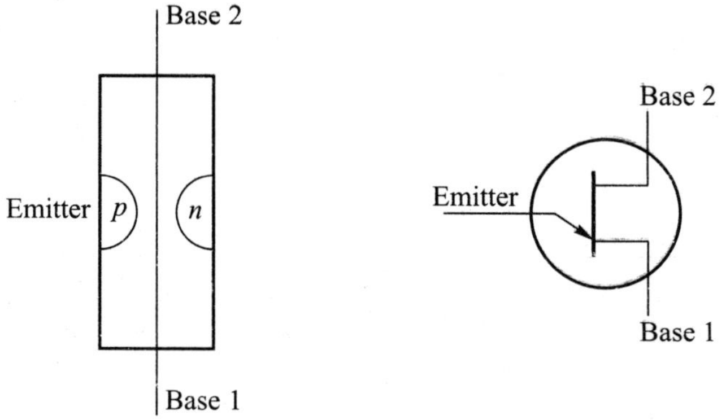

Fig. 10.1

10.2 CONSTRUCTION

It is essentially a bar of *N* type semiconductor material into which *P* type material has been diffused somewhere along its length. Contacts are then made to the device as shown; these are referred to as the emitter, base 1 and base 2 respectively. Since the device has one *p-n* junction and three leads, it is commonly called a unijunction transistor. The emitter is heavily doped having many holes. The *N* region lightly doped. This is why, the resistance between the base terminals is very high [5 to 10 kilo ohms] when emitter is open.

There are two types of unijunction transistor. The original unijunction transistor, or UJT, is a simple device that is essentially a bar of *N* type semiconductor material into which *P* type material has been diffused somewhere along its length. The 2N2646 is the most commonly used version of the UJT. The programmable unijunction transistor, or PUT, is a close cousin to the thyristor. Like the thyristor it consists of four P-N layers and has an anode and a cathode connected to the first and the last layer, and a gate connected to one of the inner layers. They are not directly interchangeable with conversional UJTs but perform a similar function. The 2N6027 is an example of such a device.

It differs from a FET that its emitter junction is much smaller than the gate surface of a FET. UJT is operated with emitter junction forward biased while FET is usually operated with the gate junction as reverse biased. It has no ability to amplify while it has ability to control a large a.c. power with a small signal. It exhibits a negative resistance characteristic and it hence it can be used as an oscillator.

The equivalent circuit shown in Fig. 10.2 has been developed to explain how the device works, and it is necessary to define the terms used in this explanation.

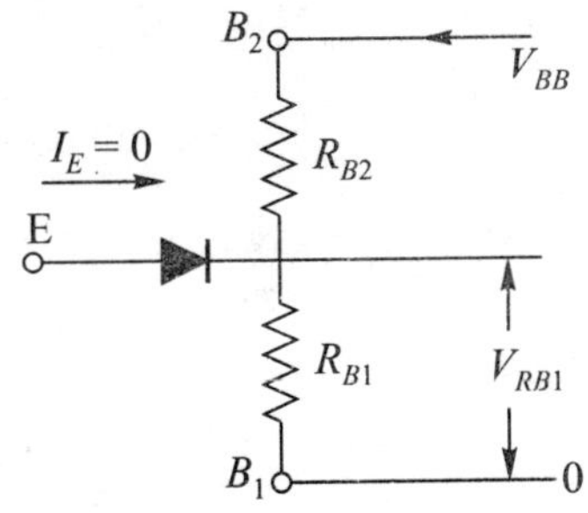

Fig. 10.2

R_{BB} is known as the <u>interbase resistance</u>, and is the sum of R_{B1} and R_{B2}.

$$R_{BB} = R_{B1} + R_{B2} \quad ...(1)$$

Note: This is only true when the emitter is open circuit.

V_{RB1} is the voltage developed across R_{B1}; this is given by the voltage divider rule:

$$V_{RB1} = \frac{R_{B1}}{R_{B1} + R_{B2}} \times V_{BB} \quad ...(2)$$

Since the denominator of equation 2 is equal to equation 1, the former can be rewritten as:

$$V_{RB1} = \frac{R_{B1}}{R_{BB}} \times V_{BB} \quad ...(3)$$

The ratio R_{B1}/R_{BB} is referred to as the <u>intrinsic standoff ratio</u> and is denoted by η (the Greek letter eta).

10.3 WORKING OF UJT

If an external voltage V_E is connected to the emitter, the equivalent circuit can be redrawn as shown in Fig. 10.3. If V_E less than V_{RB1}, the diode is reverse biased and the circuit behaves as though the emitter was open circuit. If however V_E is increased so that it exceeds V_{RB1} by at least 0.7 V, the diode becomes forward biased and emitter current I_E flows into the base 1 region. Because of this, the value of R_{B1} decreases. It has been suggested that this is due to the presence of additional charge carriers (holes) in the bar. Further increase in V_E causes the emitter current to increase which in turn reduces R_{B1} and this causes a further increase

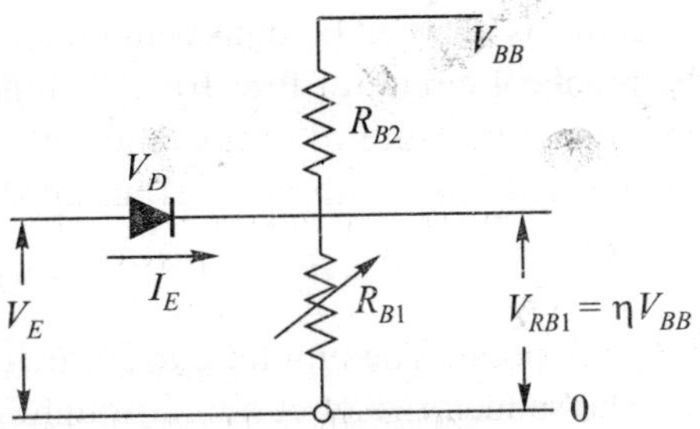

Fig. 10.3

in current. This runaway effect is termed regeneration. The value of emitter voltage at which this occurs is known as the peak voltage V_P and is given by

$$V_P = \eta V_{BB} + V_D \qquad ...(4)$$

V_D = forward voltage drop across silicon diode ($\simeq$ 0.7 V)

10.4 CHARACTERISTICS OF UJT

The characteristics of the UJT are illustrated by the graph of emitter voltage against emitter current (Fig. 10.4). As the emitter voltage is increased, the current is very small - just a few micro amps. When the peak point is reached, the current rises rapidly, until at the valley point the device runs into saturation. At this point R_{B1} is at its lowest value, which is known as the saturation resistance. At valley point, the current is increased to such an extent that no further increase in conductivity can take place. Beyond valley point, the device behaves as a conventional forward biased P-N junction diode. Between the peak point and valley point on the characteristics, the UJT exhibits a dynamic negative resistance.

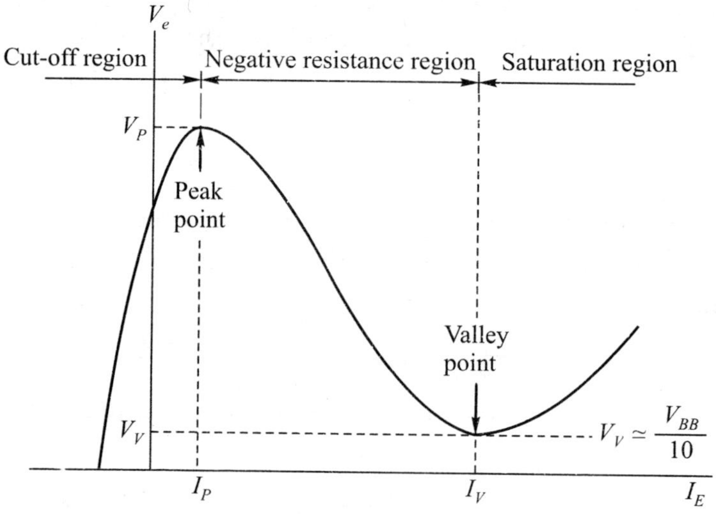

Fig. 10.4

10.5 APPLICATION OF UJT AS A RELAXATION OSCILLATOR

The simplest application of a UJT is as a relaxation oscillator, which is defined as one in which a capacitor is charged gradually and then discharged rapidly. The basic circuit is shown in Fig. 10.5; in the practical circuit of Fig. 10.6, R_3 limits the emitter current and provides a voltage pulse, while R_2 provides a measure of temperature compensation. Fig. 10.7 shows the waveforms occurring at the emitter and base 1; the first is an approximation to a saw-tooth and the second is a pulse of short duration.

The operation of the circuit is as follows: C charges through R_1 until the voltage across it reaches the peak point. The emitter current then rises rapidly, discharging C_1 through the base 1 region and R_3. The sudden rise of current through R_3 produces the voltage pulse. When the current falls to I_V the UJT switches off and the cycle is repeated. It can be shown that the time t between successive pulses is given by:

$$t = 2.3R_1C \log_{10}\frac{1}{1-\eta} \quad \text{where } \eta = 1 - e^{-t/R_1C}$$

Frequency of saw-tooth wave, $f = 1/t$ Hz, [t in seconds]

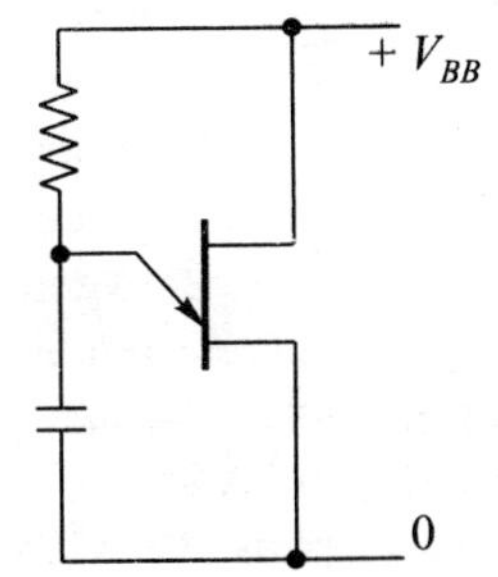

Fig. 10.5

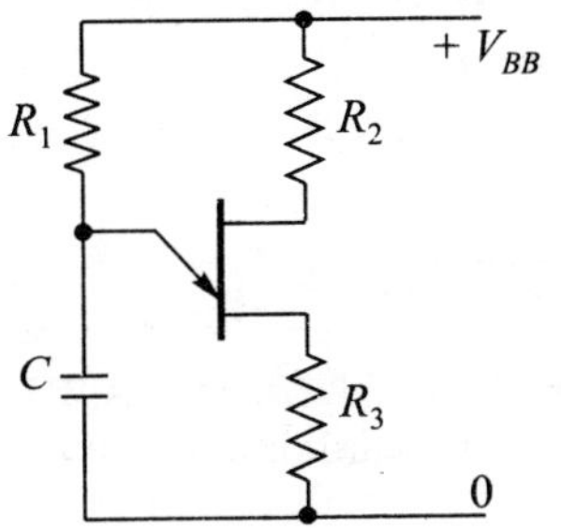

Fig. 10.6

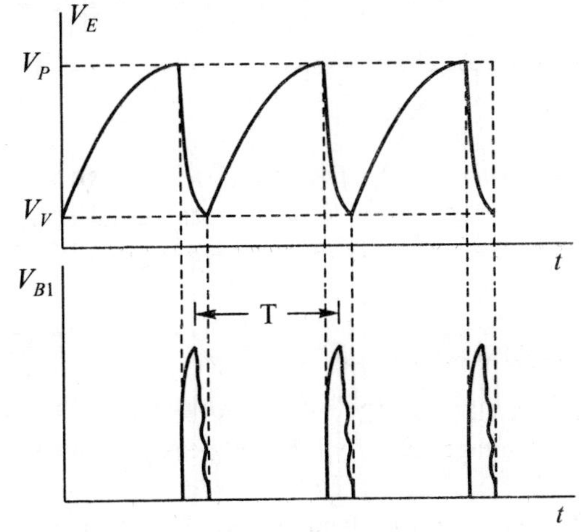

Fig. 10.7

10.6 ADVANTAGES OF UJT

The UJT has the following advantages:

1. It is a low cost device.
2. It has excellent characteristics.
3. It is a low-power absorbing device under normal operating conditions.

SOLVED EXAMPLES

Example 1: *The intrinsic stand-off ratio for a UJT is determined to be 0.6. If the interbase resistance is 10 kΩ, what are the values of R_{B1} and R_{B2}?*

Solution: $R_{BB} = 10$ kΩ, $\eta = 0.6$

$$R_{BB} = R_{B1} + R_{B2}$$

$$10 = R_{B1} + R_{B2}$$

$$\eta = \frac{R_{B1}}{R_{B1} + R_{B2}} \qquad 0.6 = \frac{R_{B1}}{10}$$

$$R_{B1} = 10 \times 0.6 = 6 \text{ k}\Omega$$

$$R_{B2} = 10 - 6 = 4 \text{ k}\Omega.$$

Example 2: *A unijunction transistor has 10 V between the bases. If the intrinsic stand off ratio is 0.65, find the value of stand off voltage. What will be the peak-point voltage if the forward voltage drop in the p-n junction is 0.7 V?*

Solution: $V_{BB} = 10$ V, $\eta = 0.65$, $V_D = 0.7$ V

Stand off ratio $= \eta V_{BB} = 0.65 \times 10 = 6.5$ V

Peak-point voltage, $V_P = \eta V_{BB} + V_D = 6.5 + 0.7 = 7.2$ V.

Example 3: *A Silicon UJT has an interbase resistance of 20 kΩ. Its η =10. Find the values of R_{B1}, R_{B2}.*

Solution:

$$R_{BB} = 20 \text{ k}\Omega$$

$$R_{B1} + R_{B2} = 20 = R_{BB}$$

$$\eta = \frac{R_{B1}}{R_{B1} + R_{B2}}$$

$$0.6 = \frac{R_{B1}}{20}$$

$$R_{B1} = 20 \times 0.6 = 12 \text{ k}\Omega$$

$$R_{B2} = 20 - 12 \text{ k}\Omega = 8 \text{ k}\Omega.$$

EXERCISES

1. A UJT has 12 V between the bases. If the intrinsic stand off ratio is 2/3, find the value of stand-off voltage. What will be the peak point voltage if the forward drop in the *pn* junction is 0.7 V? [**Ans.** 8V, 8.7 V]
2. The intrinsic stand off ratio for a UJT is 0.75 and V_{BB} = 12 V. If the forward drop in the *pn*-junction is 0.7 V, find the peak point voltage. [**Ans.** 9.7 V]
3. In a UJT, η = 0.8, V_P = 10.3 V, R_{B2} = 5 kΩ, determine R_{B1}, V_{BB}. [**Ans.** 20 kΩ, 12 V]
4. The intrinsic stand off ratio for a UJT is determined to be 0.6. If the inter-base resistance is 5 kΩ, what are the values of R_{B1}, R_{B2}? [**Ans.** R_{B1} = 3 kΩ, R_{B2} = 2 kΩ]
5. The intrinsic stand off ratio for a UJT is 0.62. If its interbase resistance is 7 kΩ, what are the values of R_{B1}, R_{B2}? [**Ans.** 4.3 kΩ, 2.6 kΩ]

QUESTIONS

A. Short Answer Questions:

1. What is UJT? Why it is so called?
2. What are the applications UJT? Explain about relaxation oscillator.
3. Draw the relaxation oscillator UJT and explain its operation.
4. Describe the characteristics of UJT.
5. Draw the equivalent circuit of a UJT and explain.
6. Define the terms 'interbase resistance' and 'intrinsic standoff ratio'.

B. Long Answer Questions:

1. Describe the construction and working of UJT. Define the term intrinsic stand off ratio. Draw the static emitter characteristics of UJT.
2. What are the main applications of UJT?
3. Describe the *V-I* characteristics of UJT.
4. Write a short note on UJT Relaxation Oscillator.
5. Describe the equivalent circuit of UJT.
6. Explain the *V-I* characteristics of UJT.
7. Explain the negative resistance characteristics of an UJT.
8. What is a UJT? Describe the *V-I* characteristics of UJT.
9. Write a short note on (*i*) UJT, (*ii*) Relaxation Oscillator.
10. Describe the construction and working of UJT.

CHAPTER 11

Silicon Controlled Rectifier (SCR)

11.1 INTRODUCTION

Silicon-controlled rectifier (or semiconductor-controlled rectifier) is a four-layer solid state device that controls current. The name "silicon controlled rectifier" or SCR is General Electric's trade name for a type of thyristor. The SCR was developed by a team of power engineers led by Gordon Hall and commercialized by Frank W. "Bill" Gutzwiller in 1957.

The SCR has appeared in the market under different names such as thyristor, thyrode and transistor. It is a unidirectional power switch and is being extensively used in switching d.c. and a.c., rectifying a.c. to give controlled d.c. output, converting d.c. to a.c., etc. Thus SCR combines the features of a rectifier and a transistor. In this chapter we shall examine the different characteristics of SCR and it's applications.

11.2 CONSTRUCTION

The Silicon-Controlled Rectifier (SCR)

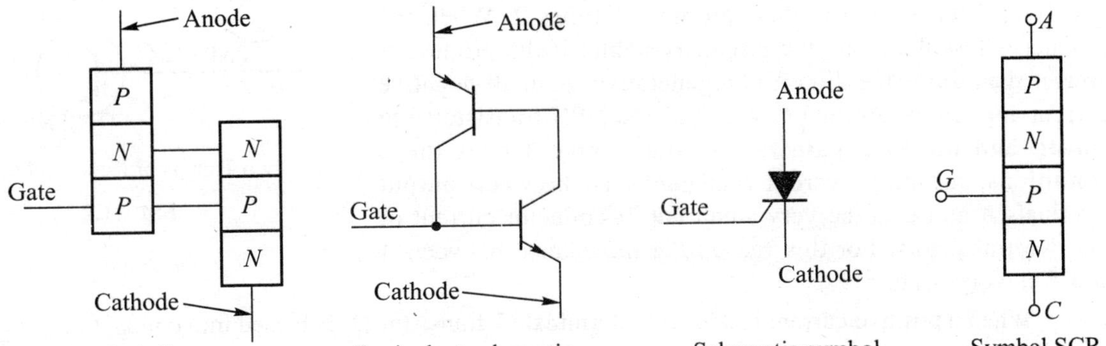

Physical diagram   Equivalent schematic   Schematic symbol   Symbol SCR

Fig. 11.1

Consider the SCR as a transistor pair, one PNP and the other NPN, connected as shown figure. It can also be treated as the essential combination of an ordinary rectifier (*p-n*) and a junction transistor (*p-n-p*) to form a pnpn device. The anode is attached to the upper *P*-layer; the cathode, *C*, is part of the lower *N*-layer; and the gate terminal, *G*, goes to the *P*-layer of the NPN triode. When a *p-n* junction is added to a junction transistor, the resulting three *p-n* junction device is called a silicon controlled rectifier. In the normal operating conditions of SCR, anode is held at high positive potential w.r.t. cathode and gate at small positive potential w.r.t. cathode. The SCR is a solid state equivalent of thyatron. The gate, anode and cathode of SCR correspond to the grid, plate and cathode of thyatron. Hence SCR sometimes can also be called as Thyristor.

11.3 WORKING

In the normal "off" state, the device restricts current to the leakage current. When the gate to cathode voltage exceeds a certain threshold, the device turns "on" and conducts current. The device will remain in the "on" state even after gate current is removed so long as current through the device remains above the holding current. Once current falls below the holding current for an appropriate period of time, the device will switch "off".

If the applied voltage increases rapidly enough, capacitive coupling may induce enough charge into the gate to trigger the device into the "on" state; this is referred to as "dv/dt triggering". This is usually prevented by limiting the rate of voltage rise across the device, perhaps by using a snubber. "dv/dt triggering" may not switch the SCR into full conduction rapidly and the partially-triggered SCR may dissipate more power than is usual, possibly harming the device.

SCRs can also be triggered by increasing the forward voltage beyond their rated breakdown voltage (also called as break over voltage), but again, this does not rapidly switch the entire device into conduction and so may be harmful so this mode of operation is also usually avoided. Also, the actual breakdown voltage may be substantially higher than the rated breakdown voltage, so the exact trigger point will vary from device to device.

11.4 TWO TRANSISTOR REPRESENTATION

In operation the collector of Q_2 drives the base of Q_1, while the collector of Q_1 feeds back to the base of Q_2. (β) 1 is the current gain of Q_1, and (β) 2 is the current gain of Q_2. The gain of this positive feedback loop is their product, 1 times 2. When the product is less than one, the circuit is stable; if the product is greater than unity, the circuit is regenerative. A small negative current applied to terminal *G* will bias the NPN transistor into cutoff, and the loop gain is less than unity. Under these conditions, the only current that can exist between output terminals *A* and *C* is the very small cutoff collector current of the two transistors. For this reason the impedance between *A* and *C* is very high.

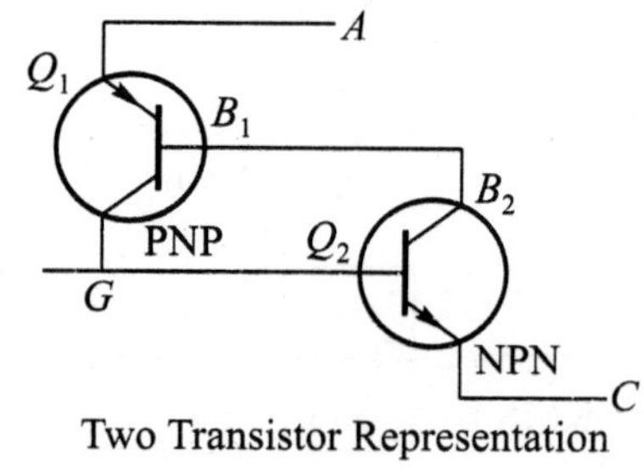

Two Transistor Representation

Fig. 11.2

When a positive current is applied to terminal *G*, transistor Q_2 is biased into conduction, causing its collector current to rise. Since the current gain of Q_2 increases with increased collector current, a point (called the break over point) is reached where the loop gain equals unity and the circuit becomes regenerative. At this point, collector current of the two transistors rapidly increases to a value limited only by the external circuit. Both transistors are driven into saturation, and the impedance between *A* and *C* is very low. The positive current applied to terminal *G*, which served to trigger the self-regenerative action, is no longer required since the collector of PNP transistor Q_1 now supplies more than enough current to drive Q_2. The circuit will remain on until it is turned off by a reduction in the collector current to a value below that necessary to maintain Conduction.

11.5 CHARACTERISTICS OF SCR

The characteristic curve for the SCR is shown in figure shown below. With no gate current, the leakage current remains very small as the forward voltage from cathode to anode is increased until the breakdown point is reached. Here the center junction breaks down, the SCR begins to conduct heavily, and the drop across the SCR becomes very low.

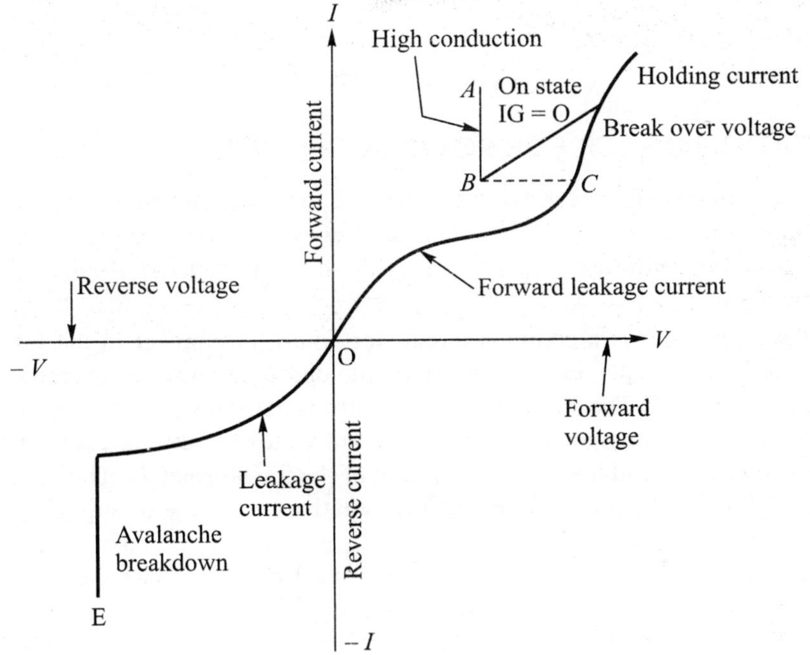

Fig. 11.3

The vertical axis +*I* and –*I* represents the device current and the horizontal axis +*V* and –*V* represents the voltage applied across the device anode to the current. The parameter I_F defines the RMS forward current that the SCR can carry in the ON state, while V_R defines the amount of voltage the unit can block in the OFF state.

A forward bias shown in figure as +*V* will result when a positive potential is applied to the anode and negative to cathode. Even after the application of a forward bias, the device remains non-conducting until the positive gate trigger voltage is applied. After the device is triggered ON it reverts to a low impedance state and current flows through the unit. The unit will remain conducting after the gate voltage has been removed. In the ON state (represented by +1), the current must be limited by load, or the damage to the SCR will result.

The reverse bias condition is represented by –*V*. A reverse bias exists when the potential applied across the SCR results in the cathode being more positive than anode. In this condition the SCR is non-conducting and the application of a trigger voltage will have no effect on the device. In the reverse bias mode, the knee of the curve is known as the Peak Inverse Voltage (PIV) or Peak Reverse Voltage (PRV) and this value cannot be exceeded or the device will breakdown and be destroyed.

Breakover Voltage: It is the minimum forward voltage, gate being open, at which SCR starts conducting heavily (*i.e.* turned ON).

Peak Reverse Voltage (PRV): It is the maximum reverse voltage (cathode positive w.r.t. anode) that can be applied to an SCR with conducting in the reverse direction.

Forward Current Rating: It is the maximum anode current that an SCR is capable of passing without destruction.

Holding Current: It is the maximum anode current, gate being open, at which SCR is turned OFF from ON conditions.

Circuit Fusing (I^2t) Rating: It is the product of square of forward surge current and the time of duration of surge.

11.6 APPLICATION OF SCR FOR POWER CONTROL

It is often necessary to control power delivered to some load such as the heating of a furnace. Series resistances or potentiometers cannot be used because they waste power in high power circuits. Under such conditions, silicon controlled rectifiers are used which are capable of adjusting the transmitted power with little waste.

The below figure shows a common circuit for controlling power in the load R_L. During the positive half-cycle of a.c. supply, end A is positive and end B is negative. Therefore, capacitor C_2 is charged through $AD_1RC_2D_4B$. The charge on the capacitor C_2 depends on the value of the potentiometer R. When the capacitor C_2 is charged through a sufficient voltage, it discharges through Zener Z. This gives a pulse to the primary and hence the secondary of the transformer T_2. This turns on SCR$_2$ which conducts currents through the load R_L. During the negative half cycle of supply, the capacitor C_1 is charged. It discharges through the zener and fires SCR$_1$ which conducts current through the load. The angle of conduction can be controlled by the potentiometer R. The greater the resistance of R, lesser is the voltage across C_1 and C_2 and hence smaller will be the time during which SCR$_1$ and SCR$_2$ will conduct in a full cycle. In this way, we control a large power of several Kw in the load R_L with the help of a small potentiometer R.

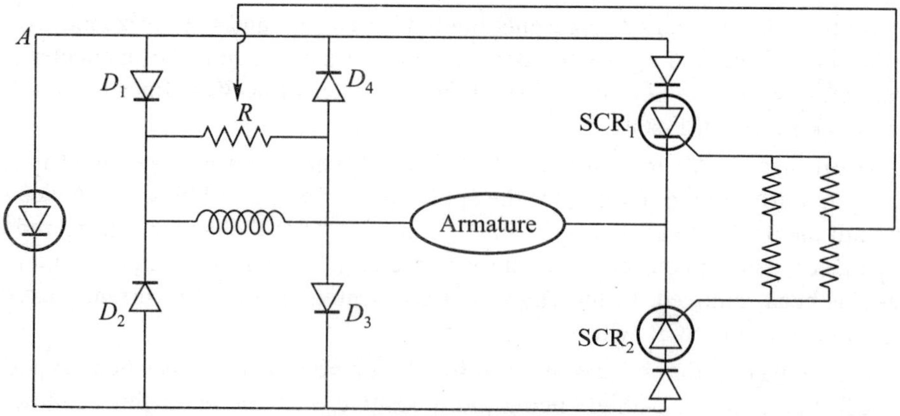

SCR for Power Control

Fig. 11.4

11.7 USES

SCRs are made with voltage ratings of up to 7500 volts, and with current ratings up to 3000 RMS amperes per device. Some of the larger ones can take over 50 kA in single-pulse operation. SCRs are used in power switching, phase control, chopper, battery chargers and inverter circuits. Industrially they are applied to produce variable DC voltages for motors (from a few to several thousand HP) from AC line voltage.

SOLVED EXAMPLES

Example 1: *An SCR in a circuit is subjected to a 50 A surge that lasts for 12 mS. Determine whether or not this surge will destroy the device. Given that circuit fusing rating is 90 A^2S.*

Solution: Circuit fusing rating = $I^2t = 50^2 \times (12 \times 10^{-3}) = 30\ A^2S$

As it is below the maximum rating of 90 A^2S, the device will not be destroyed.

Example 2: *An SCR has a circuit fusing rating of 50 A^2S. The device is being used in a circuit where it could be subjected to a 100 A surge. Determine the maximum allowable duration of such a surge.*

Solution:

$$t_{max} = \frac{I^2t}{I_S^2} = \frac{50}{100^2} = 5 \times 10^{-3}\ S = 5\ ms.$$

Example 3: *Write down the expression for the values of average voltage in a full wave SCR and find the values if $V_m = 200$ V, $\phi_1 = 30°$.*

Solution:

$$V_{dc} = \frac{V_m}{\pi}[1 + \cos\phi_1],\ \phi_1 = \text{firing angle}$$

$$V_{dc} = \frac{200}{\pi}[1 + \cos 30°]$$

$$V_{dc} = 119\ V.$$

EXERCISES

1. An SCR has a forward breakdown voltage of 150 V when a gate current of 1 mA flows in the gate circuit. If the sinusoidal voltage of 400 V peak is applied, find
   (*i*) firing angle, (*ii*) average output voltage [**Ans.** 22°, 122.6 V]
2. An SCR has a break over voltage of 400 V, and a trigger current of 10 mA. What do you infer from it? What will happen if gate current is made 15 mA?
3. An SCR has gate-cathode characteristic in the form of a straight line of slope 150 volt/amp. For trigger source voltage $V_S = 16$ V and allowable gate power dissipation of 0.8 watt, calculate R_{GS}.
   [**Ans.** 69.2 Ω]

QUESTIONS

A. Short Answer Questions:

1. What is the basic function of SCR?
2. Describe the construction of SCR and mention its characteristics.
3. Describe function of SCR.

4. Explain SCR's V-I characteristics.
5. Write the applications of SCR.
6. Describe the method of gate operation of SCR.
7. Describe the construction of SCR.
8. Explain the two-transistor analogy of an SCR.
9. Draw the forward gate characteristics of an SCR and explain it's nature.
10. Explain the turn OFF and turn ON mechanisms of an SCR.
11. Explain the power control mechanism using an SCR.
12. Describe the applications of SCR.

B. Long Answer Questions:

1. Describe the construction and working of SCR. Explain its characteristics.
2. What is an SCR? Give it's symbol and two transistor representation. Mention its applications.
3. Explain the construction and working of an SCR.
4. Describe the equivalent circuit of an SCR and its working.
5. Describe the *V-I* characteristics of SCR. Mention its applications.
6. Describe the important applications of SCR. Explain the action of an SCR as a switch. Draw the symbol of SCR.
7. Describe the *di/dt* protection and *dv/dt* protection factors to be provided to SCR.
8. Describe the gate protection of SCR. Explain the SCR function for working.

*C*HAPTER **12**

Photo Electric Devices

12.1 LIGHT DEPENDENT RESISTOR (LDR)

A photoresistor or Light Dependent Resistor or CdS Cell is a resistor whose resistance decreases with increasing incident light intensity. It can also be referred to as a photo conductor. A photoresistor is made of a high resistance semiconductor. If light falling on the device is of high enough frequency, photons absorbed by the semiconductor give bound electrons enough energy to jump into the conduction band. The resulting free electron (and its hole partner) conduct electricity, thereby lowering resistance.

A photoelectric device can be either intrinsic or extrinsic. An intrinsic semiconductor has its own charge carriers and is not an efficient semiconductor, *e.g.*, silicon. In intrinsic devices the only available electrons are in the valence band, and hence the photon must have enough energy to excite the electron across the entire band gap. Extrinsic devices have impurities, also called dopants, added whose ground state energy is closer to the conduction band; since the electrons dont't have as far to jump, lower energy photons (*i.e.*, longer wavelengths and lower frequencies) are sufficient to trigger the device. If a sample of silicon has some of its atoms replaced by phosphorus atoms (impurities), there will be extra electrons available for conduction. This is an example of an extrinsic semiconductor.

As its name implies, the light dependent resistor it is a resistive device which changes its electrical resistance from several thousand ohms in the dark to only a few hundred ohms when light falls upon it. The net effect is a decrease in resistance for an increase in illumination. Materials used as the semiconductor substrate include, Lead Sulphide (PbS), Lead Selenide (PbSe), Indium Antimonide (InSb) which detect light in the infrared range and the most commonly used of all Cadmium Sulphide (CdS), as its spectral response curve closely matches that of the human eye and can even be controlled using a simple torch as a light source. Typically it has a peak sensitivity wavelength (λ_p) of about 560 nm to 600 nm in the visible spectral range.

Cadmium Sulphide Cell (CdS)

This is most popularly used for incident radiation in the spectrum. It is formed by chemical deposition on an insulating substrate a thin layer of CdS with a small amount of silver, antimony or indium. This cell has resistance of about 1 MΩ in absolute darkness but a small resistance of only about 10 Ω with strong light. The basic construction of CdS cell is shown below.

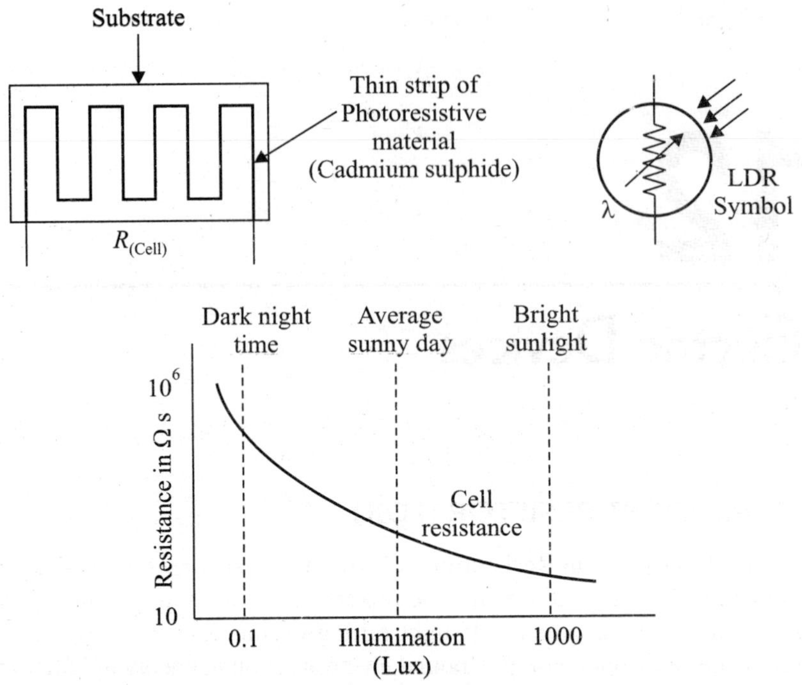

Fig. 12.1

If an external voltage source V is connected to the cell as shown in the figure, a photo-current I flows in the external circuit. With zero incident radiation (absolute darkness), the resistance of the cell is large, several MΩ and, hence the photo current I is negligibly small. With strong sunlight incident on the cell, the photo-current may be as large as 10 mA. In contrast, an ordinary semiconductor specimen gives a current of only a few tens of μA with strong sunlight. For a given cell, with definite external voltage V, the photo-current increases linearly with the intensity of illumination.

The most commonly used photo resistive device is the ORP12 Cadmium Sulphide photoconductive cell. This light dependent resistor has a spectral response of about 610 nm in the yellow to orange region of light. The resistance of the cell when unilluminated (dark resistance) is very high at about 10 MΩ's which falls to about 100 Ω's when fully illuminated (lit resistance). To increase the dark resistance and therefore reduce the dark current, the resistive path forms a zigzag pattern across the ceramic substrate. The CdS photocell is a very low cost device often used in auto dimming, darkness or twilight detection for turning the street lights "ON" and "OFF", and for photographic exposure meter type applications.

Merits

(1) High dissipation capability, one watt or more.

(2) Excellent sensitivity to visible spectrum.

(3) Low resistance on being exposed to strong sunlight.

Applications

(1) For measurement of intensity of light.

(2) For recording of modulated light intensity. *E.g.* on the sound track of motion picture film.

(3) As an on-off light sensitivity relay in control circuits.

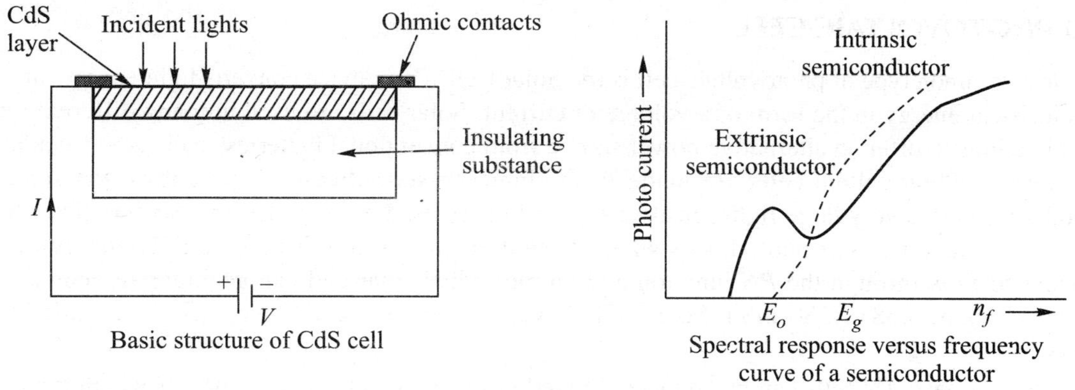

Fig. 12.2

One simple use of a Light Dependant Resistor, is as a light sensitive switch as shown below.

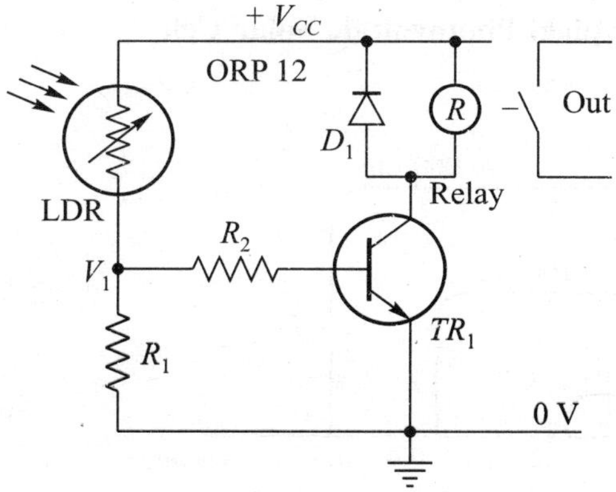

Fig. 12.3

The circuit is of a relay output light activated switch. A potential divider circuit is formed between the photoresistor, LDR and the resistor R_1. When no light is present *i.e.*, in darkness, the resistance of the LDR is very high in the Mega ohms range so zero base bias is applied to the transistor TR_1 and the relay is de-energised or "OFF". As the light level increases the resistance of the LDR starts to decrease causing the base bias voltage at V_1 to rise. At some point determined by the potential divider network formed with resistor R_1, the base bias voltage is high enough to turn "ON" the transistor TR_1 and thus activate the relay which inturn is used to control some external circuitry. As the light level falls back to darkness again the resistance of the LDR increases causing the base voltage of the transistor to decrease, turning the transistor and relay "OFF" at a fixed light level determined again by the potential divider network.

12.2 PHOTOJUNCTION DEVICES

Photojunction Devices are basically PN-Junction light detectors made from silicon semiconductors and which can detect both visible light and infrared light levels. This class of photoelectric device includes the **Photodiode** and the **Phototransistor.**

12.3 PHOTOVOLTAIC CELL

The most common type of photovoltaic cell is the Solar Cell. This device converts light energy directly into electrical energy in the form of a voltage or current. Solar cells are used in many different types of applications to offer an alternative power source from conventional batteries, such as in calculators and satellite. Photovoltaic cells are made from single crystal silicon *PN* junctions, the same as photodiodes with a very large light sensitive region but are used without the reverse bias. They have the same characteristics as photodiodes when in the dark. When illuminated the light energy causes electrons to flow through the *PN* junction and an individual solar cell can generate an open circuit voltage of about 0.58V (580 mV). Solar cells have a "Positive" and a "Negative" side just like a battery.

Individual solar cells can be connected together in series to form solar panels which increases the output voltage or connected together in parallel to increase the available current. Commercially available solar panels are rated in Watts, which is the product of the output voltage and current ($V \times I$) when fully lit.

Characteristics of a typical Photovoltaic Solar Cell

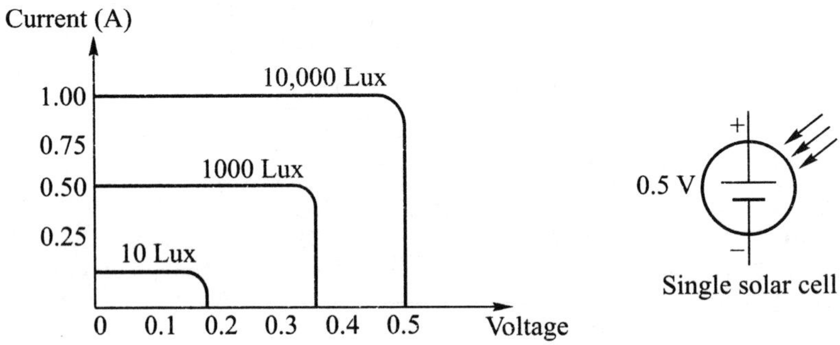

Fig. 12.4

The amount of available current from a solar cell depends upon the light intensity, the size of the cell and its efficiency which is generally very low at around 20%. To increase the overall efficiency of the cell commercially available solar cells use Polycrystalline Silicon and Amorphous Silicon, which has no crystalline structure and can generate currents of between 20 to 40 mA per cm^2. Other materials used include Gallium Arsenide, Copper Indium Diselenide and Cadmium Telluride. These different materials each have a different spectrum band response, and so can be "tuned" to produce an output voltage at different wavelengths of light.

Silicon Solar Cell

Any photovoltaic cell with sunlight incident on it may be used to generate electrical energy. Such energy converter is called a solar cell. The figure shown below gives basic structure of silicon solar cells. Effort is made to ensure that the surface area perpendicular to the cell is maximum. The metallic ring connected to the *p*-type material and the thickness of the *p*-type material are such that maximum number of photons of incident light reach junction.

Conversion efficiency η = output electrical power/input radiation power

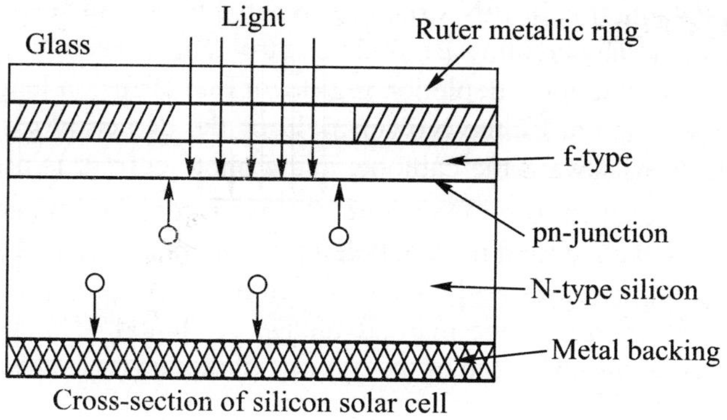

Cross-section of silicon solar cell

Fig. 12.5

12.4 PHOTODIODE

The construction of the photodiode is similar to that of a conventional *PN*-junction diode except that the diodes outer casing is transparent near to the junction so that light can fall upon the junction. LED's can also be used as photodiodes as they can both emit and detect light. All *PN*-junctions are light sensitive and can be used in a photoconductive (PC) mode with the *PN*-junction of the photodiode always "Reverse Biased" so that only the diodes leakage or dark current can flow. This reverse bias condition causes an increase of the depletion region which is the sensitive part of the junction.

A photodiode is a type of photodetector capable of converting light into either current or voltage, depending upon the mode of operation. Photodiodes are similar to regular semiconductor diodes except that they may be either exposed (to detect vacuum *UV* or *X*-rays) or packaged with a window or optical fibre connection to allow light to reach the sensitive part of the device. Many diodes designed for use specifically as a photodiode will also use a PIN junction rather than the typical *PN* junction.

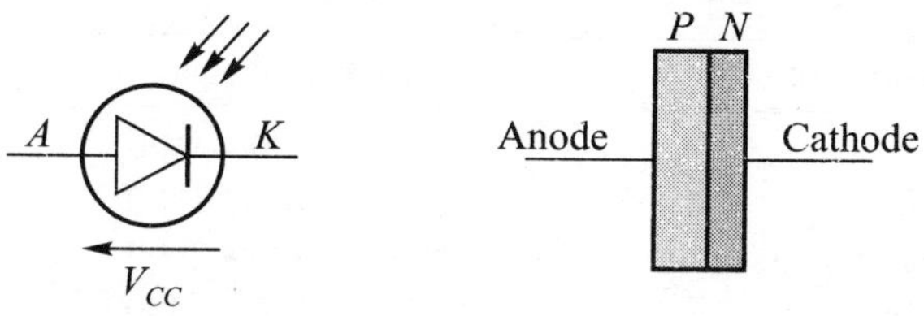

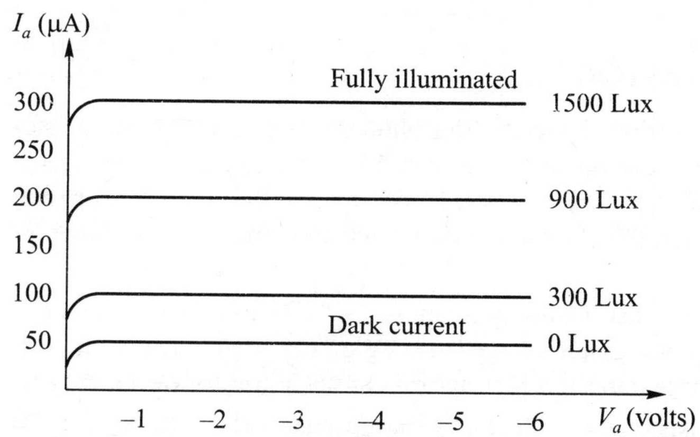

Fig. 12.6

A photodiode is a *PN* junction or PIN structure. When a photon of sufficient energy strikes the diode, it excites an electron thereby creating a mobile electron and a positively charged electron hole. If the absorption occurs in the junction's depletion region, or one diffusion length away from it, these carriers are swept from the junction by the built-in field of the depletion region. Thus holes move toward the anode, and electrons toward the cathode, and a photo current is produced.

The photodiodes dark current (0 lux) is about 10 μA for germanium and 1 μA for silicon type diodes. When light falls upon the junction more hole/electron pairs are formed and the leakage current increases. The leakage current increases as the illumination of the junction increases. Diode current is directly proportional to light intensity. One man advantage of photodiodes are their fast response to changes in light levels, but one disadvantage of this type of photo device is the relatively small current flow even when fully lit.

Photodiodes are commonly used in cameras, light meters, CD and DVD-ROM drives, TV remote controls, scanners, fax machines and copiers etc., and when integrated into operational amplifier circuits as infrared spectrum detectors for fibre optic communications, burglar alarm motion detection circuits and numerous imaging, laser scanning and positioning systems etc.

Since transistors and ICs are made of semiconductors, and contain P-N junctions, almost every active component is potentially a photodiode. Many components, especially those sensitive to small currents, will not work correctly if illuminated, due to the induced photocurrents. In most components this is not desired, so they are placed in an opaque housing. Since housings are not completely opaque to *X*-rays or other high energy radiation, these can still cause many *ICs* to malfunction due to induced photo-currents.

Material used to make a photodiode is critical to defining its properties, because only photons with sufficient energy to excite electrons across the material's bandgap will produce significant photo currents. Because of their greater bandgap, silicon-based photodiodes generate less noise than germanium-based photodiodes, but germanium photodiodes must be used for wavelengths longer than approximately 1 μm.

Materials commonly used to produce photodiodes include:

Material	Wavelength range (nm)
Silicon	190—1100
Germanium	400—1700
Indium gallium arsenide	800—2600
Lead sulfide	<1000—3500

12.5 PHOTOTRANSISTOR

An alternative photo junction device to the photodiode is the Phototransistor which is basically a photodiode with amplification and operates by exposing its base region to the light source. Phototransistors operate the same as photodiodes except that they can provide current gain and are much more sensitive than the photodiode with currents are 50—100 times greater than that of the standard photodiode.

Phototransistors consist mainly of a bipolar NPN Transistor with the collector-base *PN*-junction reverse-biased. The Phototransistor's large base region is left electrically unconnected and uses photons of light to generate a base current which inturn causes a collector/emitter current to flow.

In the NPN transistor the collector is biased positively with respect to the emitter so that the base/collectoı junction is reverse biased. Therefore, with no light on the junction normal leakage or dark current flows which is very small. When light falls on the base more electron/hole pairs are formed in this region and the current produced by this action is amplified by the transistor. The sensitivity of

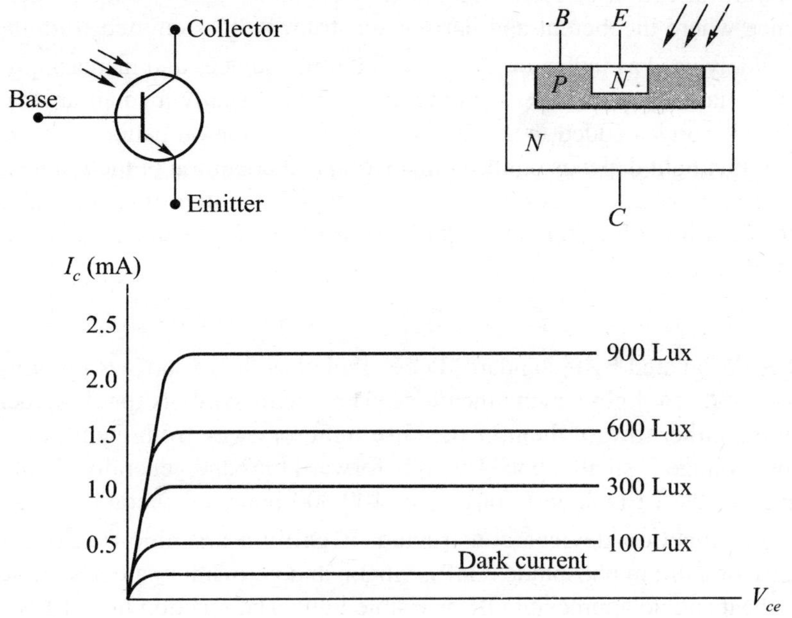

Fig. 12.7

a phototransistor is a function of the DC current gain of the transistor. Therefore, the overall sensitivity is a function of collector current and can be controlled by connecting a resistance between the base and the emitter but for very high sensitivity optocoupler type applications, Darlington phototransistors are generally used.

Photo Darlington transistors use a second bipolar NPN transistor to provide additional amplification or when higher sensitivity of a photo detector are required, but its response is slower than that of an ordinary NPN phototransistor. It consists of a normal phototransistor whose emitter output is coupled to the base of a larger bipolar NPN transistor. Because a darlington transistor configuration gives a current gain equal to a product of the current gains of two individual transistors, a photodarlington device produces a very sensitive detector.

Typical applications of Photo transistors are in opto-isolators, slotted opto switches, light beam sensors, fiber optics and TV type remote controls, etc. Infrared filters are sometimes required when detecting visible light.

Another type of photo junction semiconductor device worth a mention is the Photothyristor. This is a light activated thyristor or Silicon Controlled Rectifier, SCR that can be used as a light activated switch in a.c. applications. However their sensitivity is usually very low compared to photodiodes or phototransistors, as to increase their sensitivity to light they are made thinner around the gate junction which inturn limits the amount of current that they can switch. Then for higher current applications they are used as pilot devices in opto-couplers to switch larger more conventional thyristors.

12.6 LIGHT EMITTING DIODE (LED)

A light-emitting diode (LED) is a semiconductor diode that emits light when an electrical current is applied in the forward direction of the device, as in the simple LED circuit. The effect is a form of electroluminescence where incoherent and narrow-spectrum light is emitted from the *pn*-junction.

LEDs are widely used as indicator lights on electronic devices and increasingly in higher power applications such as flashlights and area lighting. An LED is usually a small area (less than 1 mm^2) light source, often with optics added to the chip to shape its radiation pattern and assist in reflection. The color of the emitted light depends on the composition and condition of the semiconducting material used, and can be infrared, visible, or ultraviolet. Besides lighting, interesting applications include using UV-LEDs for sterilization of water and disinfection of devices, and as grow light to enhance photosynthesis in plants.

Construction

The LED is designated by a standard diode symbol with two arrows pointing away from the cathode. The arrows indicate light leaving the diode. The circuit symbols for all optoelectronic devices have arrows pointing either toward them, if they <u>use</u> light, or away from them, if they <u>produce</u> light. The LED operating voltage is small, about 1.6 volts forward bias and generally about 10 milliamperes. The life expectancy of the LED is very long, over 100,000 hours of operation.

LEDs are *p-n* junction devices constructed of gallium arsenide (GaAs), gallium arsenide phosphide (GaAsP), or gallium phosphide (GaP). Silicon and germanium are not suitable because those junctions produce heat and no appreciable IR or visible light. The junction in an LED is forward biased and when electrons cross the junction from the *n* to the *p*-type material, the electron-hole recombination process produces some photons in the IR or visible in a process called electroluminescence. An exposed semiconductor surface can then emit light.

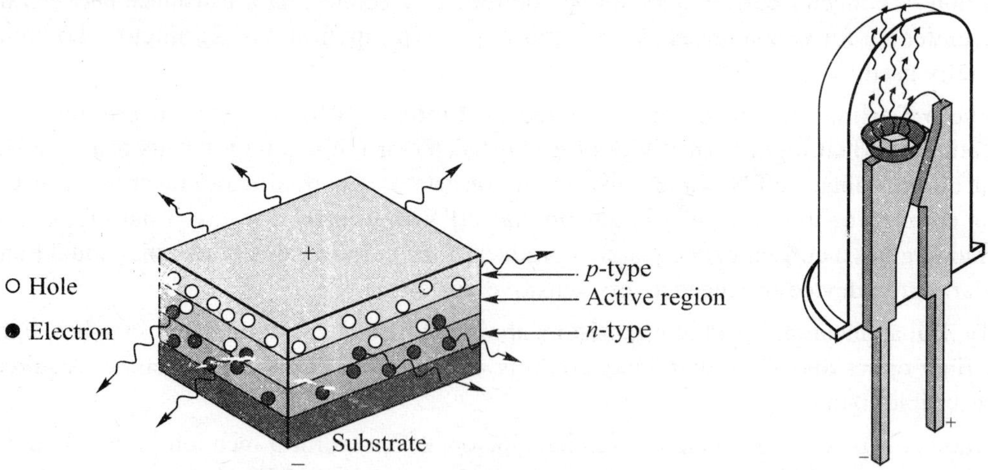

Fig. 12.8

One way to construct an LED is to deposit three semiconductor layers on a substrate. Between *p*-type and *n*-type semiconductor layers, an active region emits light when an electron and hole recombine. Considering the *p-n* combination to be a diode, then when the diode is forward biased, holes from the *p*-type material and electrons from the *n*-type material are both driven into the active region. The light is produced by a solid state process called electroluminescence.

In this particular design, the layers of the LED emit light all the way around the layered structure, and the LED structure is placed in a tiny reflective cup so that the light from the active layer will be reflected toward the desired exit direction.

LED V-I Characteristics

Before a LED can "emit" light, it needs a current to flow through it as it is a current dependent device. As the LED is to be connected in a forward bias condition across a power supply it should be Current Limited using a series resistor to protect it from excessive current flow. In most cases LEDs are operated from a low voltage DC supply, with a series resistor to limit the forward current to a suitable value from say 5 mA for a simple LED indicator to 30 mA or more where a high brightness light output is needed.

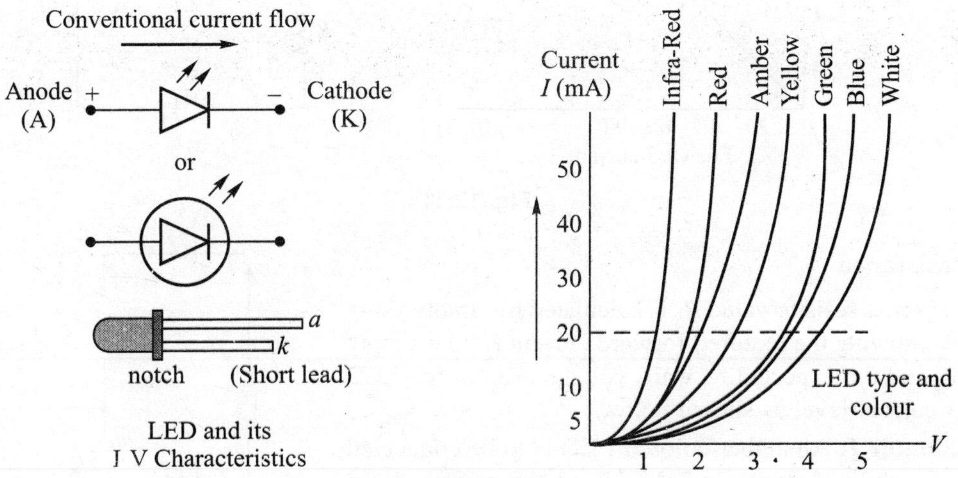

Fig. 12.9

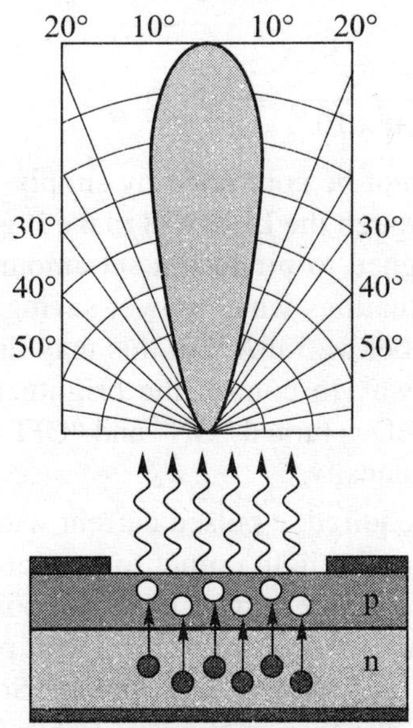

Fig. 12.10

The light output is quite linearly proportional to the current within its active region, so the light output can be precisely modulated to send an undistorted signal through a fiber optic cable. An LED is a directional light source, with the maximum emitted power in the direction perpendicular to the emitting surface. The typical radiation pattern shows that most of the energy is emitted within 20° of the direction of maximum light. Some packages for LEDs include plastic lenses to spread the light for a greater angle of visibility.

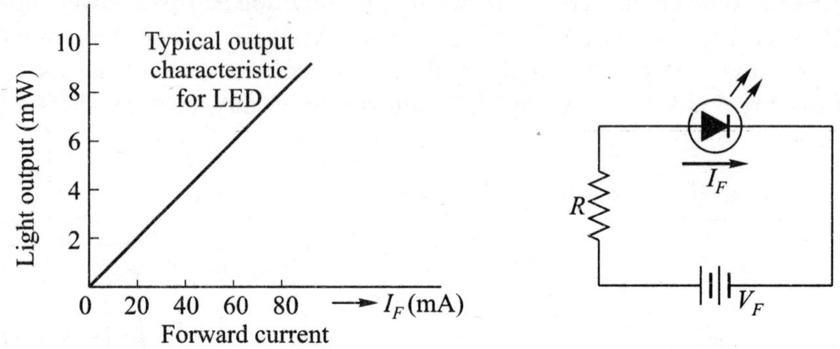

Fig. 12.11

Series Resistance

The series resistor value R_S is calculated by simply using ohm's law knowing the required forward current I_F, the supply voltage V_S and the expected forward voltage drop of the LED V_F at this current level as shown below.

Example 1: An amber colored LED is to be connected to a 5.0 V stabilized power supply. Using the circuit above calculate the series resistor required to limit the forward current to less than 10 mA.

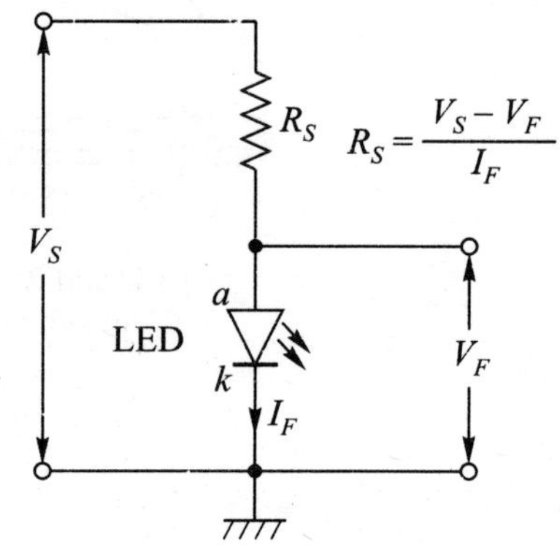

Fig. 12.12

$$R_S = \frac{V_S - V_F}{I_F} = \frac{5\text{V} - 2\text{V}}{10 \text{ mA}} = \frac{3}{10 \times 10^{-3}} = 300\ \Omega$$

The brightness of a LED cannot be controlled by simply varying the current flowing through it. Allowing more current to flow through the LED will make it glow brighter but will also cause it to dissipate more heat. LEDs are designed to produce a set amount of light operating with a forward current of about 10 to 20 mA. In situations where power savings are important, less current may be possible but reducing the current to below 5 mA, but this may dim its output to much or even turn it "OFF" completely. A much better way to control the brightness of LEDs is to use "Pulse Width Modulation" (PWM), in which the LED is turned "ON" and "OFF" continuously at varying frequencies depending upon the required light intensity.

When higher light output is required, a pulsed current with a fairly short duty cycle (ON-OFF ratio) allows the current and therefore the light output to be increased significantly during the actual pulses, while still keeping the LEDs average current level and power dissipation within its limits. The human eyes fills in the gaps between the "ON" and "OFF" light pulses, providing the pulse frequency is high enough, making it appear as a continuous light output. So pulses at a frequency of 100 Hz or more actually appear brighter to the eye than continuous light of the same average intensity.

LEDs are available in a wide range of shapes, colours and various sizes with different light output intensities available, with the most common (and cheapest to produce) being the standard 5 mm Red LED. LED's are also available in various "packages" arranged to produce both letters and numbers with the most common being that of a "Seven-Segment Display" arrangement. Nowadays, full colour flat screen LED displays are available with a large number of dedicated ICs available for driving the displays directly. Most LEDs produce a single output colour however, multi-colour LEDs are now available that can produce a variety of colours within a single device.

Displays

As well as individual colour or multi-colour LEDs, light emitting diodes can be combined together in a single package to produce displays such as bargraphs, strips, arrays and 7-segment displays. A 7-segment LED display provides a very convenient way of displaying information or digital data in the form of Numbers, Letters or even Alpha-numerical characters and as their name suggests, they consist of 7 individual LEDs (the segments), within one single display package. In order to produce the required numbers or characters from 0 to 9 and *A* to *F* respectively, on the display the correct combination of LED segments need to be illuminated. A standard 7-segment LED display generally has 8 input connections, one for each LED segment and one that acts as a common terminal or connection for all the internal segments.

There are two important types of 7-segment LED digital display.

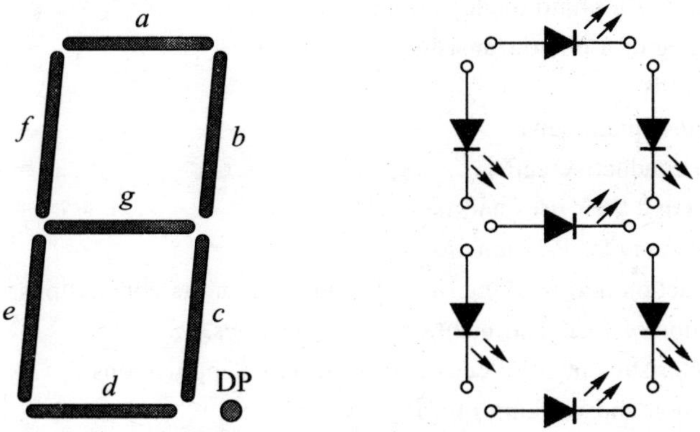

A typical 7-Segment display

Fig. 12.13

Materials

Conventional LEDs are made from a variety of inorganic semiconductor materials, producing the following colors:

- Aluminium gallium arsenide (AlGaAs)—red and infrared
- Aluminium gallium phosphide (AlGaP)—green
- Aluminium gallium indium phosphide (AlGaInP)—high-brightness orange-red, orange, yellow, and green.
- Gallium arsenide phosphide (GaAsP)—red, orange-red, orange, and yellow.
- Gallium phosphide (GaP)—red, yellow and green.
- Gallium nitride (GaN)—green, pure green (or emerald green), and blue also white (if it has an AlGaN Quantum Barrier)

- Indium gallium nitride (InGaN)-450—470 nm-near ultraviolet, bluish-green and blue
- Silicon carbide (SiC) as substrate—blue
- Silicon (Si) as substrate—blue (under development)
- Sapphire (Ab03) as substrate—blue
- Zinc selenide (ZnSe)—blue
- Diamond (C)—ultraviolet
- Aluminium nitride (AlN), aluminium gallium nitride (AlGaN), aluminium gallium indium nitride (AlGaInN)—near to far ultraviolet (down to 210 nm)

The many application of LEDs are very diverse but fall in three major category's : Visual signal application where the light goes more or less directly from the LED to the human eye, to convey a message or meaning. Illumination where LED light is reflected from object to give visual response of these objects. And finally LEDs are also used to generate light for measuring and interacting with processes that do not involve the human visual system.

EXERCISES

Short Answer Questions :

1. Explain the construction and operation of photovoltaic cell.
2. Explain the working of a photo diode and photo transistor.
3. Describe the working of a photo transistor.
4. Write about solar cells.
5. Write about a photo voltaic cells.
6. Write about photo conductive cells.
7. What are LEDs ? Give their uses and merits.
8. What do you understand by P-N junction photodiode.
9. Explain the construction and working of LED and mention its applications.
10. Explain the operation of LED and mention its applications.
11. Explain the characteristics of solar cells and mention its applications.
12. Write the construction and working of a Photovoltaic cell.
13. Write the construction and working of a solar cell.
14. Give the characteristics and working of LED.

Long Answer Questions :

1. Explain the construction, operation and characteristics of photoconductive cells.
2. Explain the construction, operation and characteristics of solar cells and LEDs.
3. Explain the construction and the working of LED. Give its merits and demerits.
4. Explain the characteristics of phototransistor. Mention its applications.
5. Describe the construction and working of LEDs.
6. What are solar cells ? Describe in detail a silicon solar cell.
7. Write a short note on Photovoltaic cells, solar cells, photo diodes and photo transistors.
8. Draw the circuit diagram of a LED and mention it's working and applications.
9. Distinguish between the photodiode, phototransistor and a solar cells.
10. Discuss the construction, working and characteristics of photo conductive cells.

FACULTY OF SCIENCE

B.Sc. I Year (New) Examination, March/April, 2009

ELECTRONICS

Paper–I : Circuit Analysis and Electronic Devices

Time : 3 Hours **Maximum Marks : 100**

*Answer **ALL** questions from Part-A and any **Four** questions from Part-B.*
*Each question carries **20 marks** in Part-A and **Five** marks in Part-B.*

PART–A

(4 × 20 = 80 Marks)

(Essay Answer Type)

1. (*a*) (*i*) Define resistance, reactance, impedance, conductance, susceptance and admittance. 6

(*ii*) Derive equations for RMS and Average value of AC voltage and define form factor. 14

Or

(*b*) (*i*) State and prove Maximum power transfer theorem. 10

(*ii*) Explain the solution of network using Node analysis method. 10

2. (*a*) (*i*) Explain the working of C.R.T. using neat diagram. 10

(*ii*) Explain the Transient response of RC circuit and derive equation for instantaneous voltage across capacitor for growth. 10

Or

(*b*) (*i*) Explain the phenomenon of resonance in series LCR circuit. 5

(*ii*) Derive equation for resonant frequency, band width and quality factor of series LCR circuit. 15

3. (*a*) (*i*) Explain depletion and diffusion capacitance of PN junction diode. 5

(*ii*) Explain the effort of Temperature on reverse saturation current. 5

(*iii*) Explain input and output characteristics of NPN transistor in CB configuration. 10

Or

(*b*) (*i*) Define the *h*-parameters of transistor in CE configuration and draw its equivalent circuit. 10

(*ii*) Explain the need for biasing of transistor and explain fixed biasing. 10

4. (*a*) (*i*) Explain FET parameters. 5

(*ii*) Explain the application of FET as voltage variable resistor. 5

(*iii*) Explain V-I characteristics of SCR with circuit diagram. 10

Or

(*b*) (*i*) Explain the construction and characteristics of LDR. 10

(*ii*) Explain the function of UJT as relaxation oscillator and derive equation for its time period. 10

PART–B
(4 × 5 = 20 Marks)
(Answer **Four** Questions)
(Short Answer Type)

5. Using mesh analysis method find currents flowing through all the resistors in the given circuit.

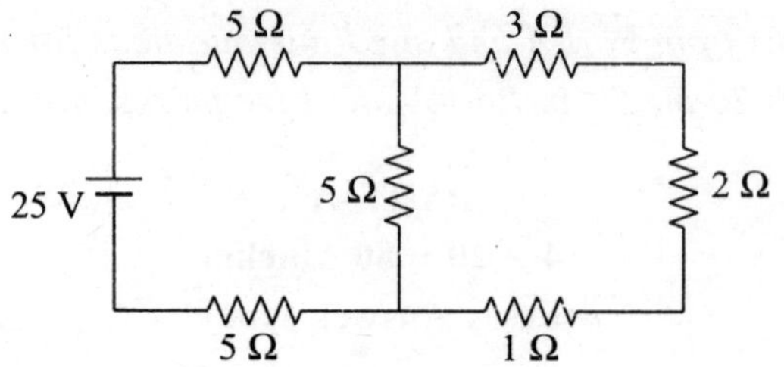

6. Prove Reciprocity theorem for the given network.

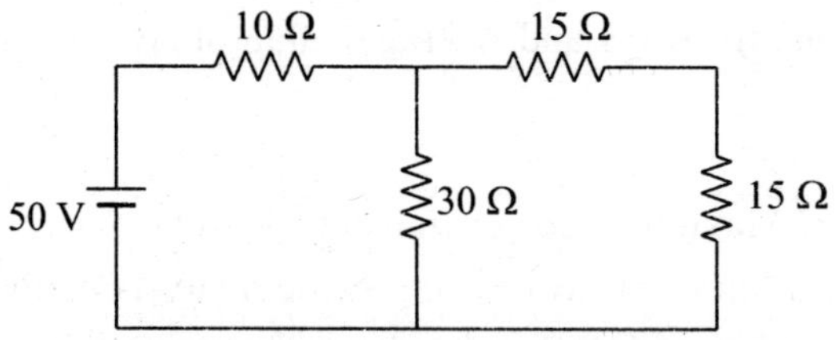

7. Explain the frequency response of RC low pass filter and define its cutoff frequency.
8. Find resonant frequency and quality factor of a given circuit.

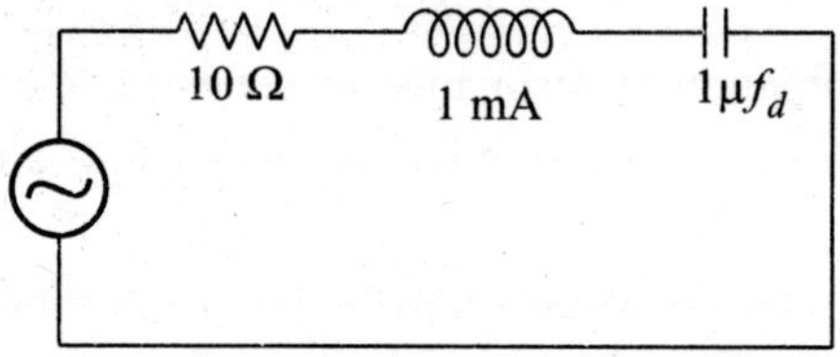

9. Explain the application of Zener diode as voltage regulator.
10. Find DC operating point of given transistor circuit (Assume Get Transistor).

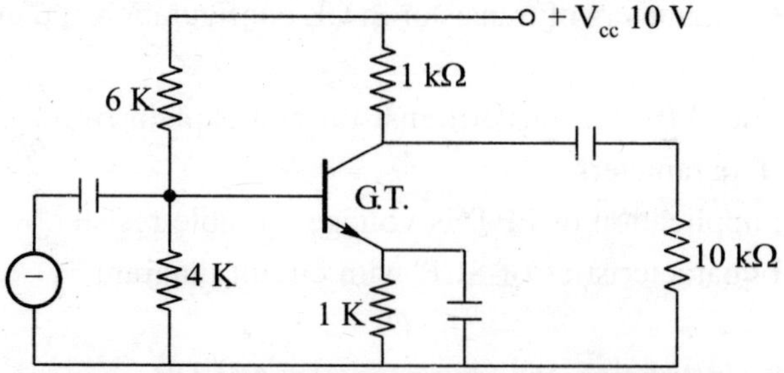

11. Explain the application of SCR for power control using suitable circuit.
12. Explain the construction of LED and describe its characteristics.